International Association of Geodesy Symposia

Chris Rizos, Series Editor
Pascal Willis, Assistant Series Editor

International Association of Geodesy Symposia

Chris Rizos, Series Editor
Pascal Willis, Assistant Series Editor

Symposium 101: Global and Regional Geodynamics
Symposium 102: Global Positioning System: An Overview
Symposium 103: Gravity, Gradiometry, and Gravimetry
Symposium 104: Sea SurfaceTopography and the Geoid
Symposium 105: Earth Rotation and Coordinate Reference Frames
Symposium 106: Determination of the Geoid: Present and Future
Symposium 107: Kinematic Systems in Geodesy, Surveying, and Remote Sensing
Symposium 108: Application of Geodesy to Engineering
Symposium 109: Permanent Satellite Tracking Networks for Geodesy and Geodynamics
Symposium 110: From Mars to Greenland: Charting Gravity with Space and Airborne Instruments
Symposium 111: Recent Geodetic and Gravimetric Research in Latin America
Symposium 112: Geodesy and Physics of the Earth: Geodetic Contributions to Geodynamics
Symposium 113: Gravity and Geoid
Symposium 114: Geodetic Theory Today
Symposium 115: GPS Trends in Precise Terrestrial, Airborne, and Spaceborne Applications
Symposium 116: Global Gravity Field and Its Temporal Variations
Symposium 117: Gravity, Geoid and Marine Geodesy
Symposium 118: Advances in Positioning and Reference Frames
Symposium 119: Geodesy on the Move
Symposium 120: Towards an Integrated Global Geodetic Observation System (IGGOS)
Symposium 121: Geodesy Beyond 2000: The Challenges of the First Decade
Symposium 122: IV Hotine-Marussi Symposium on Mathematical Geodesy
Symposium 123: Gravity, Geoid and Geodynamics 2000
Symposium 124: Vertical Reference Systems
Symposium 125: Vistas for Geodesy in the New Millennium
Symposium 126: Satellite Altimetry for Geodesy, Geophysics and Oceanography
Symposium 127: V Hotine Marussi Symposium on Mathematical Geodesy
Symposium 128: A Window on the Future of Geodesy
Symposium 129: Gravity, Geoid and Space Missions
Symposium 130: Dynamic Planet - Monitoring and Understanding…
Symposium 131: Geodetic Deformation Monitoring: From Geophysical to Engineering Roles
Symposium 132: VI Hotine-Marussi Symposium on Theoretical and Computational Geodesy
Symposium 133: Observing our Changing Earth
Symposium 134: Geodetic Reference Frames
Symposium 135: Gravity, Geoid and Earth Observation
Symposium 136: Geodesy for Planet Earth
Symposium 137: VII Hotine-Marussi Symposium on Mathematical Geodesy
Symposium 138: Reference Frames for Applications in Geosciences
Symposium 139: Earth on the Edge: Science for a Sustainable Planet

More information about this series at
http://www.springer.com/series/1345

The 1st International Workshop on the Quality of Geodetic Observation and Monitoring Systems (QuGOMS'11)

Proceedings of the 2011 IAG International Workshop, Munich, Germany, April 13 – 15, 2011

Edited by

Hansjörg Kutterer
Florian Seitz
Hamza Alkhatib
Michael Schmidt

Springer

Volume Editors
Hansjörg Kutterer
Bundesamt für Kartographie und Geodäsie
Frankfurt am Main
Germany

Florian Seitz
Technische Universität München
München
Germany

Hamza Alkhatib
Leibniz Universität Hannover
Geodätisches Institut
Hannover
Germany

Michael Schmidt
Deutsches Geodätisches Forschungsinstitut
München
Germany

Series Editors
Chris Rizos
School of Surveying
University of New South Wales
Sydney
Australia

Assistant Series Editors
Pascal Willis
Institut national de l'Information
Geographique et Forestiere
Direction Technique
Saint-Mande
France

ISSN 0939-9585
ISBN 978-3-319-10827-8 ISBN 978-3-319-10828-5 (ebook)
DOI 10.1007/978-3-319-10828-5
Springer Cham Heidelberg New York Dordrecht London

Library of Congress Control Number: 2014956880

Printed on acid-free paper

Springer is part of Springer Science+Business Media (www.springer.com)

Preface

The 1st international Workshop on the Quality of Geodetic Observation and Monitoring (QuGOMS) was devoted to the general methodology in the field of estimation and filtering with a refined uncertainty modelling emphasizing applications in engineering geodesy and Earth system observation. Geodetic multi-sensor systems and networks using terrestrial and space-borne observation techniques were thematic anchor points.

The QuGOMS workshop has been organized jointly by the study groups IC-SG2 and IC-SG3 of the Intercommission Committee on Theory (ICCT) of the International Association of Geodesy (IAG). Besides its strong relations to all IAG Commissions the workshop was also in the scope of FIG Commissions 5 and 6. Thus, it attracted scientists under the umbrellas of both IAG and FIG. The workshop took place in the rooms of the International Graduate School of Science and Engineering (IGSSE) of the Technische Universität München, Garching/Munich from 13th to 15th April 2011.

The workshop was organized in five regular sessions. To a large extent, the sessions' topics referred to the subjects of the IC-SG2 and IC-SG3 of the ICCT:

- Uncertainty modelling of geodetic data
- Theoretical studies on combination strategies and parameter estimation
- Recursive state-space filtering
- Sensor networks and multi-sensor systems in engineering geodesy
- Multi-mission approaches with view to physical processes in the Earth system

The contributed papers showed quite well the related questions and hence the close connection of methodology in the different fields of application such as global geodesy and engineering geodesy. Without doubt, it is worthwhile to continue this kind of workshop to foster scientific exchange also between scientific organizations.

There were several colleagues who contributed to the success of the workshop and of the proceedings. All editors of this volume acted also as convenors. Hamza Alkhatib coordinated the reviews of the submitted papers and communicated with the symposium editors and with Springer. Otto Heunecke took care of the contributions from engineering geodesy. Various reviewers helped to ensure a valid review process. Florian Seitz acted as local host. All in all this is gratefully acknowledged.

Frankfurt am Main, Germany
12 September 2013

Hansjörg Kutterer

Contents

Part I

Uncertainty Modeling of Geodetic Data

Modeling Data Quality Using Artificial Neural Networks

Ralf Laufer and Volker Schwieger

Abstract

Managing data quality is an important issue in all technical fields of applications. Demands on quality-assured data in combination with a more diversified quality description are rising with increasing complexity and automation of processes, for instance within advanced driver assistance systems (ADAS). Therefore it is important to use a comprehensive quality model and furthermore to manage and describe data quality throughout processes or sub-processes.

This paper focuses on the modeling of data quality in processes which are in general not known in detail or which are too complex to describe all influences on data quality. As emerged during research, artificial neural networks (ANN) are capable for modeling data quality parameters within processes with respect to their interconnections.

Since multi-layer feed-forward ANN are required for this task, a large number of examples, depending on the number of quality parameters to be taken into account, is necessary for the supervised learning of the ANN, respectively determining all parameters defining the net. Therefore the general usability of ANN was firstly evaluated for a simple geodetic application, the polar survey, where an unlimited number of learning examples could be generated easily. As will be shown, the quality parameters describing accuracy, availability, completeness and consistency of the data can be modeled using ANN. A combined evaluation of availability, completeness or consistency and accuracy was tested as well. Standard deviations of new points can be determined using ANN with sub-mm accuracy in all cases.

To benchmark the usability of ANN for a real practical problem, the complex process of mobile radio location and determination of driver trajectories on the digital road network based on these data, was used. The quality of calculated trajectories could be predicted sufficiently from a number of relevant input parameters such as antenna density and road density. The cross-deviation as an important quality parameter for the trajectories could be predicted with an accuracy of better than 40 m.

R. Laufer • V. Schwieger (✉)
Institute of Engineering Geodesy, University of Stuttgart,
Geschwister-Scholl-Str. 24D, 70174 Stuttgart, Germany
e-mail: volker.schwieger@ingeo.uni-stuttgart.de

H. Kutterer et al. (eds.), *The 1st International Workshop on the Quality of Geodetic Observation and Monitoring Systems (QuGOMS'11)*, International Association of Geodesy Symposia 140,
DOI 10.1007/978-3-319-10828-5_1,

Keywords
Data quality • Neural networks • Propagation of data quality

1 Introduction

Managing data quality requires first of all an adequate and homogenous quality model. The model which is used in the following was developed by Wiltschko (2004) and Wiltschko and Kaufmann (2005) and consists of six inherent quality characteristics. The six quality characteristics of the used quality model are shortly defined as follows:

- Availability: Measure for existence of information at a certain time and at a certain place.
- Timeliness: Measure for correlation of information with temporally changing reality.
- Completeness: Measure for existence of all relevant information to describe reality.
- Consistency: Measure for correlation of information with the information model.
- Correctness: Measure for the correlation of information with reality, timeliness assumed.
- Accuracy: Describes the correlation between determined value and real value.

Depending on the kind of data, these quality characteristics can be concretized each with several quantitative quality parameters. Quality measurement methods are needed as well to determine values for these quantitative quality parameters. With this model all relevant aspects of data quality can be sufficiently described.

Considering a data management process, the occurring input and output data can be described by quality parameters which are part of individual quality models defined for the different data. Since output data is generated from input data, their quality can be determined from the quality of the input data. However the relation between input and output data quality can be quite complex and often cannot be described in an analytical way. In some cases the dependencies are only partly known. Therefore a robust method is necessary which can handle this lack of information. As it turned out, artificial neural networks are more capable to solve this problem than other methods such as Petri nets or Monte-Carlo-simulation (Laufer 2011). An important assumption is that all inputs that influence the output significantly are captured (ideal case).

2 Data Quality Propagation with Artificial Neural Networks

In the following part, ANN will be briefly introduced and their most important advantages with respect to the other methods mentioned in the introduction are presented.

Artificial neural networks are a method which was developed around 70 years before (Zell 1997). The motivation was to copy the human brain in design and principle of operation. The networks consist of neural cells linked by artificial nerve tracts which can transmit information in an intelligent way (e.g. threshold based). Starting with very simple networks of artificial neural cells, more and more complex networks were developed over time, capable to solve more complex problems in combinatorics, pattern recognition, diagnostics and other fields of application where normal computers still get to their limits. A broad variety of applications is listed, for example, in Hagan et al. (1996). The main advantages of ANN with respect to other methods described in Laufer (2011) are:

- Massive parallel work flow (fast),
- Ability to learn and adapt,
- Ability to generalize and associate and
- High fault tolerance.

The artificial neural network can be trained in different ways depending on the kind of network and the application. In this paper only feed-forward networks and therefore supervised learning can be determined successfully. For supervised learning a sufficient amount of learning examples is necessary depending on the complexity of the process as well as the number of different input and output parameters.

In feed-forward networks, all neurons are arranged in several levels. Typically a network consists of 2–3 layers. Figure 1 shows a single artificial neuron with R inputs with the weights $w_{i,j}$ for the R different input parameters p_j of the neuron no. i. b describes the bias of the neuron. The inputs are used in the transfer function Σ to generate the net input n. The activation function f generates the net output a. The flow of information in these kinds of ANN is always fixed in one direction.

To train the network, training data sets representing the whole possible data range are necessary. Validation data sets are used to check the learning status of the net after each iteration. After finishing the training, the net performance must be evaluated independently using new data sets not yet known to the net, so-called evaluation data.

3 Propagation of Data Quality for Polar Point Determination

The polar point determination is a quite simple method to calculate Cartesian point coordinates from measured distance and direction to a new point. In the first step ANN were used

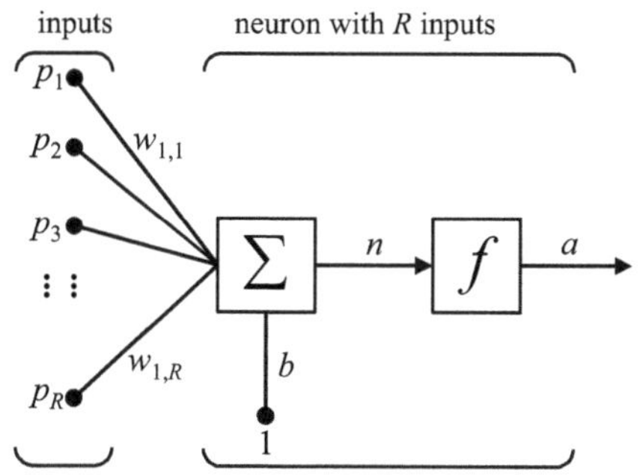

Fig. 1 Artificial neuron with R inputs

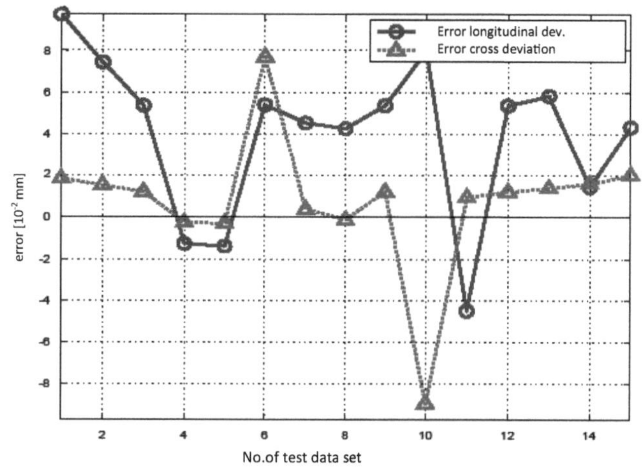

Fig. 2 Polar point determination: Propagation of accuracy (evaluation data)

to propagate the two quality parameters standard deviation (in the following abbreviated with "deviation") in longitudinal and cross direction (l and c) referring to the direction to the point. These two output parameters describing the accuracy are besides others related to the following input parameters:

- Measured distance (s)
- Standard deviation of distance (σ_s)
- Distance-related correction of measured distance (ppm)

The additional handling of availability of input data was tested in the second step. Unavailable data were flagged with the value "0". For mathematical reasons it was necessary to normalize all values (since the input parameters may vary heavily in magnitude, e.g. distances up to several km measured with a standard deviation of a few mm). For the generation of training and evaluation data, it was necessary to create a fault tree for different possible lacks in completeness (cf. Laufer 2011) for further details.

The additional handling of quality parameters describing availability demands a more complex network. For the single handling of accuracy parameters, a network with one hidden layer containing nine neurons and two output neurons (abbrev.: [9-2]) was sufficient. For additional propagation of availability a [15-15-2]-network with two hidden layers was necessary.

As shown in Fig. 2, the accuracy parameters l and c can be propagated with an uncertainty below 0.1 mm. To avoid the network memorizing the learning data sets, new example data sets not used for training or validation of the network were used to evaluate the trained network. The input data were chosen randomly out of the trained intervals.

Handling availability in addition leads to similar results: l and c were propagated with an accuracy of better than 0.1 mm, the availability with an accuracy of better than 0.03. Hence it is possible to distinguish between small but real values near zero and not unavailable data, flagged with "0".

In a third step the propagation of parameters describing completeness was determined. Therefore it was first of all necessary to arrange the input and output data in data sets, which can contain one or more single parameters. Otherwise it would not be possible to separate between a lack of completeness and a lack of availability. If a complete data set is missing, it is per definition a lack of availability. If one parameter within a data set is missing, it is a lack of completeness. To simplify matters, in this example all input parameters and all output parameters where defined as data sets.

For propagation of completeness, the input and output parameter vectors were expanded by one flag for each data set to provide information about completeness. The flag is a binary character, where the value "1" stands for "data set is complete". Whereas "0" means "data set is incomplete".

To keep the number of training data sets small and to get the ANN well trained at the same time, it is important to simulate a much higher lack in completeness during training than will occur in reality. Otherwise, in a small amount of training data, there are only a few lacks in completeness occurring. Hence the network would not be able to learn how to handle incomplete data sets.

Again, the accuracy parameters can be propagated with an accuracy of better than ±0.1 mm (Fig. 3). The completeness of the output data sets can be determined better than ±0.04. This means the completeness can be easily propagated by adding new binary parameters to the input and output vectors.

Propagation of consistency and correctness of the data can be handled as well by introducing additional binary characters. The distinction between binary values describing completeness, consistency and correctness happens automatically by the clearly defined and constant position of the parameters within the vectors.

Handling timeliness of data, on the contrary, demands time-sensitive ANN, so-called dynamic networks that can handle time-invariant data. Since the focus of our research

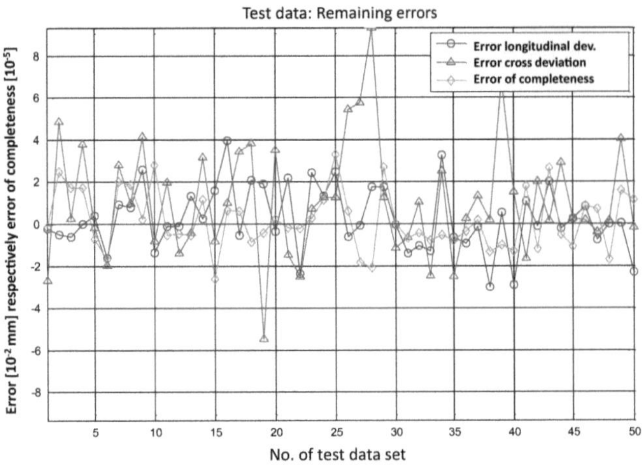

Fig. 3 Polar point determination: Propagation of accuracy and completeness (evaluation data)

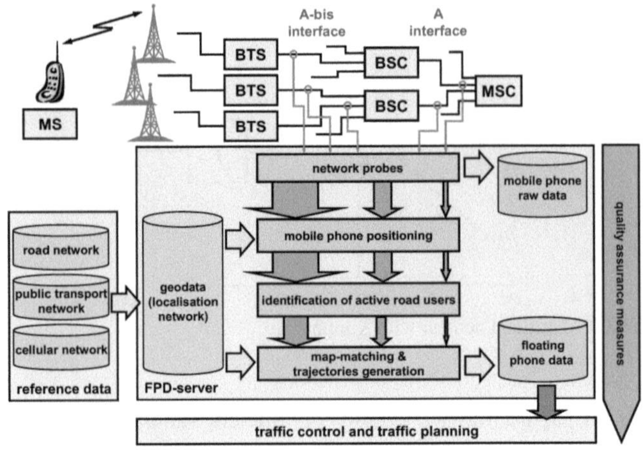

Fig. 4 Process of FPD generation (Wiltschko et al. 2007)

was on static networks, the quality character timeliness has not been investigated so far.

4 Propagation of Data Quality for Mobile Phone Positioning

The project Do-iT (data optimization for integrated telematics) with a duration of almost 4 years was funded by the Federal Ministry of Economics and Technology. The focus of the project was the generation of mobile phone trajectories, so called floating phone data (FPD), of individual motorized road users within the main road network for traffic applications (e. g. congestion detection or traffic planning). Generation of FPD is a complex process consisting of several sub-processes as outlined in Fig. 4.

At the Institute of Engineering Geodesy Stuttgart (IIGS) two different approaches regarding two different interfaces for data recording where developed. The A-interface includes data of all registered mobile phone users, not depending on the status of the phone (active or only switched on). The data on the second interface, the so-called Abis-interface ("bis" is the French word for "bonus"), contain only data of active mobile phones (phones during a call or data connection). These data always have a high spatial and temporal resolution, whereas the data from A-interface only have a medium resolution if the phone is active, otherwise a low resolution.

For practical reasons, the A-data can serve as practical example to evaluate the use of ANN for propagation of data quality. Therefore only the approach for A-bis data will be briefly described in the following.

In a first step the recorded data are sorted and prioritized in near-realtime with respect to the prospected calculability and quality of generated FPD. Static mobile phones (e.g. in buildings beside the roads), phones in trains or tram lines are to be eliminated. In a second step, single positions are generated, using beside other information, the measured signal strengths from neighboring radio cells and the theoretical signal strength propagation maps for each radio cell. The sequence of positions of each user can be converted into a sequence of road elements on the digital road network (thus a trajectory) using map-aiding methods (Czommer 2000). See Ramm and Schwieger (2008) or Do-iT (2009) to read more about the methods.

For evaluation of the generated trajectories, an empirical quality parameter describing the correctness was introduced. The parameter *cross deviation* describes the mean perpendicular deviation of a sequence of positions from the most likely route of the user on the digital road network. In Fig. 5 the situation is shown graphically. The figure is a screenshot from a project evaluation report (Do-iT 2009), therefore a reference trajectory measured by GPS is visible as well.

In the following it will be shown that the quality parameter cross deviation can be propagated sufficiently with ANN near realtime. The propagation of quality can be triggered immediately after the generation of the FPD trajectory. The following influence parameters were identified and investigated in Laufer (2011):

- Length of trajectory in m,
- Duration of trajectory in s,
- Mean antenna density in antennas per km^2,
- Difference in antenna density in antennas per km^2,
- Mean density of road network in road elements per km^2,
- Difference in density of road network in road elements per km^2,
- Number of road elements the trajectory consists of.

Besides these seven input parameters there are more parameters which have a weaker and more or less diffuse

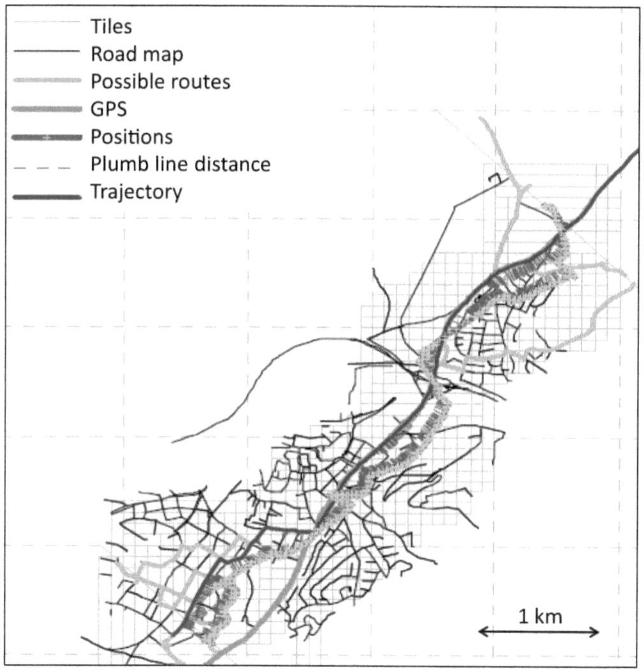

Fig. 5 Evaluation of a generated FPD trajectory with GPS

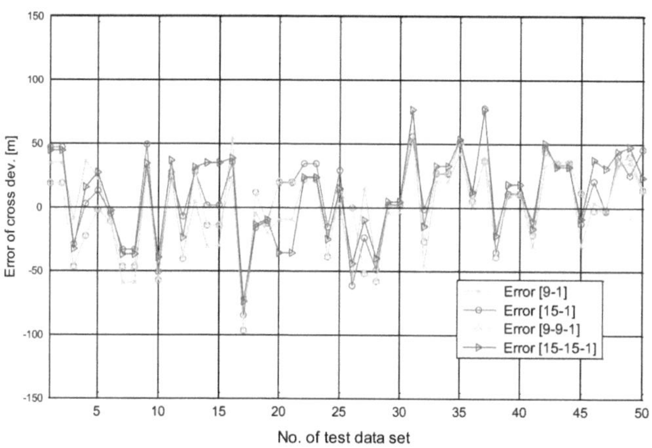

Fig. 6 Propagation of cross deviation (Laufer 2011)

influence on the quality of the FPD. These parameters were not considered within this research.

The data base for training the ANN consists of 8 recorded days. The mean cross deviation reached 180 m with a mean length of trajectories of 6.5 km and a duration of 250 s. All data was recorded in the test area of Karlsruhe and Ettlingen where denser urban road networks dominate. Only some smaller areas on the edges of the area have a more rural character.

To find the best network configuration, different net dimensions were tested as can be seen in Fig. 6 where the results for one single day (March 26th 2009) are displayed. The error of the propagated cross deviation within 50 randomly chosen test data sets does not exceed 100 m for all networks. The network [15-15-1] does perform best.

Table 1 Results for testing different net configurations and different amount of data

Net-configuration	Mean square error	Max. dev. [m]	Standard deviation [m]
9-9-1/15-15-1 best of each day	0.024	93.9	41.0
9-9-1 (all 8 days together)	0.025	103.8	41.0
15-15-1 (all 8 days together)	0.025	108.8	38.2
15-15-1 (all 8 days) 6 extrema eliminated	0.024	80.1	27.9

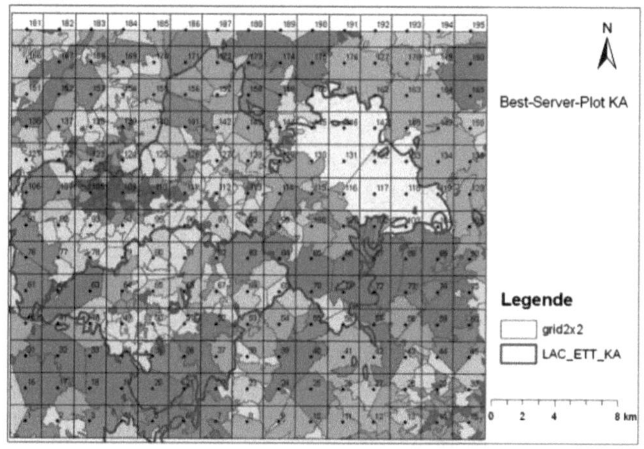

Fig. 7 Best-Server-Plot for Karlsruhe and Ettlingen

As can be seen from Table 1, one single day data is enough to get the ANN trained. The standard deviation of all 50 test data sets reaches 41 m in both cases (single days versus all 8 days together). If a network with 15 neurons in each hidden layer is used, the ANN trained with all 8 days of data together performs a bit better than with a single day and a standard deviation of 38.2 m was reached. Elimination of the six most extremal cross deviations leads to a standard deviation of only 27.9 m.

Since all results are reproducible, ANN seem to be a practical method to propagate cross deviation. Nevertheless, extremal values in cross deviation are more difficult to handle for ANN than more regular values. In a whole, ANN are capable to propagate cross deviation with a standard deviation of 30–40 m. To get a better idea of how good or bad this result is, a closer look into the process of generating FPD is necessary.

Figure 7 shows the test area of Karlsruhe with the radio cells where the Abis-data were available (inside the red surrounded area). The radio cells are very inhomogeneous in size, depending on the density of the road network and the expected amount of radio cells to handle. Since the single positions of road users are extracted from mobile phone data, particularly from the measured signal strength, the standard deviation of this method reaches 560 m for a single position (Do-iT 2009). This means that the identification of the correct route within a denser road network, as in most parts of the test area, is difficult. Furthermore the connection between the correctness of a route and the cross deviation is only based on a small number of test trajectories where GPS as reference was onboard. Therefore, the cross deviation is not a exact quantitative reference, it should be considered as an indicator to evaluate the trajectories.

Regarding these facts, the results reached with ANN are quite satisfactory. It is possible to map the dependencies between input data quality and the cross deviation as output quality parameter.

Conclusion

As could be shown for two examples, polar point determination and mobile phone positioning, ANN are capable for propagation of data quality within processes. However, the results depend on the modeling of the ANN as well as the quality of the training data. Finding the best performing network can only be done empirically by checking, for instance, the mean square error. The use of dynamic ANN provides the opportunity to handle timelines and should be part of further research. To get more experiences in using ANN and to prove their general usability for propagation of data quality more practical examples are necessary.

References

Czommer R (2000) Leistungsfähigkeit fahrzeugautonomer Ortungsverfahren auf der Basis von Map-Matching-Techniken. Bayerische Akademie der Wissenschaften, Verlag C. H. Beck, DGK, Reihe C, Nr. 535, PhD thesis

Do-iT (2009) Qualitätsbewertung Map-Matching gestützter Positionsbestimmungsverfahren auf Abis- und A-Ebene sowie Evaluierung der Positionsbestimmung durch höherwertig Ortungsverfahren. Internal project report, unpublished

Hagan MT, Demuth HB, Beale M (1996) Neural network design. PWS Publishing Company Division of Thomson Learning, USA

Laufer R (2011) Prozedurale Qualitätsmodellierung und-management für Daten – ingenieurtechnische und verkehrstechnische Anwendungen. University of Stuttgart, Institute of Engineering Geodesy, Bayerische Akademie der Wissenschaften, Verlag C. H. Beck, DGK, Reihe C, Nr. 662, PhD thesis

Ramm K, Schwieger V (2008) Mobile positioning for traffic state acquisition. J Location Based Services, Taylor & Francis, London

Wiltschko T (2004) Sichere Information durch infrastrukturgestützte Fahrerassistenzsysteme zur Steigerung der Verkehrssicherheit an Straßenknotenpunkten, VDI, Reihe 12, Nr. 570, VDI-Verlag, Düsseldorf, 2004, PhD thesis

Wiltschko T, Kaufmann T (2005) Modellierung und Bewertung von Prozessen in der Geodatenverarbeitung. In: Proceedings on AGIT Symposium und Fachmesse angewandte Geoinformatik, Universität Salzbug

Wiltschko T, Möhlenbrink W, Schwieger V (2007) Acquisition of traffic state information by mobile phone positioning. In: Proceedings on 6th European congress on ITS, Aalborg, Denmark

Zell A (1997) Simulation Neuronaler Netze, 2nd edn. München, Oldenbourg

Magic Square of Real Spectral and Time Series Analysis with an Application to Moving Average Processes

I. Krasbutter, B. Kargoll, and W.-D. Schuh

Abstract

This paper is concerned with the spectral analysis of stochastic processes that are real-valued, one-dimensional, discrete-time, covariance-stationary, and which have a representation as a moving average (MA) process. In particular, we will review the meaning and interrelations of four fundamental quantities in the time and frequency domain, (1) the stochastic process itself (which includes filtered stochastic processes), (2) its autocovariance function, (3) the spectral representation of the stochastic process, and (4) the corresponding spectral distribution function, or if it exists, the spectral density function. These quantities will be viewed as forming the corners of a square (the "magic square of spectral and time series analysis") with various connecting lines, which represent certain mathematical operations between them. To demonstrate the evaluation of these operations, we will discuss the example of a q-th order MA process.

Keywords

Moving average process • Spectral analysis • Stochastic process • Time series analysis

1 Introduction

The spectral analysis of deterministic functions and the formulation of stochastic processes belong to the well-established statistical tools in various fields within geodesy (see, e.g. Koch and Schmidt 1994; Moritz 1989; Welsch et al. 2000). We found, however, that in particular the nature of the spectral representation of stochastic processes in terms of the stochastic Fourier integral and its relationships with the autocovariance and spectral distribution (or density) function is far less well known than the details of the time-domain and Fourier analyses of deterministic functions. Our motivation for this paper is therefore to take a step towards closing this gap in understanding. We will in particular provide the reader with the key definitions of the involved stochastic processes as well as of their crucial properties (Sect. 2). Then we will state and explain the computational formulae for the spectral analysis of general real-valued covariance-stationary stochastic processes (Sect. 3). This is in contrast to the usual representation of these formulae in the mathematical statistics oriented literature (e.g. Brockwell and Davis 1991; Priestley 2004), where one generally finds only the results for complex-valued stochastic processes, which often complicates their application in practical situations. To aid the understanding of the mathematical relationships of the involved fundamental statistical quantities (stochastic process, autocovariance function, spectral representation of the process, spectral distribution or density function) we will use a corresponding graphical representation in form of a "magic square" (also in Sect. 3). We will conclude this paper with an outlook to extensions to the presented example [moving average (MA) process].

I. Krasbutter • B. Kargoll (✉) • W.-D. Schuh
Institute of Geodesy and Geoinformation, University of Bonn, Nussallee 17, 53115 Bonn, Germany
e-mail: bkargoll@geod.uni-bonn.de

H. Kutterer et al. (eds.), *The 1st International Workshop on the Quality of Geodetic Observation and Monitoring Systems (QuGOMS'11)*, International Association of Geodesy Symposia 140,
DOI 10.1007/978-3-319-10828-5_2,

2 Basic Elements of Stochastic Processes

In this chapter, we will provide a summary of basic definitions (D) and properties of the stochastic processes considered in Sect. 3.

(D1): We say that $\mathcal{X}_T = (\Omega, \mathcal{A}, P, \{\mathcal{X}_t, t \in T\})$ is a **(general) stochastic process** if and only if (iff)

- $(\Omega, \mathcal{A}, P)$ is any probability space (where Ω denotes the sample space, $\mathcal{A}$ a σ-algebra of events, and P a probability measure),
- T is any non-empty set, and
- $\mathcal{X}_t$ is a random variable defined on $(\Omega, \mathcal{A})$ for any $t \in T$.

In this paper, we will restrict our attention to real-valued and one-dimensional stochastic processes as given in the following definition.

(D2): We say that $\mathcal{X}_T$ is a **real-valued (one-dimensional) stochastic process** iff $\mathcal{X}_t : (\Omega, \mathcal{A}) \to (\mathbb{R}, \mathcal{B})$ for any $t \in T$, where $\mathcal{B}$ is the Borel sigma algebra generated by the set of all real-valued, one-dimensional, left-open and right-closed intervals.

In Sect. 3, we will use stochastic processes that have a discrete parameter set T in the time domain as well as processes with a continuous parameter set in the frequency domain. This distinction is made by the following definition.

(D3): We say that $\mathcal{X}_T$ is a

- **discrete-parameter stochastic process** (or stochastic process with discrete parameter) iff $T \subset \mathbb{Z}$. Furthermore, we call $\mathcal{X}_T$ a **discrete-time stochastic process** or **discrete-time time series** iff the elements of T refer to points in time.
- **continuous-parameter stochastic process** (or stochastic process with continuous parameter) iff $T \subset \mathbb{R}$. In addition, we call $\mathcal{X}_T$ a **continuous-frequency stochastic process** iff the elements of T refer to (angular) frequencies, in which case we will also write $T = W$.

As far as discrete-parameter stochastic processes are concerned, we will focus our attention on covariance-stationary processes (in the time domain). The precise meaning of this concept is provided as follows.

(D4): We say that $\mathcal{X}_T$ is **covariance stationary** iff

- $E\{\mathcal{X}_t\} = \mu < \infty$ (i.e. constant/finite) for any $t \in T$,
- $E\{(\mathcal{X}_t - \mu)^2\} = \sigma^2_{\mathcal{X}} < \infty$ (i.e. constant/finite) for any $t \in T$, and
- $\gamma_{\mathcal{X}}(t_1, t_2) = \gamma_{\mathcal{X}}(t_1 + \Delta t, t_2 + \Delta t)$ for any $t_1, t_2 \in T$ and any Δt with $t_1 + \Delta t, t_2 + \Delta t \in T$,

where $E\{.\}$ denotes the expectation operator and $\gamma_{\mathcal{X}}$ the autocovariance function, defined by $\gamma_{\mathcal{X}}(t_1, t_2) = E\{(\mathcal{X}_{t_1} - \mu)((\mathcal{X}_{t_2} - \mu)\}$. For a covariance-stationary stochastic process, we have that $\gamma_{\mathcal{X}}(t_1, t_2) = \gamma_{\mathcal{X}}(t_1 - t_2, 0)$ for any $t_1, t_2 \in T$ (and $0 \in T$) such that also $t_1 - t_2 \in T$; that is, we can always rewrite $\gamma_{\mathcal{X}}$ by using only a single variable argument, the second one taking the constant value 0. In light of this, we redefine the autocovariance function for covariance-stationary processes as

$$\gamma_{\mathcal{X}}(k) := \gamma_{\mathcal{X}}(k, 0) = \gamma_{\mathcal{X}}(t + k, t)$$

for any $k, t \in T$ with $t + k \in T$; the parameter k is called lag (cf. Brockwell and Davis 1991, pp. 11–12).

The fundamental instance of a covariance-stationary process and primary building block for certain other stochastic processes is white noise, defined as follows.

(D5): We say that $\mathcal{E}_T := \mathcal{X}_T$ is **(discrete-parameter) white noise** with mean 0 and variance $\sigma^2_{\mathcal{X}}$ iff

- $T \subset \mathbb{Z}$,
- $E\{\mathcal{X}_t\} = 0$ for any $t \in T$, and
- $\gamma_{\mathcal{X}}(k) = \begin{cases} \sigma^2_{\mathcal{X}} & \text{if } k = 0, \\ 0 & \text{if } k \neq 0 \end{cases}$.

Now let us consider a *non-recursive filter* C, defined by the filter equation $y_t = \sum_{k=-\infty}^{\infty} c_k u_{t-k}$ for any $t \in \mathbb{Z}$, or in lag operator notation $y_t = C(L)u_t$ with $L^k u_t := u_{t-k}$ and $C(L) = \sum_{k=-\infty}^{\infty} c_k L^k$, where $(u_t \mid t \in \mathbb{Z})$ is any filter input sequence and $(y_t \mid t \in \mathbb{Z})$ any filter output sequence (in either case of real numbers or random variables), and $(c_k \mid k \in \mathbb{Z})$ is any sequence of real-valued filter coefficients. If we view the random variables of a white noise process $\mathcal{E}_T$ as filter input to a

- *causal* (i.e. $c_k = 0$ for any $k < 0$),
- either *finite* or *infinite* (i.e. a finite or an infinite number of filter coefficients is non-zero),
- *absolutely summable* (i.e. $\sum_{k=-\infty}^{\infty} |c_k| < \infty$), and
- *invertible* (i.e. there exists an inverse filter $\overline{C}$ with filter coefficients $(\bar{c}_k \mid k \in \mathbb{N}^0)$ such that $[\overline{C}(L)C(L)]u_t = u_t$ where $\overline{C}(L) = \sum_{k=0}^{\infty} \bar{c}_k L^k$)

version of such a non-recursive filter, then we obtain the moving average process as filter output, as explained in the following definition.

(D6): If $\mathcal{E}_T$ with $T \subset \mathbb{Z}$ is (discrete) white noise with mean 0 and variance $\sigma^2_{\mathcal{E}}$, then we say that $\mathcal{L}_T : (\Omega, \mathcal{A}, P, \{\mathcal{L}_t, t \in T\})$ is a **(discrete-parameter) moving average process** of order q (or MA(q) process) (with $q \in \mathbb{N}$) iff the random variables $\mathcal{L}_t$ satisfy, for any $t \in T$, the equation

$$\mathcal{L}_t = \mathcal{E}_t + \beta_1 \mathcal{E}_{t-1} + \ldots + \beta_q \mathcal{E}_{t-q} = \beta(L)\mathcal{E}_t$$

with $\beta(L) = 1 + \beta_1 L + \ldots + \beta_q L^q$. In the limiting case of $q = \infty$, we call $\mathcal{L}_T$ an MA(∞) process.

Treating $\beta(L)$ as a complex polynomial, then, if $\beta(z) \neq 0$ for any $z \in \mathbb{C}$ with $|z| \leq 1$, then the filter β and, hence the MA(q) process, is invertible (cf. Brockwell and Davis 1991, pp. 86–87). Furthermore, whereas any MA(q) process with $q < \infty$ is covariance-stationary (cf. Priestley 2004, p. 137), the *MA*(∞) process is covariance-stationary iff the sequence $(\beta_k \mid k \in \mathbb{N}^0)$ of filter coefficients is absolutely summable (i.e. iff $\sum_{k=0}^{\infty} |\beta_k| < \infty$) (cf. Brockwell and Davis 1991, pp. 89–91).

On the other hand, as far as continuous-parameter stochastic processes are concerned, we will only deal with stochastic processes that are in a certain sense stochastically continuous and that have orthogonal increments, which is explained in the following.

(D7): We say that any continuous-parameter stochastic process $\mathcal{X}_T$ is **mean-square right-continuous** (or **mean-square continuous from the right**) at $t_0 \in T$ iff $\lim_{t \to t_0^+} E\{[\mathcal{X}_t - \mathcal{X}_{t_0}]^2\} = 0$ holds (cf. Gilgen 2006, p. 453).

(D8): We say that any continuous-parameter stochastic process $\mathcal{X}_T$ is a **stochastic process with orthogonal increments** $\Delta\mathcal{X}_{s,t} = \mathcal{X}_t - \mathcal{X}_s$ with $s, t \in T$ and $s < t$ iff $E\{\Delta\mathcal{X}_{t_1,t_2}\Delta\mathcal{X}_{t_3,t_4}\} = 0$ for any $t_1, t_2, t_3, t_4 \in T$ with $t_1 < t_2 < t_3 < t_4$ or $t_3 < t_4 < t_1 < t_2$, i.e. iff any two "non-overlapping" increments are uncorrelated.

3 Magic Square for Covariance-Stationary, Discrete-Time Processes

In the current section, we will review the mathematical operations that connect the following four fundamental quantities of a spectral analysis in the time and the frequency domain:

1. stochastic process,
2. autocovariance function,
3. the spectral representation of the stochastic process,
4. the spectral distribution function (or, if it exists, its derivative, the spectral density function).

We may view these quantities as forming the corners of a square with various connecting lines, which represent certain mathematical operations between them (see Fig. 1). We will refer to this as the Magic Square. To demonstrate the evaluation of these operations, we will consider the example of an MA(q) process. The reader should note that, with this understanding, it would be straightforward to apply the Magic Square to more complicated processes such as ARMA(p,q) processes, which would, however, exceed the limit of this paper. We begin the discussion by constructing the time domain (the "left-hand side") of the Magic Square.

3.1 Time Domain (Left-Hand Side)

We consider any stochastic process $\mathcal{X}_T$ which is

- one-dimensional and real-valued (i.e. $\Omega = \mathbb{R}^1$),
- discrete in time (with $T = \mathbb{Z}$), and
- covariance-stationary with zero mean, variance $\sigma_{\mathcal{X}}^2$ and autocovariance function $\gamma_{\mathcal{X}}$.

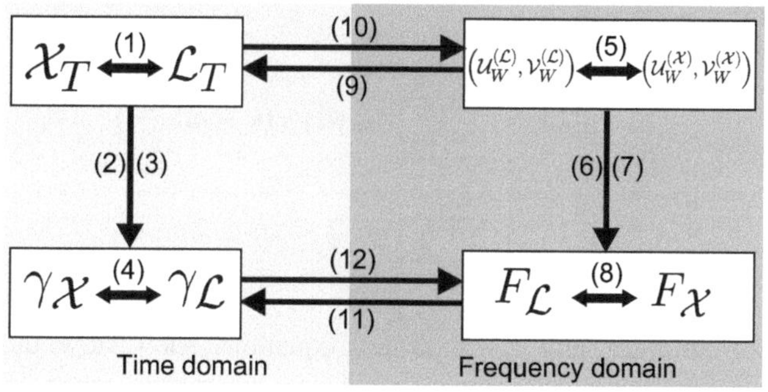

Fig. 1 Magic square for covariance-stationary discrete-time stochastic processes; *upper left* stochastic process, *lower left* autocovariance function, *upper right* spectral representation of the stochastic process, *lower right* spectral distribution function; the numbers in *brackets* indicate the mathematical operations as defined in Sects. 3.1–3.3

Furthermore, we consider any stochastic process $\mathcal{L}_T$ obtained by filtering the process $\mathcal{X}_T$, where we assume that the filter $\psi(L)$ is

- non-recursive,
- causal,
- either finite or infinite,
- absolutely summable, and
- invertible (with inverse filter $\overline{\psi}(L)$).

It follows that the process $\mathcal{L}_T$ is

- one-dimensional and real-valued (i.e. $\Omega = \mathbb{R}^1$),
- discrete in time (with $T = \mathbb{Z}$), and
- covariance-stationary (cf. Brockwell and Davis 1991, p. 84) with zero mean, variance $\sigma_{\mathcal{L}}^2$ and autocovariance function $\gamma_{\mathcal{L}}$.

The general mathematical operations within the time domain can be summarized as follows:

(1) $\mathcal{X}_T \Longleftrightarrow \mathcal{L}_T$:

$$\mathcal{L}_t = \psi(L)\mathcal{X}_t, \qquad \mathcal{X}_t = \overline{\psi}(L)\mathcal{L}_t,$$

hold for any $t \in \mathbb{Z}$. The first of these equations is an expression of the above assumption that $\mathcal{L}_T$ is a stochastic process obtained by non-recursive filtering of $\mathcal{X}_T$. The second of these equations reflects the presumed invertibility of the filter operation.

(2) $\mathcal{X}_T \Longrightarrow \gamma_{\mathcal{X}}, \mathcal{L}_T \Longrightarrow \gamma_{\mathcal{L}}$:

$$\gamma_{\mathcal{X}}(k) = E\{\mathcal{X}_{t+k}\mathcal{X}_t\}, \qquad \gamma_{\mathcal{L}}(k) = E\{\mathcal{L}_{t+k}\mathcal{L}_t\},$$

hold for any $t, k \in \mathbb{Z}$. These equations are simply an expression of the definition of the autocovariance function applied to the stochastic processes $\mathcal{X}_T$ and $\mathcal{L}_T$ with the properties stated above (cf. Priestley 2004, p. 107).

(3) $\mathcal{X}_T \Longrightarrow \gamma_{\mathcal{L}}, \mathcal{L}_T \Longrightarrow \gamma_{\mathcal{X}}$:

Substitution of (1) and shifted versions thereof into (2) yields the expressions for $\gamma_{\mathcal{L}}(k)$ and $\gamma_{\mathcal{X}}(k)$.

(4) $\gamma_{\mathcal{X}} \Longleftrightarrow \gamma_{\mathcal{L}}$:

$$\gamma_{\mathcal{L}}(k) = \sum_{m=0}^{\infty}\sum_{n=0}^{\infty} \psi_m \psi_n \gamma_{\mathcal{X}}(k - m + n),$$

$$\gamma_{\mathcal{X}}(k) = \sum_{m=0}^{\infty}\sum_{n=0}^{\infty} \overline{\psi}_m \overline{\psi}_n \gamma_{\mathcal{L}}(k - m + n)$$

hold for any $k \in \mathbb{Z}$. These equations show how the autocovariance function of a covariance-stationary stochastic process is propagated by an essentially absolutely summable filter to the autocovariance function of the filtered (covariance-stationary) process (see Proposition 3.1.2 in Brockwell and Davis 1991, p. 84).

Example: MA(q) Process

Let us consider a non-recursive, causal, finite, absolutely summable and invertible filter $\beta(L)$. If we apply such a filter to white noise $\mathcal{E}_T$ (which satisfies the conditions made for the input process $\mathcal{X}_T$), then we obtain, by definition, an invertible MA(q) process, which then satisfies the above stated properties of $\mathcal{L}_T$. Hence, we may apply the general mathematical operations stated in equations under (1)–(4) as follows.

$$\mathcal{L}_t = \beta(L)\mathcal{E}_t,$$

$$\mathcal{E}_t = \overline{\beta}(L)\mathcal{L}_t = \sum_{k=0}^{\infty} \overline{\beta}_k \mathcal{L}_{t-k}$$

The first of these equations defines the MA(q) process; the second equation yields white noise expressed in terms of the random variables of $\mathcal{L}_T$, filtered by means of a non-recursive, causal and infinite filter $\overline{\beta}$. As far as autocovariance functions are concerned, $\gamma_{\mathcal{X}} = \gamma_{\mathcal{E}}$ takes a very simple form (see the definition of white noise in Sect. 2); then, the first equation of (4) may be simplified to

$$\gamma_{\mathcal{L}}(k) = \begin{cases} \sigma_{\mathcal{E}}^2 \sum\limits_{n=0}^{q-|k|} \beta_n \beta_{n+|k|}, & \text{if}|k| \leq q \\ 0, & \text{if}|k| > q \end{cases}$$

(cf. Brockwell and Davis 1991, pp. 78–79).

3.2 Frequency Domain (Right-Hand Side)

The usual approach to a spectral representation of a stochastic process given in the time domain is to define it in the frequency domain in terms of a complex stochastic process associated with complex exponential base functions. This allows one, besides a shorter notation, to also cover the case where the stochastic process in the time domain is complex-valued. Whenever the process in the time domain is real-valued, as it is the case with the applications we have in mind, this complication is, however, unnecessary. We therefore restate the main results, given in complex notation in the literature, in terms of pairs of real stochastic processes associated with sine and cosine base functions. We find these to be closer to our natural understanding of the concept of "frequency" than complex exponentials. Thus, we will consider as the spectral representations of the processes $\mathcal{X}_T$ and $\mathcal{L}_T$ (defined in Sect. 2), in each case a tuple of two stochastic processes

$$\left(\mathcal{U}_W^{(\mathcal{X})}, \mathcal{V}_W^{(\mathcal{X})}\right) = (\Omega, \mathcal{A}, P, \{\left(\mathcal{U}_\omega^{(\mathcal{X})}, \mathcal{V}_\omega^{(\mathcal{X})}\right), \omega \in W\}),$$

$$\left(\mathcal{U}_W^{(\mathcal{L})}, \mathcal{V}_W^{(\mathcal{L})}\right) = (\Omega, \mathcal{A}, P, \{\left(\mathcal{U}_\omega^{(\mathcal{L})}, \mathcal{V}_\omega^{(\mathcal{L})}\right), \omega \in W\}),$$

which we assume to be

- one-dimensional and real-valued (i.e. $\Omega = \mathbb{R}^1$),
- frequency-continuous (with $W = [-\pi, \pi]$),
- mean-square right-continuous, and
- processes with orthogonal increments.

The relationships of these processes in the frequency domain with $\mathcal{X}_T$ and $\mathcal{L}_T$ in the time domain will become evident in Sect. 3.3. The general mathematical operations within the frequency domain are:

(5) $(\mathcal{U}_W^{(\mathcal{X})}, \mathcal{V}_W^{(\mathcal{X})}) \Longleftrightarrow (\mathcal{U}_W^{(\mathcal{L})}, \mathcal{V}_W^{(\mathcal{L})})$:

For any $\omega \in W$,

$$\mathcal{U}_W^{(\mathcal{L})}(\omega) = \int_{-\pi}^{\omega} \mathrm{Re}(H(\lambda)) d\mathcal{U}_W^{(\mathcal{X})}(\lambda) - \mathrm{Im}(H(\lambda)) d\mathcal{V}_W^{(\mathcal{X})}(\lambda),$$

$$\mathcal{V}_W^{(\mathcal{L})}(\omega) = \int_{-\pi}^{\omega} \mathrm{Im}(H(\lambda)) d\mathcal{U}_W^{(\mathcal{X})}(\lambda) + \mathrm{Re}(H(\lambda)) d\mathcal{V}_W^{(\mathcal{X})}(\lambda),$$

$$\mathcal{U}_W^{(\mathcal{X})}(\omega) = \int_{-\pi}^{\omega} \mathrm{Re}(\overline{H}(\lambda)) d\mathcal{U}_W^{(\mathcal{L})}(\omega) - \mathrm{Im}(\overline{H}(\lambda)) d\mathcal{V}_W^{(\mathcal{L})}(\lambda),$$

$$\mathcal{V}_W^{(\mathcal{X})}(\omega) = \int_{-\pi}^{\omega} \mathrm{Im}(\overline{H}(\lambda)) d\mathcal{U}_W^{(\mathcal{L})}(\omega) + \mathrm{Re}(\overline{H}(\lambda)) d\mathcal{V}_W^{(\mathcal{L})}(\lambda)$$

hold (Theorem 4.10.1 in Brockwell and Davis 1991, pp. 154–155), where $H(\omega) = \sum_{k=0}^{\infty} \psi_k e^{-ik\omega}$ with $\omega \in [-\pi, \pi]$ is the transfer function of the filter ψ, and where $\overline{H}(\omega)$ is the transfer function of the inverse filter $\overline{\psi}$ (this implies that the transfer function is generally one-dimensional, frequency-continuous and complex-valued). The relations are described by stochastic Riemann-Stieltjes-Integral, which will be explained more precisely in Sect. 3.3.

(6) $(\mathcal{U}_W^{(\mathcal{X})}, \mathcal{V}_W^{(\mathcal{X})}) \Longrightarrow F_{\mathcal{X}},\ (\mathcal{U}_W^{(\mathcal{L})}, \mathcal{V}_W^{(\mathcal{L})}) \Longrightarrow F_{\mathcal{L}}$

$$F_{\mathcal{X}}(\omega) = E\{(\mathcal{U}_W^{(\mathcal{X})}(\omega))^2\} + E\{(\mathcal{V}_W^{(\mathcal{X})}(\omega))^2\},$$

$$F_{\mathcal{L}}(\omega) = E\{(\mathcal{U}_W^{(\mathcal{L})}(\omega))^2\} + E\{(\mathcal{V}_W^{(\mathcal{L})}(\omega))^2\},$$

hold for $\omega \in W$ (Priestley 2004, pp. 250–251). These equations express the relations between the stochastic processes $\mathcal{U}_W^{(\cdot)}$, $\mathcal{V}_W^{(\cdot)}$ and the so-called spectral distribution function $F_{(\cdot)}(\omega)$. Hence, this function is real-valued and, due to $F_{(\cdot)}(-\pi) = 0$, $F_{(\cdot)}(\pi) = \gamma_{(\cdot)}(0)$, has similar properties as the probability distribution function. If the derivative $f_{(\cdot)}(\omega) = dF_{(\cdot)}(\omega)/d\omega$ exists, $f_{(\cdot)}(\omega)$ is called spectral density function and is also known as the power spectrum.

(7) $(\mathcal{U}_W^{(\mathcal{X})}, \mathcal{V}_W^{(\mathcal{X})}) \Longrightarrow F_{\mathcal{L}},\ (\mathcal{U}_W^{(\mathcal{L})}, \mathcal{V}_W^{(\mathcal{L})}) \Longrightarrow F_{\mathcal{X}}$:
Substitution of (5) into (6) yields expressions for $F_{\mathcal{L}}(\omega)$ and $F_{\mathcal{X}}(\omega)$ in terms of $\mathcal{U}_W^{(\mathcal{X})}, \mathcal{V}_W^{(\mathcal{X})}$ and $\mathcal{U}_W^{(\mathcal{L})}, \mathcal{V}_W^{(\mathcal{L})}$.

(8) $F_{\mathcal{X}} \Longleftrightarrow F_{\mathcal{L}}$:

$$F_{\mathcal{L}}(\omega) = \int_{-\pi}^{\omega} |H(\lambda)|^2 dF_{\mathcal{X}}(\lambda),$$

$$F_{\mathcal{X}}(\omega) = \int_{-\pi}^{\omega} |\overline{H}(\lambda)|^2 dF_{\mathcal{L}}(\lambda)$$

(cf. Theorem 4.4.1 in Brockwell and Davis 1991, p. 122). These equations depend on the transfer function of the corresponding filter and reflect direct relationships between the spectral density functions of $\mathcal{X}_T$ and $\mathcal{L}_T$. These integrals are the usual (deterministic) Riemann-Stieltjes integral (see also Sect. 3.3).

Example: MA(q) Process

As explained in Sect. 3.1, the input process of an MA(q) process is white noise. The spectral distribution function for white noise, given by $F_{\mathcal{E}}(\omega) = \frac{\sigma_{\mathcal{E}}^2}{2\pi}(\omega + \pi), \omega \in [-\pi, \pi]$, can be calculated from (12) in Sect. 3.3. The derivative $f_{\mathcal{E}}$ of this function, the spectral density function for white noise, clearly exists and is $f_{\mathcal{E}}(\omega) = \frac{\sigma_{\mathcal{E}}^2}{2\pi}$. To evaluate (5)–(8), we may substitute the Euler equations into $H(\omega) = \sum_{k=0}^{q} \beta_k e^{-ik\omega}$, we may rewrite this as

$$H(\omega) = \sum_{k=0}^{q} \beta_k \cos(k\omega) - i \sum_{k=1}^{q} \beta_k \sin(k\omega).$$

Hence, (8) can be rewritten for an MA(q) process as

$$F_{\mathcal{L}}(\omega) = \frac{\sigma_{\mathcal{E}}^2}{2\pi} \int_{-\pi}^{\omega} \left(\sum_{k=0}^{q} \beta_k \cos(k\lambda)\right)^2 + \left(\sum_{k=1}^{q} \beta_k \sin(k\lambda)\right)^2 d\lambda.$$

3.3 Transitions Between the Time and Frequency Domain

In pursuing a spectral analysis of time-series, one establishes a link between discrete-time covariance-stationary stochastic processes and continuous-frequency mean-square right-continuous stochastic processes with orthogonal increments in form of a stochastic integral, which is very similar to the connection of continuous deterministic functions and the Fourier transform via the Fourier integral. This link can be explained in four steps:

(a) First we have to familiarize ourselves with the usual (i.e. deterministic) form of the *Riemann-Stieltjes integral*. The key idea here is that one seeks to integrate some function f (the integrand) with respect to some other function g (the integrator) over some domain of integration; this is achieved by defining a Riemann-Stieltjes sum with respect to some partition of the domain of integration and then to determine its "limit" (i.e. the integral value) as the partition becomes infinitely fine (cf. Bartle 1976, Chap. 29).

(b) The next step is to replace the deterministic integrator by some continuous-parameter stochastic process with parameter set T. Then, one defines a stochastic Riemann-Stieltjes sum with respect to some partition of the interval T and subsequently determines its "limit in mean square" as the partition becomes infinitely fine; thus, the integral value becomes a random variable (cf. Priestley 2004, pp. 154–155).

(c) Then, one replaces the general integrator process by a continuous-frequency mean-square right-continuous stochastic process with orthogonal increments (which may be viewed as the variables of a "stochastic Fourier transform") and the general integrand by some complex exponential or sine/cosine with discrete-time parameter t. Then, the time-variable random integral variables constitute a discrete-time stochastic process (cf. Brockwell and Davis 1991, Sects. 4.6–4.8).

(d) Finally, we have to distinguish two cases: Either some discrete-time covariance-stationary stochastic process $\mathcal{X}_T$ is given and one has to find a corresponding continuous-frequency mean-square right-continuous process with orthogonal increments as its spectral representation, or one defines a process in the frequency domain and seeks its time-domain representation (cf. Brockwell and Davis 1991, Sect. 4.9).

In the following, we will treat the two cases described in (d) by formulating the mathematical operations from the frequency domain into time domain and vice versa. In addition, the mathematical relationships between the autocovariance and spectral distribution functions will be explained. We will, however, not mention certain obvious transition relationships that can be obtained via simple substitution.

(9) $\mathcal{X}_T \Longleftarrow (\mathcal{U}_W^{(\mathcal{X})}, \mathcal{V}_W^{(\mathcal{X})})$, $\mathcal{L}_T \Longleftarrow (\mathcal{U}_W^{(\mathcal{L})}, \mathcal{V}_W^{(\mathcal{L})})$:

$$\begin{Bmatrix} \mathcal{X}_t \\ \mathcal{L}_t \end{Bmatrix} = \int_{-\pi}^{\pi} \cos(\omega t) \begin{Bmatrix} d\mathcal{U}_W^{(\mathcal{X})}(\omega) \\ d\mathcal{U}_W^{(\mathcal{L})}(\omega) \end{Bmatrix} + \sin(\omega t) \begin{Bmatrix} d\mathcal{V}_W^{(\mathcal{X})}(\omega) \\ d\mathcal{V}_W^{(\mathcal{L})}(\omega) \end{Bmatrix}$$

hold for any $t \in \mathbb{Z}$ and $\omega \in W$. These equations may be viewed as the stochastic counterparts to the Fourier integral (written in terms of sine/cosine base functions) of some deterministic function; the reader will find these equations in terms of complex exponentials, for instance, in Brockwell and Davis (1991, Theorem 4.8.2, pp. 145–147) or Priestley (2004, pp. 246–252).

(10) $\mathcal{X}_T \Longrightarrow (\mathcal{U}_W^{(\mathcal{X})}, \mathcal{V}_W^{(\mathcal{X})})$, $\mathcal{L}_T \Longrightarrow (\mathcal{U}_W^{(\mathcal{L})}, \mathcal{V}_W^{(\mathcal{L})})$:

$$\begin{Bmatrix} \mathcal{U}_W^{(\mathcal{X})}(\omega) - \mathcal{U}_W^{(\mathcal{X})}(\nu) \\ \mathcal{V}_W^{(\mathcal{X})}(\omega) - \mathcal{V}_W^{(\mathcal{X})}(\nu) \end{Bmatrix} \overset{m.s.}{\longleftarrow} \frac{1}{2\pi} \times \sum_{t=-\infty}^{\infty} \mathcal{X}_t \int_{\nu}^{\omega} \begin{Bmatrix} \cos(t\lambda)d\lambda \\ \sin(t\lambda)d\lambda \end{Bmatrix},$$

$$\begin{Bmatrix} \mathcal{U}_W^{(\mathcal{L})}(\omega) - \mathcal{U}_W^{(\mathcal{L})}(\nu) \\ \mathcal{V}_W^{(\mathcal{L})}(\omega) - \mathcal{V}_W^{(\mathcal{L})}(\nu) \end{Bmatrix} \overset{m.s.}{\longleftarrow} \frac{1}{2\pi} \times \sum_{t=-\infty}^{\infty} \mathcal{L}_t \int_{\nu}^{\omega} \begin{Bmatrix} \cos(t\lambda)d\lambda \\ \sin(t\lambda)d\lambda \end{Bmatrix},$$

hold for any $t \in \mathbb{Z}$ and $\omega, \nu \in W$. These equations show the reversed operation given by (9), so that one obtains the increments $\mathcal{U}_W^{(\cdot)}(\omega) - \mathcal{U}_W^{(\cdot)}(\nu)$, $\mathcal{V}_W^{(\cdot)}(\omega) - \mathcal{V}_W^{(\cdot)}(\nu)$ and not $U_W^{(\cdot)}(\omega)$ or $V_W^{(\cdot)}(\omega)$ themselves (Theorem 4.9.1 in Brockwell and Davis 1991, pp. 151–152). Here, $\overset{m.s.}{\longrightarrow}$ denotes convergence in mean square.

(11) $\gamma_{\mathcal{X}} \Longleftarrow F_{\mathcal{X}}, \gamma_{\mathcal{L}} \Longleftarrow F_{\mathcal{L}}$

$$\begin{Bmatrix} \gamma_{\mathcal{X}}(k) \\ \gamma_{\mathcal{L}}(k) \end{Bmatrix} = \int_{-\pi}^{\pi} \cos(k\omega) \begin{Bmatrix} dF_{\mathcal{X}}(\omega) \\ dF_{\mathcal{L}}(\omega) \end{Bmatrix},$$

hold for $k \in \mathbb{Z}, \omega \in W$ and describe the mathematical relationships between a given spectral distribution and the autocovariance function (known as Wold's theorem, a discrete version of the Wiener-Khintchine Theorem), see Brockwell and Davis (1991, Corollary 4.3.1, p. 119) or Priestley (2004, pp. 222–226). The described Fourier transform is reduced to a cosine transform due to the fact that the autocovariance function of any real stochastic process is even (see Priestley 2004, p. 214).

(12) $\gamma_{\mathcal{X}} \Longrightarrow F_{\mathcal{X}}, \gamma_{\mathcal{L}} \Longrightarrow F_{\mathcal{L}}$

$$\begin{Bmatrix} F_{\mathcal{X}}(\omega) \\ F_{\mathcal{L}}(\omega) \end{Bmatrix} = \begin{Bmatrix} \gamma_{\mathcal{X}}(0) \\ \gamma_{\mathcal{L}}(0) \end{Bmatrix} \frac{\omega + \pi}{2\pi} + \frac{1}{\pi} \sum_{k=1}^{\infty} \begin{Bmatrix} \gamma_{\mathcal{X}}(k) \frac{\sin k\omega}{k} \\ \gamma_{\mathcal{L}}(k) \frac{\sin k\omega}{k} \end{Bmatrix}$$

hold for $k \in \mathbb{Z}, \omega \in W$ and is the inverse operation to (11); see Brockwell and Davis (1991, Theorem 4.9.1, pp. 151–152) and Priestley (2004, pp. 222–226).

Example: MA(q) Process

In the previous sections the main results in the time and frequency domain for an MA(q) process were presented. The above mentioned equations in this section can be used to verify these results.

Conclusion and Outlook

In this paper we demonstrated certain aspects of the Magic Square, which connects a covariance-stationary stochastic process with its autocovariance function, its spectral representation, and the corresponding spectral distribution or density function (if it exists). To keep the presentation short, we focussed on the example of a moving average process and its transition from the time into the frequency domain. The application of more complex (and more widely used) stochastic processes in the time domain such as autoregressive moving average processes would be an obvious extension of this scenario, which we will deal with in the future. Furthermore, it would be valuable to explore the principles behind the transition from the frequency into the time domain by specifying suitable spectral processes and to find their time-domain representations.

References

Bartle RG (1976) The elements of real analsysis, 2nd edn. Wiley, New York

Brockwell PJ, Davis RA (1991) Time series: theory and methods, 2nd edn. Springer, New York

Gilgen H (2006) Univariate time series in geosciences—theory and examples. Springer, Berlin

Koch KR, Schmidt M (1994) Deterministische und stochastische Signale. Dümmler, Bonn

Moritz H (1989) Advanced physical geodesy. Wichmann, Karlsruhe

Priestley MB (2004) Spectral analysis and time series. Elsevier Academic, Amsterdam

Welsch, W, Heunecke O, Kuhlmann H (2000) Handbuch Ingenieurgeodäsie, Auswertung geodätischer Überwachungsmessungen. Wichmann, Heidelberg

Describing the Quality of Inequality Constrained Estimates

L. Roese-Koerner, B. Devaraju, W.-D. Schuh, and N. Sneeuw

Abstract

A key feature of geodetic adjustment theory is the description of stochastic properties of the estimated quantities. A variety of tools and measures have been developed to describe the quality of ordinary least-squares estimates, for example, variance-covariance information, redundancy numbers, etc. Many of these features can easily be extended to a constrained least-squares estimate with equality constraints. However, this is not true for inequality constrained estimates.

In many applications in geodesy the introduction of inequality constraints could improve the results (e.g. filter and network design or the regularization of ill-posed problems). This calls for an adequate stochastic modeling accompanying the already highly developed estimation theory in the field of inequality constrained estimation. Therefore, in this contribution, an attempt is made to develop measures for the quality of inequality constrained least-squares estimates combining Monte Carlo methods and the theory of quadratic programming. Special emphasis is placed on the derivation of confidence regions.

Keywords

Confidence regions • Convex optimization • Inequality constrained least-squares • Monte Carlo method • Stochastic modeling

1 Introduction and Motivation

In many applications in geodesy some bounds or restrictions on the parameters are known in advance. Truncating the parameter space by formulating this knowledge as inequality constraints often helps to improve the results. One can think of the estimation of non-negative variance components for example or constraints on the power spectral density in the design of decorrelation filters (Roese-Koerner et al. 2012a).

However, besides the process of actually determining a solution of a problem, it is also important to give a measure of its accuracy. In presence of inequality constraints, it is no longer possible to project the uncertainty of the observations to the parameters by applying the law of error propagation. This is, because there is no analytical relationship between observations and parameters. Even if a variance-covariance (VCV) matrix could be obtained, it would not yield a realistic error description, as one has to deal with truncated probability density functions (PDFs).

Up to now, there have been different approaches for a quality description in presence of inequality constraints. For example Liew (1976) proposed to first identify all constraints, which are exactly satisfied (by solving the problem). Afterwards, these constraints are treated as equality constraints, all other constraints are discarded and the VCV

L. Roese-Koerner (✉) • W.-D. Schuh
Institute of Geodesy and Geoinformation, University of Bonn, Bonn, Germany
e-mail: lutz.roese-koerner@geod.uni-bonn.de; schuh@geod.uni-bonn.de

B. Devaraju • N. Sneeuw
Institute of Geodesy, University of Stuttgart, Stuttgart, Germany
e-mail: devaraju@gis.uni-stuttgart.de; sneeuw@gis.uni-stuttgart.de

H. Kutterer et al. (eds.), *The 1st International Workshop on the Quality of Geodetic Observation and Monitoring Systems (QuGOMS'11)*, International Association of Geodesy Symposia 140,
DOI 10.1007/978-3-319-10828-5_3,

matrix of the equality constrained problem is computed. A disadvantage of this method is, that inactive constraints (e.g. constraints, which do not hold with equality) are neglected and do not constrain the confidence region.

Geweke (1986) and Zhu et al. (2005) treated inequalities as prior information in a Bayesian sense and truncated the probability density functions. To obtain a valid PDF, the function has to be scaled, which could be thought of as distributing the probability mass in the infeasible region over the whole function, which might be not realistic. Furthermore, the numerical evaluation of the PDF is computationally expensive in the multivariate case.

In this paper, we aim at giving a quality description for Inequality Constrained Least-Squares (ICLS) problems using Monte Carlo methods. In contrast to the idea of scaling, we want to obtain a PDF of the estimated parameters, which is identical to the PDF of an Ordinary Least-Squares (OLS) estimate inside the feasible region and where all probability mass in the infeasible region is projected onto its boundaries.

The paper is organized as follows. In Sect. 2 we define the ICLS problem and provide references on several solvers. Our proposed method is described in Sect. 3, as is the derivation of confidence regions and a brief description of a sensitivity analysis for the constraints. In Sect. 4 a case study is carried out to illustrate the application of our approach. The insights that have been gained are summarized in Sect. 5.

2 Background

First, the well-known linear OLS estimation model is extended to a linear ICLS estimation. We assume a deterministic model of the form

$$\mathbf{y} + \mathbf{v} = \mathbf{A}\mathbf{x}, \tag{1}$$

with vector of observations $\mathbf{y}$, vector of residuals $\mathbf{v}$, $(n \times m)$ design matrix $\mathbf{A}$ and vector of unknown parameters $\mathbf{x}$. n is the number of observations, m the number of parameters. The design matrix is assumed to have full rank m and all quantities are assumed to be real valued. The (possibly fully populated) VCV matrix $\mathbf{Q}$ of the observations is assumed to be known. The aim is to minimize the quadratic form

$$\Phi(\mathbf{x}) = \mathbf{v}^T \mathbf{Q}^{-1} \mathbf{v}. \tag{2}$$

Clearly, this aim can be achieved, by applying the usual OLS estimator

$$\hat{\mathbf{x}} = (\mathbf{A}^T \mathbf{Q}^{-1} \mathbf{A})^{-1} \mathbf{A}^T \mathbf{Q}^{-1} \mathbf{y}. \tag{3}$$

Throughout this paper, symbols marked with a hat refer to unconstrained quantities whereas tildes refer to quantities of a constrained solution.

We now introduce p inequality constraints in a matrix vector notation:

$$\mathbf{B}^T \mathbf{x} \leq \mathbf{b}. \tag{4}$$

$\mathbf{B}$ is a $(m \times p)$ matrix of constraints and $\mathbf{b}$ the corresponding right-hand side. As it is not known in advance which constraints will lead to changes in the parameters (i.d. are exactly satisfied or "active") the usual techniques of equality constrained estimation can not be applied. However, many algorithms have been developed to solve such an ICLS problem of the form

$$\text{minimize} \quad \Phi(\mathbf{x}) = \mathbf{v}^T \mathbf{Q}^{-1} \mathbf{v} \tag{5a}$$

$$\text{subject to} \quad \mathbf{B}^T \mathbf{x} \leq \mathbf{b} \tag{5b}$$

(which is often referred to as Quadratic Program (QP) as we want to minimize a quadratic objective function $\Phi(\mathbf{x})$ with respect to some linear constraints). Most of the solvers can be subdivided into two classes: Simplex Methods and Interior Point Methods.

In each iteration of a Simplex method a search direction is computed and projected onto a subset of the constraints. If at least one constraint is active in the solution, the optimal point will be at the boundary of the feasible set (the set where all constraints are satisfied). Therefore, one follows the borderline of the feasible set until the optimal solution is reached. If it is not on the boundary, in the last iteration the projection is neglected, resulting in a step into the interior of the feasible set. Examples for solvers of this type are the Active Set Method (cf. Gill et al. 1981, pp. 167–173) or Dantzigs Simplex Algorithm for Quadratic Programming (Dantzig 1998, pp. 490–498).

Interior Point Methods on the other hand, substitute the original—possibly hard to solve—problem by a sequence of easier to solve ones. Then a so called "central path" through the interior of the feasible region is followed until the optimal solution is reached. Examples are the Logarithmic Barrier method or primal-dual methods (cf. Boyd and Vandenberghe 2004, pp. 568–571 respectively pp. 609–613).

Other approaches also include the idea of aggregating all inequality constraints into one complex equality constraint (Peng et al. 2006) or transforming (5) into a Linear Complementarity Problem (cf. Koch 2006, pp. 24–25), which can be solved using Lemke's Algorithm (cf. Fritsch 1985).

As we want to focus on the quality description, we will not pursue the process of actually solving an ICLS problem but refer to the above mentioned authors. All results within this paper were computed using the Active Set Method.

3 MC-QP Method

As a VCV matrix is no longer representative in the inequality constrained case, we want to directly propagate the probability density of the observations. Especially in cases, where either no analytical solution is known or it would computationally be very expensive to obtain, Monte Carlo (MC) methods often are used. For the ICLS problem, MC integration seems to be perfectly suited as the probability distribution of the observations is assumed to be known but there is no analytical relationship between observations and parameters.

So the idea is to draw M samples of the observations, solve the resulting QPs and use them to obtain an empirical probability density function of the estimated parameters.

3.1 Propagation of the Probability Density

Assuming that we have a fully populated VCV matrix $\mathbf{Q}$ of the observations, we want to draw M samples $\mathbf{s}_{\mathcal{Y}}^{(i)}$ of the observations, which follow the normal distribution

$$\mathcal{Y} \sim N(E\{\mathcal{Y}\}, \mathbf{Q}). \tag{6}$$

$E\{\mathcal{Y}\}$ denotes the expectation operator of the random variable $\mathcal{Y}$, the random counterpart of the deterministic variable $\mathbf{y}$. As estimator of $E\{\mathcal{Y}\}$ we use an unconstrained OLS estimate

$$\hat{\mathbf{y}} = \mathbf{A}\hat{\mathbf{x}} \tag{7a}$$

$$= \mathbf{A}(\mathbf{A}^T\mathbf{Q}^{-1}\mathbf{A})^{-1}\mathbf{A}^T\mathbf{Q}^{-1}\mathbf{y}. \tag{7b}$$

The process of sampling from the above described distribution can be done as follows (cf. Koch 2007, p. 197): First $\mathbf{Q}$ is factorized into the product of two upper right triangular matrices $\mathbf{R}$ using the Cholesky factorization

$$\mathbf{Q} = \mathbf{R}^T\mathbf{R}. \tag{8}$$

Afterwards, M standard normal distributed samples $\mathbf{s}_{\mathcal{E}}^{(i)}$ are drawn from the distribution

$$\mathcal{E} \sim N(\mathbf{0}, \mathbf{I}), \tag{9}$$

with identity matrix $\mathbf{I}$ and transformed to

$$\mathbf{s}_{\mathcal{Y}}^{(i)} = \hat{\mathbf{y}} + \mathbf{R}^T\mathbf{s}_{\mathcal{E}}^{(i)}, \quad i = 1, \ldots, M. \tag{10}$$

These samples are used as input for the quadratic program (5), which is solved using the Active Set Method, producing M solution $\mathbf{s}_{\tilde{\mathcal{X}}}^{(i)}$. Hence, we can achieve an empirical joint density function of the parameters $\tilde{\mathcal{X}}$ by computing and normalizing the histogram of the solutions.

One can think of this approach as computing M different instances of the problem, resulting in M different objective functions with M different minima. However, as the constraints remain unchanged, all minima inside the feasible region coincide with the solution of an OLS estimation. All solutions outside the feasible region on the contrary are projected to the closest point (in the metric, given through the objective function) that fulfills all constraints. As all these points will be at the boundary of the feasible set, this will lead to the accumulation of probabilities, described in Sect. 1.

The task of solving M different QPs is computational demanding. However, as with the solution of the original problem a good initial solution is known, QP algorithms will converge fast.

3.2 Confidence Regions (HPD Regions)

As with the MC-QP approach we not only obtain a point estimate but a whole empirical PDF, we can easily compute confidence intervals. In Chen and Shao (1999) and the Supplement 1 to the "Guide to the expression of uncertainty in measurement" (GUM, ISO 2008, pp. 5 and 30) the confidence interval of a Monte Carlo estimate (called highest probability density (HPD) region) is defined as the shortest interval, which contains $1 - \alpha$ percent of the data (with $1-\alpha$ being the level of significance). This definition extends naturally to the n-dimensional case:

The (1-α)-confidence region Ω of an n-dimensional problem is defined as the smallest region containing $1-\alpha$ percent of the data

$$P\{\mathbf{x}|\Omega\} = 1 - \alpha. \tag{11}$$

This region can be computed by simply sorting the values of the n-dimensional histogram described in Sect. 3.1 by value, starting with the largest one. Afterwards, the cumulative sum is computed until $1 - \alpha$ percentage is reached. All bins of the histogram that added up to $1 - \alpha$ percentage form the confidence region. One has to be aware that this region is not necessarily connected, due to the accumulation of probability mass at the boundary of the feasible region (Chen and Shao 1999, p. 84).

Figure 1 illustrates such confidence intervals for a one dimensional example for $M \rightarrow \infty$. The PDF of the OLS estimate with $E\{\mathcal{X}\} = 0$ is plotted in light gray. The α percent, which are not included in the confidence interval (shaded areas) are symmetrically distributed at both tails of the distribution ($\frac{\alpha}{2}$ at each side). This symmetry is destroyed when introducing the inequality constraint $x \leq 1$ and performing an ICLS estimate.

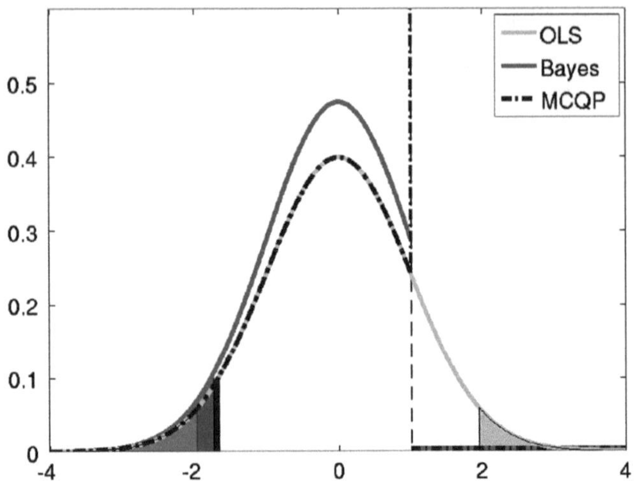

Fig. 1 Probability density functions and confidence intervals of different estimates for a one dimensional problem: OLS (*light gray*), Bayesian ICLS approach (*gray*) and MC-QP method (*dark gray, dash-dotted*). The *dashed line* represents the inequality constraint $x \leq 1$ and the minimal value of the objective function is reached for $x = 0$. The parts not contained in the confidence intervals are the *shaded regions* below the *curves*. In contrast to an OLS estimate the confidence region of an ICLS estimate can be asymmetric, which is why the OLS PDF has symmetric *shaded regions*, while those of Bayes and MC-QP are one-sided

The PDF of the MC-QP estimate is indicated by the dash-dotted dark gray line which coincides with the OLS estimate in the feasible region and accumulates all the probability mass in the infeasible region at its boundary. Thus, as the confidence interval contains the values which are most likely, the whole α percent not included in the confidence interval are at the left tail of the PDF (depending on the constraint). So the confidence interval is bounded on one side by the constraint and on the other side by the α percent that are "most unlikely". As can been seen in Fig. 1, this results in a smaller confidence interval.

However, this is not true for the Bayesian estimate (gray curve). The symmetry is destroyed here as well, but the scaling of the PDF leads to a shift of the beginning of the $(1 - \alpha)$-percent-interval and therefore to a bigger interval compared to the MC-QP method.

3.3 Influence of the Constraints

So far, we investigated the distribution of the estimated parameters and their confidence region. However, it might be also of interest, to determine the influence of the constraints onto the parameters. This can be done either on a global level, determining if the overall change in the result due to the constraints is significant or at a local level, investigating the individual influence of each constraint on each parameter. Due to the limited space, we will only very briefly discuss the different options for such a sensitivity analysis and provide some references for further reading. A more detailed discussion could for example be found in Roese-Koerner et al. (2012b).

On a global scale, one can perform a hypothesis testing. Here, the sum of squared residuals of the unconstrained OLS estimate is compared with the sum of squared changes through the constraints (Koch 1981). Another global measure is the ratio of the probability mass inside and outside the feasible region (measured by checking in each Monte Carlo iteration if at least one constraint is active or not).

To analyse the local influence, the Lagrangian

$$L(\mathbf{x}, \mathbf{k}) = \Phi(\mathbf{x}) + \mathbf{k}^T(\mathbf{B}^T\mathbf{x} - \mathbf{b}) \tag{12}$$

of the ICLS problem (5) is needed. It is the sum of the original objective function $\Phi(\mathbf{x})$ and the rearranged constraints multiplied with the Lagrange multipliers $\mathbf{k}$. The Lagrange multipliers of the optimal solution can be determined and give a measure for the "activeness" of a constraint. This can be used to quantify the influence of each constraint on each parameter (cf. Boyd and Vandenberghe 2004, p. 252).

4 Case Study

In this case study, the MC-QP method is applied to derive stochastic information of some quantities of a very simple ICLS problem: the estimation of a line of best fit with a constrained minimal intercept. We have intentionally chosen a simple problem to focus on the methodological aspects of the MC-QP method and on the comparison of different confidence regions.

Assume the following uncorrelated and normal distributed observations $\mathbf{y}$ to be measured at the supporting points $\mathbf{t}$:

$$\mathbf{y} = \begin{bmatrix}-4.0\ 0.0\ 1.5\ 3.0\ 2.0\end{bmatrix}^T, \qquad \mathbf{t} = \begin{bmatrix}1\ 2\ 3\ 4\ 5\end{bmatrix}^T.$$

The deterministic model reads

$$y_i + v_i = x_1 t_i + x_2. \tag{13}$$

The parameter space should be constrained, so that only intercepts of at least -3.5 are accepted:

$$x_2 \geq -3.5 \qquad \Longleftrightarrow \qquad \underbrace{\begin{bmatrix}0 & -1\end{bmatrix}}_{\mathbf{B}^T} \underbrace{\begin{bmatrix}x_1 \\ x_2\end{bmatrix}}_{\mathbf{x}} \leq \underbrace{\begin{bmatrix}3.5\end{bmatrix}}_{\mathbf{b}}. \tag{14}$$

Therefore, we have an ICLS problem in the form of (5) and can apply the MC-QP method. The unconstrained OLS solution reads

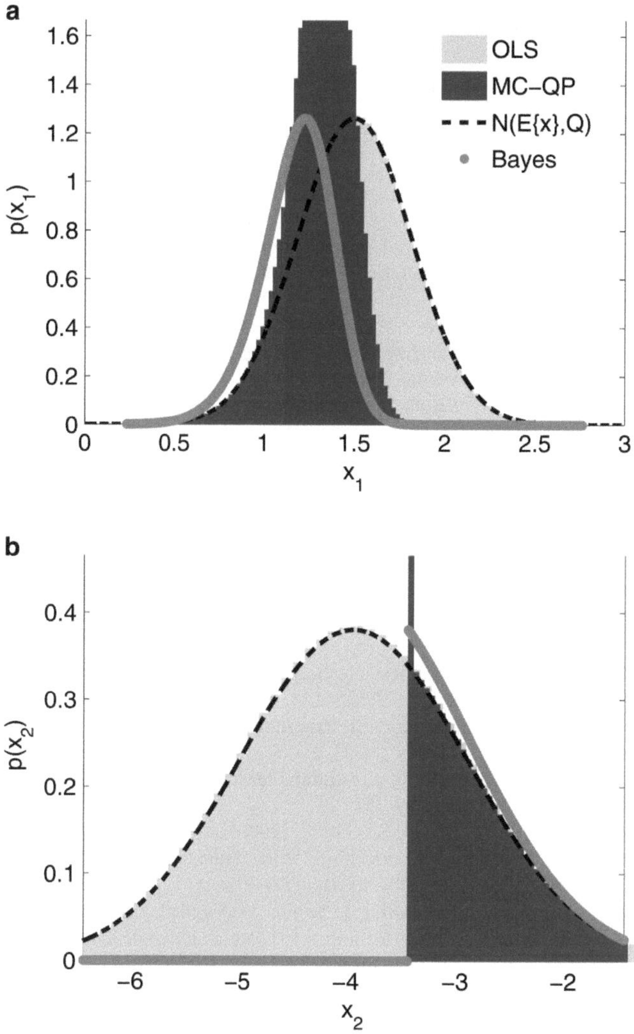

Fig. 2 Empirical marginal densities of the parameters after $M = 10,000,000$ Monte Carlo iterations. The PDF of the Monte Carlo estimates are plotted in *light gray* (OLS) and *dark gray* (ICLS), the *dashed line* is the analytical PDF of the OLS estimate and the PDF of the Bayesian approach in plotted in *gray*. (**a**) Estimates of the marginal density of x_1, (**b**) estimates of the marginal density of x_2

$$\hat{\mathbf{x}} = \begin{bmatrix} 1.5 \\ -4.0 \end{bmatrix} \tag{15}$$

and the ICLS solution, which was obtained using the Active Set Method, reads

$$\tilde{\mathbf{x}} = \begin{bmatrix} 1.3636 \\ -3.5000 \end{bmatrix}. \tag{16}$$

Comparison of the marginal densities of parameter x_2, which are illustrated in Fig. 2b, shows, that the MC-QP estimate (dark gray bars) is nearly identical to the OLS estimate (light gray bars) inside the feasible region. All probability mass

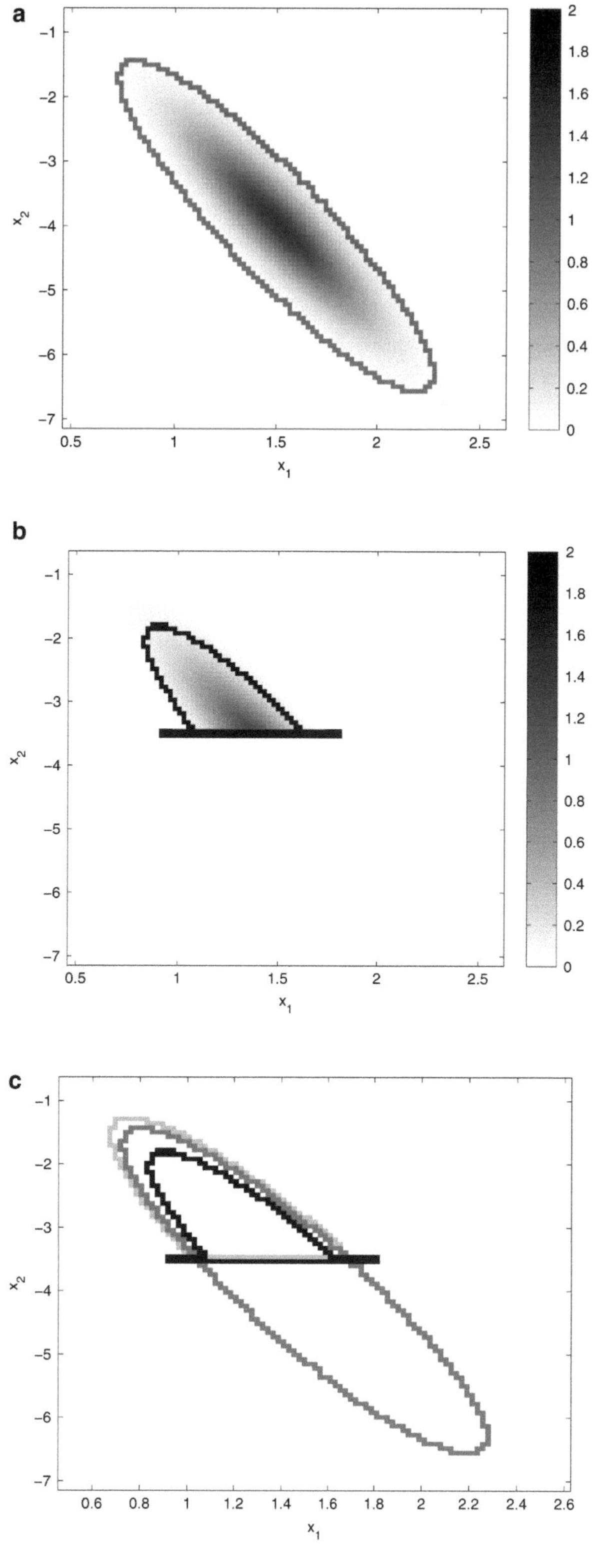

Fig. 3 Empirical joint densities and confidence regions of the parameters after $M = 10,000,000$ Monte Carlo iterations. (**a**) OLS: joint PDF and confidence region, (**b**) MC-QP: joint PDF and confidence region, (**c**) confidence regions of Bayesian (*light gray*), OLS (*darker gray*) and MC-QP approach (*black*)

left of the constraint is projected onto the boundary of the feasible set, resulting in a peak at $x_2 = -3.5$. The Bayesian estimate (gray curve) is a scaled version of the analytical PDF of the OLS estimate (dashed curve) inside the feasible region.

The more "peaky" form and the shift of the maximum of the MC-QP estimate of parameter x_1 (Fig. 2a) results from correlations between the parameters. The shift is even stronger in the Bayesian approach. Figure 3a illustrates the joint densities of the OLS estimate and the corresponding 95 % confidence region. The joint PDF of the MC-QP estimate is shown in Fig. 3b and is identical to the OLS estimate inside the feasible region. Here, the accumulation on the boundary can be seen as well. In Fig. 3c the confidence regions of the different estimates are compared. As discussed in Sect. 3.2, the confidence region of the MC-QP estimate (black) becomes smallest due to the accumulation of the probability mass on the boundary. On the contrary, applying the Bayesian method (light gray) leads to a bigger confidence region due to the scaling of the PDF. In this case study, the confidence region of the Bayes estimate is even bigger than the one of the OLS estimate (darker gray) because a huge part of the PDF is truncated.

5 Summary and Outlook

The proposed MC-QP method allows a stochastic description of ICLS estimates but it is computationally expensive to apply. It was shown that the introduction of inequality constraints within this framework leads to smaller confidence regions.

The possibilities of a sensitivity analysis (which were only mentioned briefly here) as well as the determination of the influence of constraints on correlations between parameters are to be addressed in future work.

References

Boyd S, Vandenberghe L (2004) Convex optimization. Cambridge University Press, Cambridge

Chen MH, Shao QM (1999) Monte Carlo estimation of Bayesian credible and HPD intervals. J Comput Graph Stat 8(1):69–92

Dantzig G (1998) Linear programming and extensions. Princeton University Press, Princeton

Fritsch D (1985) Some additional informations on the capacity of the linear complementarity algorithm. In: Grafarend E, Sansò F (eds) Optimization and design of geodetic networks. Springer, Berlin/Heidelberg/New York/Tokyo, pp 169–184

Geweke J (1986) Exact Inference in the inequality constrained normal linear regression model. J Appl Econ 1(2):127–41

Gill P, Murray W, Wright M (1981) Practical optimization. Academic Press, London

ISO (2008) Evaluation of measurement data—Supplement 1 to the "Guide to the expression of uncertainty in measurement"—Propagation of distributions using a Monte Carlo method, Joint Committee for Guides in Metrology, Bureau International des Poids et Mesures, Geneva

Koch A (2006) Semantische Integration von zweidimensionalen GIS-Daten und Digitalen Geländemodellen. Ph.D. thesis, Universität Hannover

Koch KR (1981) Hypothesis testing with inequalities. Bollettino di geodesia e scienze affini 2:136–144

Koch KR (2007) Introduction to bayesian statistics, 2nd edn. Springer, Berlin

Liew CK (1976) Inequality constrained least-squares estimation. J Am Stat Assoc 71(355):746–751

Peng J, Zhang H, Shong S, Guo C (2006) An aggregate constraint method for inequality-constrained least squares problems. J Geodesy 79:705–713. Doi:10.1007/s00190-006-0026-z

Roese-Koerner L, Krasbutter I, Schuh WD (2012a) A constrained quadratic programming technique for data-adaptive design of decorrelation filters. In: Sneeuw N, Novak P, Crespi M, Sanso F, Sideris MG (eds) VII Hotine-Marussi Symposium on Mathematical Geodesy, Rome, 2009, Springer, International Association of Geodesy Symposia, vol 137, pp 165–170. DOI 10.1007/978-3-642-22078-4_25

Roese-Koerner L, Devaraju B, Sneeuw N, Schuh WD (2012b) A stochastic framework for inequality constrained estimation. J Geodesy 86(11):1005–1018. Doi:10.1007/s00190-012-0560-9

Zhu J, Santerre R, Chang XW (2005) A Bayesian method for linear, inequality-constrained adjustment and its application to GPS positioning. J Geodesy 78:528–534. Doi:10.1007/s00190-004-0425-y

GNSS Integer Ambiguity Validation Procedures: Sensitivity Analysis

J. Wang and T. Li

Abstract

Global Navigation Satellite Systems (GNSS) have been widely used for many precise positioning and navigation applications. In satellite-based precise positioning, as carrier phase measurements are used, the determination of correct integer carrier phase ambiguities is a key issue in obtaining accurate and reliable positioning results. Therefore much effort has been investigated in developing a robust quality control procedure which can effectively validate the ambiguity resolution results. Such a quality control procedure has been traditionally based on the so-called F-ratio and R-ratio tests. A major shortcoming of these two ratio tests is that their probability distributions, which are normally considered to be Fisher distributed, are still unknown, which precludes the possibility to evaluate the confidence level for the ambiguity validation test. To overcome such a shortcoming, an alternative ambiguity validation test based on the so-called W-ratio has been proposed, which allows for a more rigorous quality control procedure under the assumption that the fixed integer ambiguities are constant. The W-ratio test has been widely used by many researchers. Like any other quality control procedures, there are assumptions to be made, for example, it is assumed that both functional and stochastic models are correct, in the W-ratio test. This paper presents a sensitivity analysis for the new ambiguity validation test based on the W-ratio as well as the other two ratio tests. The analysis will cover the sensitivities of the ratio tests to undetected gross errors (outliers), stochastic models, and geometry strengths relating to a variety of satellite constellations, such as GPS, GPS/GLONASS integration with real data sets. While the performances of different ratio tests are analyzed in terms of the so-called ambiguity "correct" rates based on the ground truth integer ambiguity values.

Keywords

Ambiguity Validation test • F-ratio • GNSS • R-ratio • W-ratio

1 Introduction

Precise kinematic relative GNSS-based positioning requires the reliable determination of the carrier phase integer ambiguities. The carrier phase measurements are ambiguous, with the ambiguity—the integer number of signal wavelengths between satellite and antenna—being an unknown value a priori. Hence the determination of the integer ambiguities, commonly referred to as ambiguity resolution (AR), is the most critical data analysis step for precise GNSS-based positioning. With fixed integer ambiguities, the carrier phases can be used as unambiguous precise range measurements.

The ambiguity resolution process consists of two steps, namely: ambiguity estimation and ambiguity validation.

J. Wang • T. Li (✉)
School of Surveying and Geospatial Engineering, The University of New South Wales, Sydney, NSW 2052, Australia
e-mail: tao.li@unsw.edu.au

H. Kutterer et al. (eds.), *The 1st International Workshop on the Quality of Geodetic Observation and Monitoring Systems (QuGOMS'11)*, International Association of Geodesy Symposia 140,
DOI 10.1007/978-3-319-10828-5_4, © Springer International Publishing Switzerland 2015

In ambiguity search, integer ambiguity parameters are initially treated as real (continuous) parameters. The real-valued (float) ambiguity parameters, together with other unknown parameters such as the coordinates of the roving receiver, can be estimated using a least-squares or Kalman filtering algorithm. The float solution of the real-valued ambiguity estimates and their associated statistics is then used to construct a search window, which is assumed to contain the correct integer ambiguities. The process of searching all possible integer ambiguity combinations within the search window is then performed using a search criterion based on the minimisation of the quadratic form of the least-squares residuals (Teunissen 1993, 1995). It is important to note that most ambiguity research techniques provide ambiguity estimation within a few tens of milliseconds, and such a performance normally satisfies most applications. The best integer ambiguity combination from the search that results in the minimum quadratic form of the least-squares residuals will be considered as the most likely (best) solution. The next step for the ambiguity resolution is to apply a so-called discrimination (validation) test to ensure that the most likely integer ambiguity combination is statistically better than the second best combination, as defined by the second minimum quadratic form of the least-squares residuals.

Traditionally ambiguity validation test procedures have been based on the so-called F-ratio and R-ratio test, (Frei and Beutler 1990; Euler et al. 1991). An alternative ambiguity validation (discrimination) test procedure has been proposed by Wang et al. (1998), in order to overcome the drawback of the validation test using the F-ratio and R-ratio (in which the probability distribution are unknown). This procedure is based on the ratio (called W-ratio) of the difference between the minimum and second minimum quadratic forms of the least-squares residuals and its standard deviation. Recent research has shown that these ratios could be treated as integer aperture estimators (Teunissen 2003), but the critical values are computed through intensive simulations (Li and Wang 2012).

A comparative study of the major ambiguity validation procedures has been carried out (Wang et al. 2000), showing that the ambiguity discrimination tests based on F-and W-ratios are generally close to the success probability of ambiguity resolution. However, the results should be different if undetected outlier and/or unmodelled systematic errors are remaining in the raw observations. In addition, a different stochastic modelling method and geometry of the satellite constellation may also affect the ambiguity test. Thus, this contribution will focus on a sensitivity analysis for the ambiguity validation tests.

2 Ambiguity Estimation and Validation Procedure

2.1 Ambiguity Estimation

In the case of using least-squares, the so-called Gauss–Markov model for linearised GNSS (single- or double-differenced) measurements is written as (Wang et al. 1998):

$$l = A_c x_c + A_k x_k + v \quad (1)$$

$$D = \sigma^2 Q = \sigma^2 P^{-1} \quad (2)$$

where is the $n \times 1$ *measurement vector,* n is the number of measurements; is the $n \times 1$ vector of the random errors; is the $m \times 1$ double-differenced ambiguity parameter vector, and m is the number of ambiguity parameters; is the $t \times 1$ vector of all other unknown parameters, and t is the number of all other unknowns (except ambiguities); A_k is the design matrix for the ambiguity parameters; A_c is the design matrix for the other unknown parameters; D is the covariance matrix; Q is the cofactor matrix; P is the weight matrix; and σ^2 is the a priori variance factor, assumed as 1.

Based on the principle of least-squares (minimum), the estimates of the unknowns in Eq. (1) can be obtained:

$$\widehat{x} = Q_{\widehat{x}} A^{\mathrm{T}} P l \quad (3)$$

with:

$$Q_{\widehat{x}} = \left(A^T P A\right)^{-1} \quad (4)$$

where $\widehat{x} = (\widehat{x}_c,\ \widehat{x}_k)^T$ and $A = (A_c,\ A_k)$. $Q_{\widehat{x}}$ is the cofactor matrix of the estimated vector, which can be represented by the following partitioned matrices:

$$Q_{\widehat{x}} = \begin{bmatrix} Q_{\widehat{x}_c} & Q_{\widehat{x}_c \widehat{x}_k} \\ Q_{\widehat{x}_k \widehat{x}_c} & Q_{\widehat{x}_k} \end{bmatrix}. \quad (5)$$

Furthermore, from Eqs. (1), (3) and (5), the least-squares residuals are obtained as follows:

$$\widehat{v} = l - A\widehat{x} = Q_{\widehat{v}} P l \quad (6)$$

where $Q_{\widehat{v}} = Q - A Q_{\widehat{x}} A^T$ is the cofactor matrix of the residuals. With the estimated residual vector $\widehat{v}$ and weight matrix P, the *a posteriori* variance cofactor can be estimated as:

$$\widehat{s}^2 = \frac{\Omega_0}{f} \quad (7)$$

where $\Omega_0 = \widehat{v}^T P \widehat{v} = l^T P Q_{\widehat{v}} P l = l^T P l - l^T P A \widehat{x}$, and $f = n - m - t$.

The above solution is referred to as the float solution, with the real-valued ambiguity estimates and their associated statistics. The next step is to take into account the fact that the ambiguities should be integer valued. This means that the integer ambiguities can be obtained by applying an appropriate searching method, e.g. the LAMBDA method (Teunissen 1995). The best integer ambiguity combination that results in the minimum quadratic form of the least-squares residuals will be considered as the most likely (*best*) solution. Normally, the first two best ambiguity combinations are identified for validation purposes.

2.2 Ambiguity Validation with W-Ratio Test

The W-ratio is defined as (Wang et al. 1998):

$$W = \frac{d}{\sqrt{Var(d)}} \tag{8}$$

where

$$d = \Omega_s - \Omega_m \tag{9}$$

$$Var(d) = \delta^2 Q_d \tag{10}$$

$Q_d = 4 \cdot (K_s - K_m)^T Q_{\widehat{xk}}^{-1} (K_s - K_m)$ is the cofactor of d, and δ^2 is the variance factor. K_m and K_s are the best and second best integer candidates. Two different variance cofactors can be chosen (Wang et al. 1998). If the *a posteriori* variance cofactor $\widehat{s}^2$ is used, the W-ratio is referred as W_s and it has a truncated Student's t-distribution; otherwise if a priori variance σ^2 is used, W-ratio is referred as W_a, which has a truncated normal distribution.

3 Sensitivity Analysis for the Validation Tests

3.1 Description of Test Data Sets and Data Processing

To analyse the sensitivity of the carrier phase integer ambiguity validation tests based on the W-ratio, F-ratio, and R-ratio, a variety of numerical tests were carried out. All of the test data are given in Table 1.

Data Set A: The outliers of 0.5, 1.0, 2.0 m were simulated for pseudo-ranges while the outliers of 3, and 5.0 cm for carrier phases, respectively. Only one outlier was simulated in either pseudo-ranges or carrier phases for each data set at any simulation test.

Table 1 Real data sets for the tests

Data set	SV no.	Data span (s)	Obs. type	Remark
A (GPS)	6	2,402	Dual	06.02.2010, Sydney
B(GPS+GLO)	8(5)	600	Single	25.08.2010, Sydney
C(GPS+GLO)	8(5)	600	Single	25.08.2010, Sydney

Data Set B: Single-frequency data from eight GPS satellites and five GLONASS satellites were utilized for analysing the effect of realistic stochastic modelling.

Data Set C: Single-frequency data from eight GPS satellites and five GLONASS satellites were used to gain an insight into the benefits of using a combination of GPS and GLONASS for ambiguity resolution.

Each data set was processed by least-squares on an epoch by epoch basis. The preset standard deviations for both L1 and L2 pseudo-ranges are 0.158 m, whereas those for L1 and L2 carrier phase are 0.0032 m. The best and second best ambiguity vector (ambiguity estimation) was obtained using the LAMBDA method.

3.2 Impact of Undetected Outliers

Tests were carried out with the Data Set A using the epoch-by-epoch solution mode to study the influence of undetected outliers (gross errors) on the ambiguity validation test. Table 2 shows the averaged R-, F- and W-ratios, and their corresponding success rate, as well as a posteriori variance values $\widehat{s}^2$, with different magnitudes pseudo-range outliers. In addition, correct ambiguity validation rates were obtained. Since the best integer ambiguity combinations identified from the search process are identical to the true integer ambiguity, the rates could be easily computed by actual test results using R-, F- and W-ratios with the true values. In all the following tests, a fail-rate of 0.001 for R-ratio test which is based on the Integer Aperture estimator (Teunissen 2009) was used, and the conventional critical value of 2.0 for F-ratio, and W-ratio with 99 % confidence level (one-tail) were used.

It can be seen from Table 2 that the statistics of the validation tests were getting worse when looking over the test statistics, which, as expected, become smaller in proportion to the increase of the intentionally added outliers. It is very hard to validate the best ambiguity combination with the R-, F- and W-ratio values when an outlier of 2 m was added. One possible reason would be understood from the estimated a posteriori variance factors shown in the last column, which should be close to unity according to the least-squares estimation theory (Cross 1983). However, if the variance factor is significantly different from unity (it may be rejected by an appropriate hypothesis test), it is suspected

Table 2 Averaged statistic values regarding ambiguity validation with respect to undetected pseudo-range outliers (epoch by epoch solution)

SV	Outlier (m)	R	F	Wa	Ws	$\hat{s}^2$
2	0.5	6.43	5.33	7.15	5.98	1.43
	1	2.85	2.52	4.84	2.88	2.82
	2	1.20	1.16	1.18	0.41	8.26
5	0.5	9.35	7.44	7.61	6.78	1.26
	1	3.83	3.29	5.72	3.62	2.50
	2	1.26	1.21	1.41	0.51	7.67
10	0.5	6.52	5.51	7.40	6.10	1.47
	1	2.87	2.56	5.07	2.92	3.01
	2	1.18	1.14	1.02	0.34	9.04
12	0.5	6.30	5.34	6.88	6.33	1.18
	1	2.74	2.48	4.58	3.17	2.09
	2	1.45	1.35	1.90	0.79	5.80
29	0.5	9.42	7.52	8.23	7.33	1.26
	1	4.32	3.67	6.51	4.22	2.38
	2	1.35	1.29	1.84	0.70	6.94
Outlier free		13.05	10.16	8.33	8.78	0.90
SV	Outlier (m)	Success rate (%)				
		R	F	Wa	Ws	
2	0.5	97.2	96.3	97.3	94.9	
	1	73.8	68.5	69	43.6	
	2	1.5	0.1	0.8	0.0	
5	0.5	99.5	99.3	99.0	98.0	
	1	92.7	88.7	91.4	66.4	
	2	0.8	0.0	1.3	0.0	
10	0.5	99.6	99.3	97.3	96.8	
	1	82	74.7	85.3	44.2	
	2	1.3	0.5	1.5	0.0	
12	0.5	98.8	92.8	99.0	98.3	
	1	68	61.7	63.4	46.2	
	2	12.7	6.5	10.1	0.67	
29	0.5	99.1	98.8	98.9	97.6	
	1	94.5	92.3	91.7	78.4	
	2	1.2	0.0	3.2	0.0	
Outlier free		99.9	99.9	99.9	99.8	

that outliers (gross errors) exist in the measurements or there is a problem with the fidelity of stochastic and functional model (Cross 1983). Since we have already assumed that the outliers are undetected, it can be concluded that the large undetected outliers in the pseudo-range measurements cause the model non-fidelity, and hence lead to unrealistic a posteriori variance estimation, that makes the R-, F- and W-ratios extremely small. Fortunately, such a large error considered in the tests could be detected by the unit variance hypothesis test with critical value 2.7 with 99.0 % confidence.

Table 3 is a summary of the ambiguity validation test results with respect to the undetected outliers in carrier phase observations for each satellite. It can be concluded from the results that the correct ambiguity can still be validated even with a couple of centimetre outliers in one carrier phase measurement.

Table 3 Averaged statistics values on the ambiguity validation with respect to undetected carrier phase observation outliers (epoch by epoch solution)

SV	Outlier (cm)	R	F	Wa	Ws	$\hat{s}^2$
2	3	2.68	2.57	4.47	5.02	0.89
	5	1.22	1.21	1.3	1.48	0.88
5	3	2.28	2.21	4.12	4.53	0.91
	5	1.25	1.24	1.42	1.56	0.91
10	3	2.47	2.39	5.01	5.33	0.94
	5	1.21	1.21	1.36	1.36	1.00
12	3	3.8.0	3.52	5.36	5.83	0.92
	5	1.85	1.78	2.62	2.85	0.92
29	3	2.48	2.40	4.39	4.72	0.93
	5	1.49	1.47	2.00	2.13	0.94
Outlier free		13.05	10.16	9.25	8.78	0.90
SV	Outlier (cm)	Success rate (%)				
		R	F	Wa	Ws	
2	3	64.7	63.6	74.3	75.1	
	5	1.8	1.3	10.1	11.2	
5	3	59.2	56.7	68.8	70.2	
	5	5.3	0.7	11.9	12.4	
10	3	65.9	64.4	77.1	77.6	
	5	1.1	0.4	10.3	9.2	
12	3	73.3	72.2	82	81.1	
	5	31.4	26.7	38.1	39.9	
29	3	62.8	60.0	66.5	67.9	
	5	10.3	23.3	11.0	10.2	
Outlier free		99.9	99.9	99.9	99.8	

3.3 Impact of Stochastic Modeling

In order to analyze the impact of a stochastic model on ambiguity resolution, two types of stochastic models were considered here namely, the ‘Preset’ and ‘Realistic’ models. Note that a weight matrix is equally weighted in the ‘Preset’ model. For the ‘Realistic’ model, a realistic measurement noise covariance matrix estimation method (Wang 2000) was utilized.

Data set B was processed by the mathematical model used in Wang (2000), which applies double difference separately in GPS and GLONASS measurements. By applying the realistic stochastic model, the ambiguity dilution of precision (ADOP) values (Teunissen and Odijk 1997) drop significantly comparing with the pre-set stochastic model, which can be seen from Fig. 1. From the 23rd epoch on, the ADOP

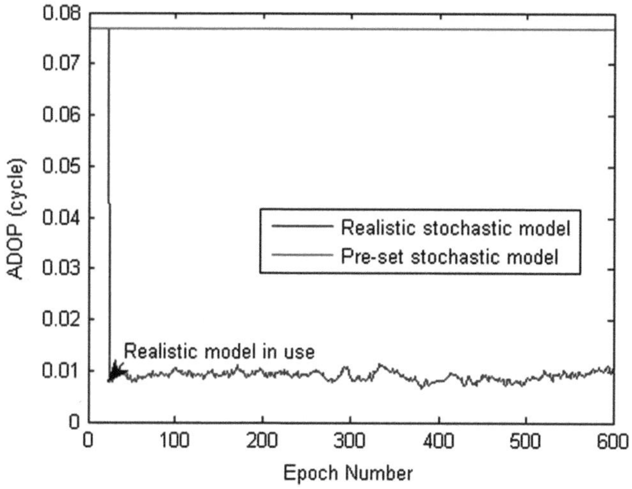

Fig. 1 ADOP values for both pre-set and realistic stochastic model

Table 4 Ambiguity validation statistics by pre-set and realistic models

Model	R	F	Wa	Ws	Truth
Pre-set	546/600	414/600	427/600	332/600	600/600
Realistic	577/577	577/577	577/577	577/577	577/577

value with the realistic model fluctuates around 0.01 cycles, whereas the ADOP values for pre-set model stays in around 0.075. Apparently there is no consideration for the changing of measurement scenario for the pre-set stochastic model.

Table 4 shows the test statistics for ambiguity validation results. Even though both models can provide us with correct integer ambiguities in each epoch, the validation results are quite different. The pre-set model can only validate successful ambiguities by up to 546 out of 600 by R-ratio test with pre-defined fail-rate 0.001, and by F-ratio and W-ratio, the accepted numbers become less. However, with the realistic model, for each validation method, they can all determine the ambiguities correctly in each epoch, and at the same time, all the resolved ambiguities are correctly accepted.

3.4 Satellite Configurations

With the increasing number of satellites for GLONASS, there is more flexibility for the augmentation of GPS applications. To analyze the impact of GLONASS in ambiguity resolution, data set C was processed firstly with GPS only, and then with the integration of GPS and GLONASS.

Figure 2 shows the ADOP values with regard to GPS only and GPS+GLONASS processed by with the pre-set stochastic model, respectively. Obviously, with 5 more GLONASS satellites integrated, there is a huge improvement in the satellite geometry, and the ADOP values become fairly small, around 0.01 cycles, see the red-dash line. In Table 5, the

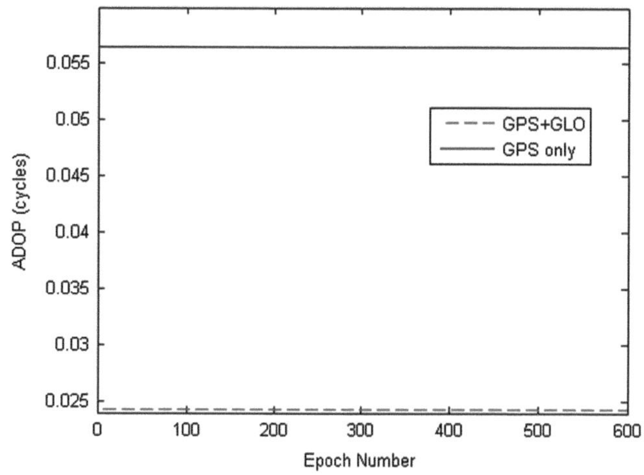

Fig. 2 ADOP values in case of GPS only and GPS+GLONASS

Table 5 Ambiguity validation statistics with different satellite constellations, 1 for GPS, 2 for GPS+GLONASS

Constellations		R	F	Wa	Ws	Correct rate (%)
1	Mean value	3.24	3.06	2.83	3.06	84.1
	Validation rates	46.7 %	43.2 %	53.1 %	30.4 %	
2	Mean value	6.23	5.82	14.13	11.60	100
	Validation rates	100 %	100 %	100 %	100 %	

Table 6 Ambiguity validation statistics with different satellite constellations, 1 for GPS, 2 for GPS+GLONASS

Constellations			R	F	Wa	Ws	Correct rate
1	Accepted	Correct	264	256	312	180	504/599
		Incorrect	16	3	6	2	
	Rejected	Correct	240	248	192	324	
		Incorrect	79	92	89	93	
2	Accepted	Correct	599	599	599	599	599/599
		Incorrect	0	0	0	0	
	Rejected	Correct	0	0	0	0	
		Incorrect	0	0	0	0	

mean values for different ambiguity validation methods as well as the validation results are presented. For the GPS only solutions, the correct rate is 84.1 %, which was obtained by comparing with the ground truth ambiguities in each epoch. The ambiguity validation methods can only accept up to 53.1 % of epochs by Wa ratio, and for the other methods, they are too conservative. When integrating GPS and GLONASS satellites, the mean values for all the validation methods increase dramatically, and the corresponding validation results are identical to the correct rate. This reveals the fact that the integration of GPS and GLONASS allows more chances to resolve the ambiguity correctly.

In Table 6, the numbers of acceptance and rejection are given. It is shown that, in case of poor satellite geometry,

Type I error and Type II error are inevitable for all ambiguity validation methods.

Concluding Remarks

Ambiguity validation is one of the most important steps in ambiguity resolution. In this contribution, we have performed a series of data processing to investigate the sensitivity of the ambiguity validation tests with respect to not only errors remaining in the raw measurements, but different stochastic models and satellite configurations.

The numerical results have shown that the validation test statistics, such as R-, F- and W-ratios, are sensitive to the magnitude of remaining measurement errors. It should also be emphasised that existence of outliers in pseudo-range measurements significantly degrades the performance of the validation tests. It has also been shown that appropriate stochastic modelling is critical, not only to improve the estimation precision, but also to enhance the performance of the ambiguity validation tests. An implementation of an online stochastic model reflects the real measurement scenario, and then greatly improves the ambiguity resolution performance.

Finally, it has been demonstrated that the inclusion of GLONASS improves the validation results. On the other hand, the analysis with an integrated GPS/GLONASS system has shown that on the fly ambiguity resolution based on single-frequency measurements could be possible, provided there are a large number of satellite measurements.

References

Cross PA (1983) Advanced least squares applied to position fixing. Working Paper No. 6, Department of Surveying, Polytechnic of East London, 205 pp

Euler HJ, Schaffrin B (1991) On a measure for the discernability between different ambiguity solutions in the static-kinematic GPS mode. In: IAG symposia no 107, kinematic systems in geodesy, surveying, and remote sensing. Springer, Berlin, pp 285–295

Frei E, Beutler G (1990) Rapid static positioning based on the fast ambiguity resolution approach FARA: theory and first results. Manuser Geod 15(4):325–356

Li T, Wang J (2012) Some remarks on GNSS integer ambiguity resolution methods. Surv Rev 44(326):320–328

Teunissen PJG (1993) A new method for fast carrier phase ambiguity estimation. In: IEEE position, location and navigation symposium (PLANS). Las Vegas, 11–15 April, pp 562–573

Teunissen PJG (1995) The least-squares ambiguity decorrelation adjustment: a method for fast GPS integer ambiguity estimation. J Geodesy 70(1–2):65–82

Teunissen PJG (2003) Integer aperture GNSS ambiguity resolution. Artif Satellites 38(3):79–88

Teunissen PJG (2009) The GNSS ambiguity ratio-test revisited: a better way of using it. Surv Rev 41(312):138–151

Teunissen PJG, Odijk D (1997) Ambiguity dilution of precision: definition, properties and application. In: Proceedings of the institute of navigation (ION 1997). Kansas City, pp 891–899

Wang J (2000) Stochastic modelling for RTK GPS/GLONASS positioning. J US Inst Navigat 46(4):297–305

Wang J, Stewart MP, Tsakiri M (1998) A discrimination test procedure for ambiguity resolution on-the-fly. J Geophys Res 72:644–653

Wang J, Stewart MP, Tsakiri M (2000) A comparative study of the integer ambiguity validation procedures. Earth Planets Space 52(10):813–817

Optimal Design of Deformation Monitoring Networks Using the Global Optimization Methods

M. Yetkin and C. Inal

Abstract

Geodetic networks are very important tools that can be used to monitor crustal movements or the deformation of structures. However, a geodetic network must be designed to sufficiently meet some network quality requirements such as accuracy, reliability, sensitivity and economy. This is the subject of geodetic network optimization. Traditional methods have been used for solving geodetic optimization problems. On the other hand, some evolutionary algorithms such as the particle swarm optimization algorithm have been started to be recently used. These methods are inspired by optimization and adaptation processes that are encountered in the nature. They are iterative procedures for quickly and efficiently solving complex optimization problems. They may provide global optimum solution or at least near-optimum solutions to problems. In this paper, the use of the shuffled frog-leaping algorithm for the optimal design of a deformation monitoring network is studied. The aim is to design and optimize a geodetic network in terms of high reliability.

Keywords

First-order design • Geodetic network optimization • Nature-inspired optimization algorithms • Network reliability • Stochastic search methods

1 Introduction

Mathematically, geodetic network optimization can be defined as minimizing or maximizing a given objective function that represents the desired network quality criteria such as reliability. The reliability of geodetic networks, which was originated from Baarda (1968), can be defined as the ability of a network to detect and resist against blunders in the observations Pelzer (1980) and (Seemkooei 2001a). Geodetic network optimal design can help in identifying and eliminating gross errors in the observations as well as in minimizing the effects of undetectable gross errors existing in the observations. Thus, in order to avoid a misinterpretation of the undetected gross errors as deformation phenomena, a monitoring network should be designed with high reliability (Kuang 1996). The importance of optimal design of deformation monitoring networks has also been investigated in Niemeier (1982).

Different geodetic network optimization problems may be classified into different orders. There are zero-order design (ZOD) problem, first-order design (FOD) problem, second-order design (SOD) problem and third-order design (THOD) problem (Grafarend 1974; Grafarend and Sanso 1985). The FOD is the choice of the optimum locations for the stations (Berné and Baselga 2004). In other words, an optimum network configuration that will satisfy the desired network quality criteria can be designed. For example, the optimal locations of the reference stations in a deformation monitoring network may be determined (Kuang 1996). This may be

M. Yetkin (✉)
Department of Geomatics Engineering, Izmir Katip Celebi University, 35620 Izmir, Turkey
e-mail: mevluty82@gmail.com

C. Inal
Department of Geomatics Engineering, Selcuk University, 42250 Konya, Turkey

H. Kutterer et al. (eds.), *The 1st International Workshop on the Quality of Geodetic Observation and Monitoring Systems (QuGOMS'11)*, International Association of Geodesy Symposia 140,
DOI 10.1007/978-3-319-10828-5_5,

an application of FOD in deformation monitoring networks. The main aim of the present contribution is to optimize a geodetic network that is established for deformation monitoring in the sense of high reliability using a stochastic optimization method. The selected optimization algorithm is the shuffled frog leaping algorithm (Elbeltagi et al. 2005; Amiri et al. 2009). The optimal distribution of network points can be determined using this method. As known, the network reliability depends on both the configuration matrix A (the geometry of the network) and the weight matrix of observations P (the accuracy of the observations).

2 The Optimal Design Problem

There are three main components of an optimization problem: the objective function, optimization variables and the solution algorithm. The location of network points and/or weights of observables can be found in a way that the pre-set quality criteria are satisfied for the network. To accomplish this, the design criteria must be expressed as a function of station coordinates. In this paper the reliability of network is taken into account as the quality criterion.

The function to be optimized is known as the objective function. The optimal design of a geodetic network can determine the optimal distribution of network stations such that the network is designed with high reliability. At the optimal design stage we have to consider that the network must ensure that the detection of blunders as completely as possible and the effects of the undetected ones on the estimated parameters should be minimum. A general reliability criterion in this case can be given as

$$min\,(r_i) = max. \tag{1}$$

That means that the minimum redundancy numbers of observables is maximized. r_i is the redundancy number of ith observation and the corresponding diagonal element of the redundancy matrix R (Kuang 1996). R is computed

$$\mathrm{R} = \mathrm{I} - \mathrm{A}\left(\mathrm{A}^{\mathrm{T}}\mathrm{PA}\right)^{-1}\mathrm{A}^{\mathrm{T}}\mathrm{P} \tag{2}$$

The optimal design problems can be solved by means of various optimization methods. Traditional methods are the trial and error method and the analytical methods such as linear programming. In this paper, the implementation of a meta-heuristic method called the shuffled frog leaping algorithm (SFLA) in the solution of the classical geodetic network design problem is investigated. SFLA aims to emulate and model the behavior of frogs searching for food laid on stones randomly located in a swamp and combines the advantages of the genetic-based memetic algorithm and the social behavior-based particle swarm optimization algorithm. Our preliminary results on its application in geodetic network optimization are presented in this paper.

3 The Shuffled Frog Leaping Algorithm

Modern metaheuristic algorithms such as genetic algorithm, ant colonies, particle swarm optimization, bee algorithms, firefly algorithms and so forth to find optimal solutions to optimization problems have been developed (Yang 2008). Optimization algorithms could be divided into two groups: deterministic and stochastic. Deterministic techniques produce the same set of solutions if the iterative search process starts with the same initial point. They are almost all local search techniques and may fail in finding global optima. Nevertheless, stochastic algorithms may produce different solutions even if the same starting point is used. The advantage of the stochastic algorithms is their ability to jump out of local optima thanks to their stochastic component. Generally stochastic algorithms are meta-heuristic (Yang 2010). A recent kind of nature-inspired meta-heuristic algorithm is called the shuffled frog leaping algorithm (Eusuff and Lansey 2003; Elbeltagi et al. 2005; Eusuff et al. 2006; Rahimi-Vahed and Mirzaei 2007; Amiri et al. 2009). Some of iteratively applied stochastic search heuristics have also been applied to geodetic networks (Dare and Saleh 2000; Berné and Baselga 2004; Yetkin et al. 2009; Yetkin 2013).

In the SFLA, the population consists of a set of solutions. The solutions (frogs) are partitioned into subsets called as memeplexes. Each memeplex perform a local exploration. Within each memeplex, the individual frogs hold ideas about the search space that can be infected by the ideas of other frogs through a memetic evolution in which the parameters of optimization problem are updated in order to reach to the optimal solution. After a defined number of memetic evolution (local search) step, ideas are passed among memeplexes in a shuffling process. The local exploration and the shuffling processes continue until defined convergence criteria are satisfied (Elbeltagi et al. 2005).

An initial population of F frogs is created randomly. For D-dimensional problems (D variables), *i*th frog is represented as $X_i = (x_1, x_2, \ldots, x_D)_i$. For example, the coordinates of the network stations are the variables in the geodetic FOD problem. Afterwards, the frogs are ranked in a descending order according to their performance value that is selected as the reliability criterion in this paper. Then, the whole population is partitioned into m memeplexes, each containing n frogs, i.e. $\mathrm{F} = \mathrm{m} \times \mathrm{n}$. In this process, the first frog goes to the first memeplex, the second frog goes to the second memeplex, frog m goes to the mth memeplex, frog $\mathrm{m} + 1$ goes back to the first memeplex, etc. (Elbeltagi et al. 2005).

As mentioned above, within each memeplex a memetic evolution called frog-leaping algorithm is performed to improve the quality of the solutions, i.e. each memeplex is evolved. The frogs with the best and the worst fitnesses are termed X_b and X_w, respectively. Also, the frog with the global best fitness is identified as X_g, then the position of the frog with the worst fitness in each cycle is improved using the following formulae:

$$S_i = \text{rand} \times (X_b - X_w) \quad (3)$$

$$X_{new} = X_{current} + S_i \quad (4)$$

where S_i is the change in *i*th frog position; $X_{current}$ is the current position of the frog with the worst fitness; X_{new} is the new position of the frog with the worst fitness; and rand is a random number between 0 and 1. If this process generates a better solution, it replaces the worst frog. Otherwise, the computations in Eqs. (3) and (4) are repeated but X_g replaces X_b. If no improvement becomes possible in this latter case, then a new solution is randomly generated to replace the worst frog with another frog having any arbitrary fitness (Elbeltagi et al. 2005; Amiri et al. 2009).

The memetic evolutionary steps within each memeplex are repeated for a specific number of iterations. The evolved memeplexes are shuffled for global exploration. The frogs are sorted in descending order of their fitness. The memeplexes are formed. Then, the memetic evolutions within each memeplex are applied. The global exploration is repeated until the solutions are converged.

Application of the SFLA meta-heuristic strategy is basically summarized as follows

1. Create a random population of F solutions (frogs).
2. Calculate the fitness for each individual. Sort the population F in descending order of their fitness.
3. Divide F into m memeplexes. There are n solutions (frogs) in each memeplex ($F = m \times n$).
4. Evolve each memeplex. The number of evolving iterations for each memeplex is N.
5. Shuffle the evolved memeplexes and return to step 2 until a pre-defined number of function evaluations is reached.

4 An Example

This example deals with the optimal first order design of a geodetic network by employing the SFLA. Let us consider a monitoring network to be observed by GPS techniques as shown in Fig. 1. The datum of the network is provided by minimum constraints; point R_1 has been considered as a fixed station. R_1, R_2 and R_3 are reference points. O_1 and O_2 are object points. The locations of R_1 and object points are not allowed to change and their local coordinates are given in Table 1. On the other hand, the locations of R_2 and R_3 are allowed to change in a stable area to find the most reliable network. The problem is to determine the exact position of the two reference points. As is well known, the reliability of a network is dependent upon both the configuration matrix A and the weight matrix **P**.

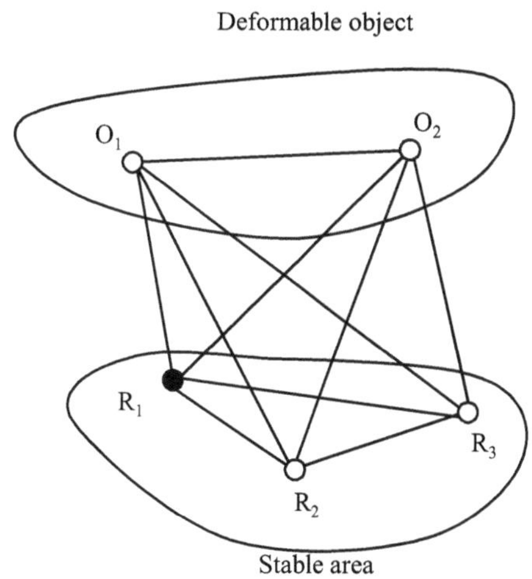

Fig. 1 The monitoring network

Table 1 The coordinates of fixed points

Fixed points	X (m)	Y (m)
R_1	500	1,000
O_1	750	2,500
O_2	3,000	2,250

The typical distance-dependent precision for the independent baselines is given as

$$\sigma\,(\text{baseline}) = 5\ \text{mm} + 1\ \text{ppm} \quad (5)$$

GPS receivers' standard technical specifications in relative positioning can be used for computation of standard deviations of the baseline components. The A matrix that might be used at the design stage is the typical of the distance derivatives due the fact that the stochastic model only considers planimetric distances (Berné and Baselga 2004, p. 52).

A deformation monitoring network consists of reference and object points. To optimize the network configuration, one may assume that the locations of the reference points are allowed to vary to some extent (Kuang 1996). However, as is well known, reference stations should be established in stable places. Therefore, the search domain for the locations of the reference stations must be determined according to this important rule. The boundaries for the search domains of R_2 and R_3 are the same and given as $x_{min} = 0$ m, $x_{max} = 5000$ m, $y_{min} = 0$ m and $y_{max} = 2000$ m. We will start with random 50 initial solutions. These initial solutions can be generated using a random number generator. The best and the worst of initial solutions are shown in Table 2.

Table 2 The worst and the best of initial solutions

min. r_i	X_2	Y_2	X_3	Y_3
0.06105	708.96	224.69	1055.3	732.73
0.33998	1850.9	1,520	722.05	190.03

Table 3 The optimization results

min r_i	X_2	Y_2	X_3	Y_3
0.42929	1177.1	1743.2	1303.5	1841.8

Table 4 Statistical analysis of different runs

Run	Min r_i
1	0.40961
2	0.41915
3	0.40647
4	0.42929
5	0.40540
6	0.41720
7	0.41821
8	0.41288
9	0.42251
10	0.39943
11	0.41990
Best	0.42929
Worst	0.39943
Median	0.41720

The SFLA is used as described above to search optimal point locations such that the minimum redundancy number can be maximized. The main parameters of SFLA are: number of frogs is 50; number of memeplexes is 5; number of generation for each memeplex before shuffling is 20; and number of shuffling iterations is 1,000. The obtained optimal solution is given in Table 3. Monumentation of the network stations can be made according to the optimal station coordinates listed in this table.

It is noted that the SFLA is probabilistic since it involves random numbers. For pseudo random number generators, once the random seed is determined, the resulting random stream is fixed. If different initial seeds are used while running the algorithm, very different solutions might be obtained. Therefore, we may run the algorithm several times. Then the result with the highest minimum redundancy number will be selected as the final solution. Table 4 shows the statistical analysis of different runs.

Conclusions

A new stochastic optimization approach to the optimal first-order design of geodetic networks has been examined. The optimization procedure gives the optimal locations for any network point after appropriate optimization criterion is given. In the present case the optimization criterion was the maximization of the minimum redundancy number. As is well known, this is the general criterion for internal reliability (Seemkooei 2001b). The stochastic optimization algorithms may be extensively used in geodetic network optimization problems. Such methods can provide sufficient solutions. Thus, optimal networks for deformation monitoring or other geodetic purposes can be obtained.

This paper suggests the use of the SFLA for optimization of a deformation monitoring network. Optimal design using the SFLA needs neither linearization nor differentiation of the objective function. Even tough optimization problems can be easily solved by means of simple mathematical computations. The diversity of the solutions is provided by random number generation.

Acknowledgment This study was funded by the Scientific Research Projects Grant of Selcuk University (Project No: 11701016).

References

Amiri B, Fathian M, Maroosi A (2009) Application of shuffled frog leaping algorithm on clustering. Int J Adv Manuf Technol 45:199–209
Baarda W (1968) A testing procedure for use in geodetic networks. Neth. Geod. Com. Publ. On Geodesy, New Series, No. 5. Delft, Netherlands
Berné JL, Baselga S (2004) First-order design of geodetic networks using the simulated annealing method. J Geod 78:47–54
Dare P, Saleh H (2000) GPS network design: logistics solution using optimal and near-optimal methods. J Geod 74:467–478
Elbeltagi E, Hegazy T, Grierson D (2005) Comparison among five evolutionary-based algorithms. Adv Eng Inform 19:43–53
Eusuff MM, Lansey KE (2003) Optimizing of water distribution network. J Water Resour Plann Manage 129(3):210–225
Eusuff MM, Lansey KE, Pasha F (2006) Shuffled frog leaping algorithm: a memetic metaheuristic for discrete optimisation. J Eng Opt 38(2):129–154
Grafarend EW (1974) Optimization of geodetic networks. Bolletino di Geodesia a Science Affini 33(4):351–406
Grafarend EW, Sanso F (1985) Optimisation and design of geodetic network analysis and optimal design. Springer, Berlin
Kuang SL (1996) Geodetic network analysis and optimal design. Ann Arbor, Ann Arbor
Niemeier W (1982) Design, diagnosis and optimization of monitoring networks in engineering surveying. The Centennial Convention of CIS, April 19–23, Ottowa
Pelzer H (1980) Some criteria for the accuracy and the reliability of geodetic networks, DGK, B, No: 252, Munich, pp 55–67
Rahimi-Vahed A, Mirzaei AH (2007) A hybrid shuffled frog leaping algorithm for a mixed model assembly line sequencing problem. Comput Ind Eng 53:642–666
Seemkooei AA (2001a) Comparison of reliability and geometrical strength criteria in geodetic networks. J Geod 75:227–233
Seemkooei AA (2001b) Strategy for designing geodetic network with high reliability and geometrical strength. J Surv Eng 127(3):104–117
Yang XS (2008) Nature-inspired metaheuristic algorithms. Luniver, Cambridge

Yang XS (2010) Firefly algorithm, stochastic test functions and design optimization. Int J Bio-inspired Comput 2(2):78–84
Yetkin M (2013) Metaheuristic optimisation approach for designing reliable and robust geodetic networks. Surv Rev 45(329):136–140
Yetkin M, Inal C, Yigit CO (2009) Use of the particle swarm optimization algorithm for second-order design of levelling networks. J Appl Geod 3:171–178

Part II

Theoretical Studies on Combination Strategies and Parameter Estimation

Towards the Combination of Data Sets from Various Observation Techniques

M. Schmidt, F. Göttl, and R. Heinkelmann

Abstract

Nowadays, heterogeneous data sets are often combined within a parameter estimation process in order to benefit from their individual strengths and favorable features. Frequently, the different data sets are complementary with respect to their measurement principle, the accuracy, the spatial and temporal distribution and resolution, as well as their spectral characteristics.

This paper gives first a review on various combination strategies based on the Gauss-Markov model; special attention will be turned on the stochastic modeling of the input data, e.g. the influence of correlations between different sets of input data. Furthermore, the method of variance component estimation is presented to determine the relative weighting between the observation techniques.

If the input data sets are sensitive to different parts of the frequency spectrum the multi-scale representation might be applied which basically means the decomposition of a target function into a number of detail signals each related to a specific frequency band. A successive parameter estimation can be applied to determine the detail signals.

Keywords

Combination strategies • Gauss-Markov model • Operator-software impact • Variance component estimation • Multi-scale representation

1 Introduction

To achieve the goals of the Global Geodetic Observing System (GGOS) a combination of complementary sensor and observation systems has to be applied (Rummel et al. 2005). This combination should be performed within a parameter estimation process in order to benefit from their individual strengths and favorable features. Data sets could be complementary, e.g. with respect to their measurement principle, accuracies, spatial and temporal distribution and resolution or their spectral characteristics. Various combination strategies can be applied, e.g. the rigorous combination within the *Gauss-Markov model* or the model of *variance component estimation* (Koch 1999).

In reference frame estimation geometric space-geodetic techniques such as VLBI (Very Long Baseline Interferometry), GNSS (Global Navigation Satellite Systems) and SLR (Satellite Laser Ranging) are usually combined; special attention has to be turned on the datum definition and the consideration of constraints; see e.g. Seitz (2015). High-resolution geoid determination is performed by combining the modern gravimetric space-geodetic techniques, i.e. CHAMP (CHAllenging Minisatellite Payload), GRACE (Gravity Recovery and Climate Experiment) and GOCE (Gravity field and steady-state Ocean Circulation Explorer)

M. Schmidt (✉) • F. Göttl
Deutsches Geodätisches Forschungsinstitut (DGFI),
Alfons-Goppel-Str. 11, 80539 München, Germany
e-mail: schmidt@dgfi.badw.de

R. Heinkelmann
Helmholtz Centre Potsdam, GFZ German Research Centre for Geosciences, Telegrafenberg, 14473 Potsdam, Germany

H. Kutterer et al. (eds.), *The 1st International Workshop on the Quality of Geodetic Observation and Monitoring Systems (QuGOMS'11)*, International Association of Geodesy Symposia 140,
DOI 10.1007/978-3-319-10828-5_6,

with terrestrial, airborne and shipborne measurements as well as data from altimetry missions. This can be achieved within a rigorous combination, mostly by considering *prior information* in form of a given gravity model (see e.g. van Loon and Kusche 2007) or by *least-squares collocation* (see e.g. Kühtreiber and Abd-Elmotaal 2007). Since GRACE provides accurate satellite data for the long wavelength part (larger 500 km) of the gravity field, GOCE however for the wavelength part between 100 and 500 km a *spectral combination* (see e.g. Gitlein et al. 2007) or a multi-scale representation (see e.g. Haagmans et al. 2002) can be applied for high-resolution geoid modeling.

Recently data combination procedures entered other fields of geodetic interest such as in hydrosphere or atmosphere modeling; see e.g. Dettmering et al. (2011). A crucial point is always the proper choice of the complete covariance matrix of the observations. A detailed discussion on this topic is, for instance, given by van Loon (2008).

The outline of this paper is the following. After introducing the general Gauss-Markov model in Sect. 2, we present in Sect. 3 altogether six models for combining data sets from various observations techniques. As an example we demonstrate in Sect. 4 how a full covariance matrix can be constructed for time series processed in different analysis centers (ACs) from the same original measurements. The different ACs are treated as here like different observation techniques. In Sect. 5 finally we discuss the multi-scale representation for providing a concept for a multi-scale (spectral) combination.

2 Gauss-Markov Model

The Gauss-Markov model is defined as

$$\mathbf{y} + \mathbf{e} = \mathbf{X}\boldsymbol{\beta} \quad \text{with} \quad D(\mathbf{y}) = \sigma^2\mathbf{P}^{-1} = \sigma^2\mathbf{Q}\,, \tag{1}$$

wherein $\mathbf{y}$ is the $n \times 1$ vector of the observations, $\mathbf{e}$ the related $n \times 1$ vector of the observation errors, $\mathbf{X}$ the $n \times u$ given coefficient (design) matrix, $\boldsymbol{\beta}$ the $u \times 1$ vector of the unknown parameters, σ^2 the unknown variance factor, $\mathbf{P}$ the $n \times n$ given positive definite weight matrix of the observations and $\mathbf{Q} = \mathbf{P}^{-1}$ the $n \times n$ given cofactor matrix, furthermore $n > u$ holds; see e.g. Koch (1999). In case of rank$\mathbf{X} = u$ we denote the model (Eq. (1)) as *Gauss-Markov model of full rank*. However, if rank$\mathbf{X} = r < u$ holds, the model (Eq. (1)) is called *Gauss-Markov model not of full rank* and the rank deficiency amounts $d = u - r$.

The *least squares method* yields the normal equation system

$$\mathbf{N}\boldsymbol{\beta} = \mathbf{b}\,, \tag{2}$$

wherein we introduced the normal equation matrix $\mathbf{N} = \mathbf{X}^T\mathbf{P}\mathbf{X}$ as well as the "right-hand side" vector $\mathbf{b} = \mathbf{X}^T\mathbf{P}\mathbf{y}$. In case of full rank, i.e. rank$\mathbf{N} = u$ we obtain the solution

$$\hat{\boldsymbol{\beta}} = \mathbf{N}^{-1}\mathbf{b}\,. \tag{3}$$

In case of a rank deficiency, i.e. rank$\mathbf{N} = r < u$ the solution reads

$$\overline{\boldsymbol{\beta}} = \mathbf{N}^{-}\mathbf{b}\,, \tag{4}$$

wherein $\mathbf{N}^{-}$ means a generalized inverse of $\mathbf{N}$, e.g. the pseudoinverse $\mathbf{N}^{+}$.

In the two cases the covariance matrices of the estimations $\hat{\boldsymbol{\beta}}$ and $\overline{\boldsymbol{\beta}}$ are given as

$$D(\hat{\boldsymbol{\beta}}) = \sigma^2\mathbf{N}^{-1}\,, \tag{5}$$

$$D(\overline{\boldsymbol{\beta}}) = \sigma^2\mathbf{N}^{-}\,. \tag{6}$$

An estimation of the variance factor can be derived from the *maximum-likelihood method*; see e.g. Koch (1999).

3 Combination Models

Whereas the vector $\mathbf{y}$ introduced in the Gauss-Markov model (Eq. (1)) can be assumed to be the observation vector of a single technique, we discuss in the following several combination models (CMs) within the *multi-technique* case. To be more specific, we combine the observation vectors $\mathbf{y}_p$ with $p = 1, \ldots, P$ from altogether P techniques such as GNSS, VLBI or SLR. Note, that we identify a set of GNSS observation sites or GRACE ACs (see Sect. 4) also with different observation techniques or groups. In the CMs we distinguish between different stochastic approaches for the observation vectors and different kinds of unknown deterministic parameters.

CM 1: First we assume that for each technique p the vector $\boldsymbol{\beta}$ of the unknown parameters is the same; i.e. we reformulate the Gauss-Markov model (Eq. (1)) as

$$\mathbf{y}_p + \mathbf{e}_p = \mathbf{X}_p\boldsymbol{\beta} \quad \text{with} \quad C(\mathbf{y}_p, \mathbf{y}_q) = \sigma^2\mathbf{Q}_{p,q} \tag{7}$$

for $q, p = 1, \ldots, P$, wherein $\mathbf{y}_p$ is the $n_p \times 1$ vector of the observations and $\mathbf{e}_p$ the $n_p \times 1$ vector of the observation errors. Furthermore, $\mathbf{X}_p$ is the $n_p \times u$ given design matrix and $\mathbf{Q}_{p,q} = \mathbf{Q}_{q,p}^T$ the $n_p \times n_q$ given cofactor matrix between the observation vectors $\mathbf{y}_p$ and $\mathbf{y}_q$; the other quantities have been already defined in the context of Eq. (1). Next we rewrite CM 1 (Eq. (7)) and obtain the formulation

$$\begin{bmatrix} \mathbf{y}_1 \\ \mathbf{y}_2 \\ \cdot \\ \mathbf{y}_P \end{bmatrix} + \begin{bmatrix} \mathbf{e}_1 \\ \mathbf{e}_2 \\ \cdot \\ \mathbf{e}_P \end{bmatrix} = \begin{bmatrix} \mathbf{X}_1 \\ \mathbf{X}_2 \\ \cdot \\ \mathbf{X}_P \end{bmatrix} \boldsymbol{\beta}$$

with

$$D(\begin{bmatrix} \mathbf{y}_1 \\ \mathbf{y}_2 \\ \cdot \\ \mathbf{y}_P \end{bmatrix}) = \sigma^2 \begin{bmatrix} \mathbf{Q}_{1,1} & \mathbf{Q}_{1,2} & \dots & \mathbf{Q}_{1,P} \\ \mathbf{Q}_{2,1} & \mathbf{Q}_{2,2} & \dots & \mathbf{Q}_{2,P} \\ & \dots\dots\dots\dots & & \\ \mathbf{Q}_{P,1} & \mathbf{Q}_{P,2} & \dots & \mathbf{Q}_{P,P} \end{bmatrix} . \quad (8)$$

Note, we require for the total number $n = \sum_{p=1}^{P} n_p$ of observations that the inequality $n > u$ holds. The complete design matrix determines the total rank of the model. CM 1 (Eq. (8)) could be transferred into the Gauss-Markov model (Eq. (1)) by defining the $n \times 1$ vectors $\mathbf{y} = \left[\mathbf{y}_1^T, \mathbf{y}_2^T, \dots, \mathbf{y}_p^T\right]^T$ and $\mathbf{e} = \left[\mathbf{e}_1^T, \mathbf{e}_2^T, \dots, \mathbf{e}_p^T\right]^T$ as well as the corresponding coefficient matrix $\mathbf{X}$ and weight matrix $\mathbf{P}$. Thus, the estimation of the unknown parameter vector $\boldsymbol{\beta}$ is obtained from Eqs. (3) or (4), respectively.

CM 2: In our second approach we assume that the observation vectors of the different techniques are mutually uncorrelated, i.e. we set $\mathbf{Q}_{p,q} = \mathbf{0}$ for $p \neq q$ in CM 1 (Eq. (8)). With $\mathbf{P}_{p,p} = \mathbf{Q}_{p,p}^{-1}$ we obtain with $\mathbf{N}_p = \mathbf{X}_p^T \mathbf{P}_{p,p} \mathbf{X}_p$ and $\mathbf{b}_p = \mathbf{X}_p^T \mathbf{P}_{p,p} \mathbf{y}_p$ the normal equation system

$$\left(\sum_{p=1}^{P} \mathbf{N}_p \right) \boldsymbol{\beta} = \sum_{p=1}^{P} \mathbf{b}_p \,. \quad (9)$$

Due to the neglect of the cofactor matrices $\mathbf{Q}_{p,q}$ for $p \neq q$, i.e. the correlations between the observation vectors $\mathbf{y}_p$ and $\mathbf{y}_q$, the estimated variances of the estimated unknown parameter vector

$$\hat{\boldsymbol{\beta}} = \left(\sum_{p=1}^{P} \mathbf{N}_p \right)^{-1} \sum_{p=1}^{P} \mathbf{b}_p \quad (10)$$

are usually too optimistic. Note, in Eq. (10) we require $\text{rank}(\sum_{p=1}^{P} \mathbf{N}_p) = u$.

CM 3: In the third approach we separate for each technique the $u \times 1$ vector $\boldsymbol{\beta} = [\boldsymbol{\beta}_p^T, \boldsymbol{\beta}_c^T]^T$ into a $u_p \times 1$ technique dependent subvector $\boldsymbol{\beta}_p$ of so-called *local* parameters and a technique independent $u_c \times 1$ subvector $\boldsymbol{\beta}_c$ of *common* or *global* parameters. Under this assumption the extended version of CM 1 (Eq. (7)) reads

$$\mathbf{y}_p + \mathbf{e}_p = \left[\mathbf{X}_{p,p} \ \ \mathbf{X}_{p,c}\right] \begin{bmatrix} \boldsymbol{\beta}_p \\ \boldsymbol{\beta}_c \end{bmatrix}$$

$$\text{with} \quad C(\mathbf{y}_p, \mathbf{y}_q) = \sigma^2 \mathbf{Q}_{p,q} \,, \quad (11)$$

wherein $\mathbf{X}_{p,p}$ and $\mathbf{X}_{p,c}$ are given $n_p \times u_p$ and $n_p \times u_c$ coefficient block matrices. Reformulating this model yields

$$\begin{bmatrix} \mathbf{y}_1 \\ \mathbf{y}_2 \\ \cdot \\ \mathbf{y}_P \end{bmatrix} + \begin{bmatrix} \mathbf{e}_1 \\ \mathbf{e}_2 \\ \cdot \\ \mathbf{e}_P \end{bmatrix} = \begin{bmatrix} \mathbf{X}_{1,1} & \dots & \mathbf{0} & \mathbf{X}_{1,c} \\ \mathbf{0} & \dots & \mathbf{0} & \mathbf{X}_{2,c} \\ & \dots\dots\dots & & \\ \mathbf{0} & \dots & \mathbf{X}_{P,P} & \mathbf{X}_{P,c} \end{bmatrix} \begin{bmatrix} \boldsymbol{\beta}_1 \\ \boldsymbol{\beta}_2 \\ \cdot \\ \boldsymbol{\beta}_P \\ \boldsymbol{\beta}_c \end{bmatrix}$$

with

$$D(\begin{bmatrix} \mathbf{y}_1 \\ \mathbf{y}_2 \\ \cdot \\ \mathbf{y}_P \end{bmatrix}) = \sigma^2 \begin{bmatrix} \mathbf{Q}_{1,1} & \mathbf{Q}_{1,2} & \dots & \mathbf{Q}_{1,P} \\ \mathbf{Q}_{2,1} & \mathbf{Q}_{2,2} & \dots & \mathbf{Q}_{2,P} \\ & \dots\dots\dots & & \\ \mathbf{Q}_{P,1} & \mathbf{Q}_{P,2} & \dots & \mathbf{Q}_{P,P} \end{bmatrix} . \quad (12)$$

CM 4: In this case we introduce for each technique an unknown individual variance component σ_p^2 and an unknown covariance components $\sigma_{p,q}$ for the covariance matrices between two individual techniques. Thus, the generalized version of CM 3 (Eq. (11)) reads

$$\mathbf{y}_p + \mathbf{e}_p = \left[\mathbf{X}_{p,p} \ \ \mathbf{X}_{p,c}\right] \begin{bmatrix} \boldsymbol{\beta}_p \\ \boldsymbol{\beta}_c \end{bmatrix}$$

$$\text{with} \quad C(\mathbf{y}_p, \mathbf{y}_q) = \sigma_{p,q} \mathbf{Q}_{p,q} \quad (13)$$

with $\sigma_{p,q} = \sigma_{q,p}$ and $\sigma_{p,p} = \sigma_p^2$. Reformulating the model above yields the *linear model with unknown variance and covariance components*

$$\begin{bmatrix} \mathbf{y}_1 \\ \mathbf{y}_2 \\ \cdot \\ \mathbf{y}_P \end{bmatrix} + \begin{bmatrix} \mathbf{e}_1 \\ \mathbf{e}_2 \\ \cdot \\ \mathbf{e}_P \end{bmatrix} = \begin{bmatrix} \mathbf{X}_{1,1} & \dots & \mathbf{0} & \mathbf{X}_{1,c} \\ \mathbf{0} & \dots & \mathbf{0} & \mathbf{X}_{2,c} \\ & \dots\dots\dots & & \\ \mathbf{0} & \dots & \mathbf{X}_{P,P} & \mathbf{X}_{P,c} \end{bmatrix} \begin{bmatrix} \boldsymbol{\beta}_1 \\ \boldsymbol{\beta}_2 \\ \cdot \\ \boldsymbol{\beta}_P \\ \boldsymbol{\beta}_c \end{bmatrix}$$

with

$$D(\begin{bmatrix} \mathbf{y}_1 \\ \mathbf{y}_2 \\ \cdot \\ \mathbf{y}_P \end{bmatrix}) = \begin{bmatrix} \sigma_1^2 \mathbf{Q}_{1,1} & \sigma_{1,2} \mathbf{Q}_{1,2} & \dots & \sigma_{1,P} \mathbf{Q}_{1,P} \\ \sigma_{1,2} \mathbf{Q}_{2,1} & \sigma_2^2 \mathbf{Q}_{2,2} & \dots & \sigma_{2,P} \mathbf{Q}_{2,P} \\ & \dots\dots\dots\dots\dots & & \\ \sigma_{1,P} \mathbf{Q}_{P,1} & \sigma_{2,P} \mathbf{Q}_{P,2} & \dots & \sigma_P^2 \mathbf{Q}_{P,P} \end{bmatrix} ; \quad (14)$$

for the solution see, for instance, Koch (1999).

CM 5: In our next approach we again assume that the observation vectors of the different techniques are mutually uncorrelated, i.e. $\mathbf{Q}_{p,q} = \mathbf{0}$ for $p \neq q$ in CM 4 (Eq. (14)). Thus, we obtain from CM 4 (Eq. (13)) the *linear model with unknown variance components*

$$\mathbf{y}_p + \mathbf{e}_p = \left[\mathbf{X}_{p,p} \ \ \mathbf{X}_{p,c}\right] \begin{bmatrix} \boldsymbol{\beta}_p \\ \boldsymbol{\beta}_c \end{bmatrix}$$

$$\text{with} \quad C(\mathbf{y}_p, \mathbf{y}_q) = \delta_{p,q} \sigma_{p,q} \mathbf{Q}_{p,q} \,, \quad (15)$$

wherein the delta symbol $\delta_{p,q}$ is defined as $\delta_{p,q} = 1$ for $p = q$ and $\delta_{p,q} = 0$ for $p \neq q$. Note, that e.g. the computation of the International Terrestrial Reference Frame (ITRF) is based on CM 5 (Eq. (15)); see Seitz (2015).

As a specialization of CM 5 (Eq. (15)) we remove the local parameter vectors $\boldsymbol{\beta}_p$ and obtain with $\boldsymbol{\beta}_c = \boldsymbol{\beta}$ the linear model with unknown variance component components as

$$\mathbf{y}_p + \mathbf{e}_p = \mathbf{X}_p\,\boldsymbol{\beta} \quad \text{with } C(\mathbf{y}_p, \mathbf{y}_q) = \delta_{p,q}\,\sigma_{p,q}\,\mathbf{Q}_{p,q} \tag{16}$$

as an extension of CM 1 (Eq. (7)). Applying e.g. the least squares method to CM 5 (Eq. (16)) we obtain the normal equation system

$$\left(\sum_{p=1}^{P} \frac{1}{\sigma_p^2}\mathbf{N}_p\right)\boldsymbol{\beta} = \sum_{p=1}^{P} \frac{1}{\sigma_p^2}\mathbf{b}_p\,. \tag{17}$$

In case that the system is of full rank, the estimation of the unknown vector $\boldsymbol{\beta}$ reads

$$\hat{\boldsymbol{\beta}} = \left(\sum_{p=1}^{P} \frac{1}{\sigma_p^2}\mathbf{N}_p\right)^{-1}\left(\sum_{p=1}^{P} \frac{1}{\sigma_p^2}\mathbf{b}_p\right). \tag{18}$$

The variance components σ_p^2 can be estimated from the residuals

$$\hat{\mathbf{e}}_p = \mathbf{X}_p\,\hat{\boldsymbol{\beta}} - \mathbf{y}_p \tag{19}$$

according to $\hat{\sigma}_p^2 = (\hat{\mathbf{e}}_p^T\,\mathbf{P}_{p,p}\,\hat{\mathbf{e}}_p)/r_p$ by means of the partial redundancy r_p (Koch 1999). Consequently, the estimation (18) has to be performed iteratively. An efficient calculation of the variance components is given by a fast Monte-Carlo implementation of the iterative maximum-likelihood variance component estimation as described by Koch and Kusche (2001).

CM 6: In this approach we assume that *prior information* for the expectation vector $E(\boldsymbol{\beta}) = \boldsymbol{\mu}_\beta$ and the covariance matrix $D(\boldsymbol{\beta}) = \mathbf{Q}_\beta$ of the unknown parameter vector $\boldsymbol{\beta}$ are available. To be more specific we introduce the additional Gauss-Markov model

$$\boldsymbol{\mu}_\beta + \mathbf{e}_\beta = \boldsymbol{\beta} \quad \text{with} \quad D(\boldsymbol{\mu}_\beta) = \sigma_\beta^2\,\mathbf{Q}_\beta \tag{20}$$

wherein $\mathbf{e}_\beta$ is the $u \times 1$ vector of the errors for the prior information and σ_β^2 an unknown variance factor of the pseudo observation vector $\boldsymbol{\mu}_\beta$. By combining the P models (Eq. (16)) and the additional model (Eq. (20)) we obtain the *linear model with unknown variance components and prior information* as

$$\begin{bmatrix}\mathbf{y}_1\\ \mathbf{y}_2\\ \cdot\\ \mathbf{y}_P\\ \boldsymbol{\mu}_\beta\end{bmatrix} + \begin{bmatrix}\mathbf{e}_1\\ \mathbf{e}_2\\ \cdot\\ \mathbf{e}_P\\ \mathbf{e}_\beta\end{bmatrix} = \begin{bmatrix}\mathbf{X}_1\\ \mathbf{X}_2\\ \cdot\\ \mathbf{X}_P\\ \mathbf{I}\end{bmatrix}\boldsymbol{\beta} \quad \text{with}$$

$$D\left(\begin{bmatrix}\mathbf{y}_1\\ \mathbf{y}_2\\ \cdot\\ \mathbf{y}_P\\ \boldsymbol{\mu}_\beta\end{bmatrix}\right) = \begin{bmatrix}\sigma_1^2\mathbf{Q}_{1,1} & \mathbf{0} & \dots & \mathbf{0} & \mathbf{0}\\ \mathbf{0} & \sigma_2^2\mathbf{Q}_{2,2} & \dots & \mathbf{0} & \mathbf{0}\\ & \dots\dots\dots\dots\dots & & & \\ \mathbf{0} & \mathbf{0} & \dots & \sigma_P^2\mathbf{Q}_{P,P} & \mathbf{0}\\ \mathbf{0} & \mathbf{0} & \dots & \mathbf{0} & \sigma_\beta^2\mathbf{Q}_\beta\end{bmatrix}. \tag{21}$$

The extended normal equations for the unknown parameter vector $\boldsymbol{\beta}$ including the unknown variance components σ_p^2 and σ_β^2 are given as

$$\left(\sum_{p=1}^{P} \lambda_p\,\mathbf{N}_p + \mathbf{P}_\beta\right)\hat{\boldsymbol{\beta}} = \sum_{p=1}^{P} \lambda_p\,\mathbf{b}_p + \mathbf{P}_\beta\,\boldsymbol{\mu}_\beta \tag{22}$$

with $\lambda_p = \sigma_\beta^2/\sigma_p^2$. Since the matrix $\mathbf{P}_\beta = \mathbf{Q}_\beta^{-1}$ is positive definite and the P matrices $\mathbf{N}_p$ at least positive semidefinite, the normal equation matrix $(\sum_{p=1}^{P} \lambda_p\,\mathbf{N}_p + \mathbf{P}_\beta)$ is regular and Eq. (22) can be solved for

$$\hat{\boldsymbol{\beta}} = \left(\sum_{p=1}^{P} \lambda_p\mathbf{N}_p + \mathbf{P}_\beta\right)^{-1}\left(\sum_{p=1}^{P} \lambda_p\mathbf{b}_p + \mathbf{P}_\beta\boldsymbol{\mu}_\beta\right), \tag{23}$$

see Koch (2000) or Koch and Kusche (2001).

4 Numerical Example

Assuming a non-rigid Earth body the motion of the Celestial Intermediate Pole (CIP) is calculable from the differential equation

$$p(t) + \frac{i}{\Omega_{CW}}\frac{dp(t)}{dt} = \chi(t)\,, \tag{24}$$

wherein Ω_{CW} means the Chandler frequency and $i = \sqrt{-1}$; for more details see e.g. Gross (2007). Equation (24) states that the complex-valued motion of the CIP, $p(t)$, is driven by the equatorial *excitation function* $\chi(t) = \chi_1(t) + i\,\chi_2(t)$. According to $\chi_j(t) = \chi_j^{mass}(t) + \chi_j^{motion}(t)$, the components $\chi_j(t)$ with $j = 1, 2$ can each be separated into a matter term $\chi_j^{mass}(t)$ and a motion term $\chi_j^{motion}(t)$. Whereas the latter is caused by moving masses the matter term is the consequence of mass changes within the Earth system and can be calculated by the degree two spherical harmonic coefficients $\Delta C_{2,1}(t)$ and $\Delta S_{2,1}(t)$ of the gravitational potential

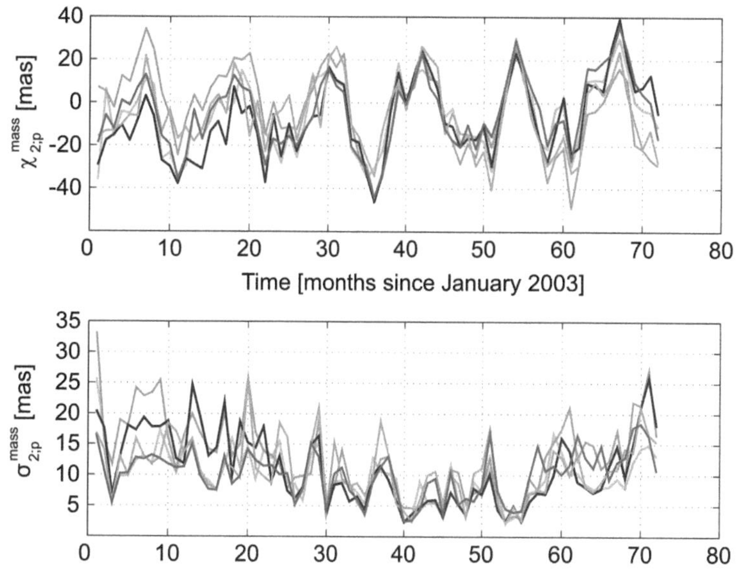

Fig. 1 *Top*: monthly time series of the excitation function $\chi^{mass}_{2;p}(t_k)$ calculated from the degree 2 spherical harmonic coefficients $\Delta S_{2,1;p}(t_k)$ from $P = 5$ GRACE ACs, namely CSR (*cyan*), GFZ (*dark blue*), JPL (*blue*), IGG (*orange*) and DEOS (*magenta*) at times t_k with $k = 1, \ldots 72$ within the time interval between January 2003 and December 2008; *bottom*: empirical standard deviations $\sigma^{mass}_{2;p}(t_k)$ of the time series; all data in mas

according to $\chi^{mass}_1(t) = \chi^{mass}_1(\Delta C_{2,1}(t))$ and $\chi^{mass}_2(t) = \chi^{mass}_2(\Delta S_{2,1}(t))$.

Several processing or analysis centers (ACs) of the GRACE K-band measurements, e.g. GFZ (GeoForschungsZentrum Potsdam), CSR (Center of Space Research, University of Texas, Austin, USA), JPL (Jet Propulsion Laboratory, Pasadena, USA), IGG (Institute for Geodesy and Geoinformation, University of Bonn) and DEOS (Delft Institute of Earth Observation and Space Systems), provide time-variable gravitational potential models. Indicating the different ACs by the index $p = 1, \ldots, P$ with P the total number of ACs we decompose $\chi^{mass}_j(t) =: \chi^{mass}_{j;p}(t)$ according to

$$\chi^{mass}_{j;p}(t) = \chi^{mass}_j(t) + \Delta\chi^{mass}_{j;p}(t) \qquad (25)$$

into an *AC independent* term $\chi^{mass}_j(t)$ and an *AC dependent* correction term $\Delta\chi^{mass}_{j;p}(t)$. The first term stays for the fact that all gravitational potential models are derived from the same GRACE input data. However, different parameterizations, software packages, background models, standards and procedures (e.g. outlier detection), etc. cause AC dependent influences which can be summarized as so-called operator-software impact (OSI) parameter; see e.g. Kutterer et al. (2009) or Fang (2007). Thus, the term $\Delta\chi^{mass}_{j;p}(t)$ means the *OSI deviation* of the processed value $\chi^{mass}_{j;p}(t)$ from the "true" value $\chi^{mass}_j(t)$. Figure 1 shows exemplarily 5 monthly time series $\chi^{mass}_{2;p}(t_k)$ with $p = 1, \ldots, P = 5$ at discrete times t_k for a time span of 6 years between January 2003 and December 2008, wherein the index $k = 1, \ldots, K = 72$ indicates the 72 months starting with January 2003. As can be seen from the bottom panel of Fig. 1 the empirical monthly standard deviations $\sigma^{mass}_{2;p}(t_k)$ calculated via the relation

$$\sigma^{mass}_{j;p}(t_k) = \frac{1}{\sqrt{P-1}} \left(\sum_{\substack{q=1\\q\neq p}}^{P} \left(\chi^{mass}_{j;q}(t_k) - \chi^{mass}_{j;p}(t_k) \right)^2 \right)^{1/2}$$

are for $j = 2$ in the range between 10 and 15 mas. Usually the time series $\chi^{mass}_{j;p}(t_k)$ are considered as uncorrelated signals, although they are derived from the same GRACE raw measurements. As a consequence of this the accuracies of estimations from a combination of these time series are usually too optimistic.

For simplification we assume in the following that no systematic offsets or trends exist between the time series shown in Fig. 1 and that the OSI deviations $\Delta\chi^{mass}_{j;p}(t_k)$ could be interpreted as random variables with expectation values $E(\Delta\chi^{mass}_{j;p}(t_k)) = 0$. By introducing the $K \times 1$ observation vectors $\mathbf{y}_p = \mathbf{y} + \Delta\mathbf{y}_p = (\chi^{mass}_{j;p}(t_k))$ we reformulate CM 1 (Eq. (7)) as

$$\mathbf{y}_p + \mathbf{e}_p = \mathbf{I}_K \boldsymbol{\beta} \quad \text{with} \quad C(\mathbf{y}_p, \mathbf{y}_q) = \sigma^2 \mathbf{Q}_{p,q} \qquad (26)$$

for $q, p = 1, \ldots, P$ with the $K \times K$ unit matrix $\mathbf{X}_p = \mathbf{I}_K$, the $K \times 1$ vector $\boldsymbol{\beta} = (\chi^{mass}_j(t_k))$ of the unknown excitation values $\chi^{mass}_j(t_k)$.

In the *traditional approach*, i.e. neglecting the cofactor matrices $\mathbf{Q}_{p,q}$ for $p \neq q$, the Gauss-Markov model (Eq. (26)) is reduced to CM 2 with the normal Eqs. (9). To be more specific, we define

$$\mathbf{Q}_{p,q} = \delta_{p,q}\, \mathbf{Q}_{y,y} = \delta_{p,q} \left(\frac{1}{P} \sum_{p=1}^{P} \mathbf{Q}_{y_p,y_p} \right) \qquad (27)$$

with $\mathbf{Q}_{y_p,y_p} = \mathrm{diag}\left[(\sigma^{mass}_{j;p}(t_1))^2, \ldots, (\sigma^{mass}_{j;p}(t_K))^2 \right]$.

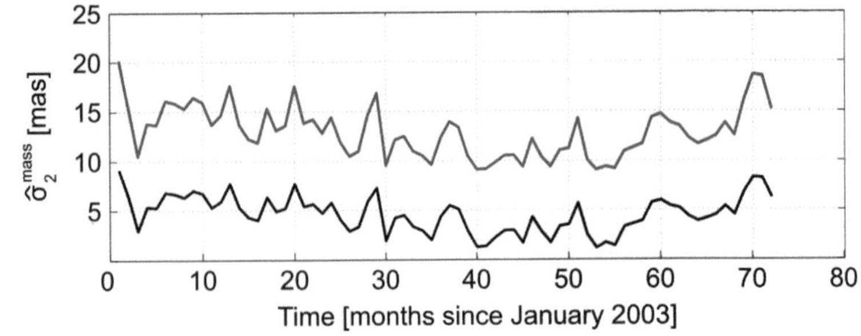

Fig. 2 Estimated standard deviations $\hat{\sigma}_2^{mass;trad}(t_k)$ and $\hat{\sigma}_2^{mass;osi}(t_k)$ from the traditional approach (*black curve*) and from the alternative approach (*red curve*)

Since the normal equation matrix is regular, the estimation $\hat{\boldsymbol{\beta}} =: \hat{\boldsymbol{\beta}}_{trad} = (\hat{\chi}_j^{mass;trad}(t_k))$ is obtained from Eq. (10).

In the *alternative approach* we consider the decomposition (25) and define for CM 1 (Eq. (7)) the covariance matrix $\mathbf{Q}_{p,q}$ as

$$\mathbf{Q}_{p,q} = \mathbf{Q}_{y,y} + \delta_{p,q}\,\mathbf{Q}_{\Delta y_p,\Delta y_p}\,, \tag{28}$$

wherein the matrix $\mathbf{Q}_{y,y}$ was already introduced by Eq. (27). According to Kutterer et al. (2009) we define the OSI deviation $\mathbf{Q}_{\Delta y_p,\Delta y_p}$ of the cofactor matrix as

$$\mathbf{Q}_{\Delta y_p,\Delta y_p} = \alpha^2\,\mathbf{Q}_{y_p,y_p}\,. \tag{29}$$

The calculation of the OSI parameter α is explained in detail by Fang (2007) and Heinkelmann et al. (2011) and is not repeated here. With a given value α the estimation $\hat{\boldsymbol{\beta}} =: \hat{\boldsymbol{\beta}}_{osi} = (\hat{\chi}_j^{mass;osi}(t_k))$ can be computed from the normal equation system of the Gauss-Markov model (Eq. (26)) considering Eqs. (28) and (29).

Due to the chosen cofactor matrices for the traditional and the alternative approach according to Eqs. (26)–(29) the estimated unknowns $\hat{\chi}_j^{mass;trad}(t_k)$ and $\hat{\chi}_j^{mass;osi}(t_k)$ (not shown here) are identical. However, as expected the estimated standard deviations are quite different as can be seen from Fig. 2. Whereas the black curve in Fig. 2, representing the standard deviations $\hat{\sigma}_2^{mass;trad}(t_k)$ of the estimated parameters according to the traditional approach, simulates unrealistic high accuracies, the range of the estimated standard deviations $\hat{\sigma}_2^{mass;osi}(t_k)$ of the alternative approach agrees well with the empirical accuracies shown in the bottom panel of Fig. 1. Thus, the results of the OSI approach seem to be more realistic.

The presented approach is just one way to consider the OSI deviations. Several other strategies are outlined by Kutterer et al. (2009), Fang (2007) or Heinkelmann et al. (2011) in detail. These procedures include variance component estimations according to CM 4 (Eq. (13)) and CM 5 (Eq. (15)).

5 Multi-scale Combination

One promising modern tool for the representation and the combination of input data from different observation techniques is the *multi-scale representation (MSR)*. The MSR means viewing on a signal under different resolutions (microscope effect). In other words the MSR provides approximations of the signal under different resolution levels. These approximations are representable as series expansions in so-called scaling functions; furthermore the differences between the approximations of two adjacent levels are called detail signals, also representable by series expansions, but this time in so-called wavelet functions.

In case of using scaling and wavelet functions as base functions for modeling the target function, e.g. the gravity field, two ways can be used for combining data from different observation techniques, namely (1) the combination of data sets from different observation techniques on the highest resolution level or (2) the estimation of the target function on different resolution levels due to the distribution and the sensitivity of the data from different techniques.

The MSR is explained in detail in many publications, e.g., by Schmidt (2007, 2012). Here we give a brief summary for the two-dimensional (2-D) case.

In the 2-D multi-scale approach we model the target function F as $F(\mathbf{x}) \approx F_J(\mathbf{x}) = F_J(x,y)$ depending on the 2-D position vector $\mathbf{x} = [x,y]^T$ on the highest resolution level J as

$$F_J(x,y) = \sum_{k_1=0}^{K_J-1}\sum_{k_2=0}^{K_J-1} d_{J;k_1,k_2}\,\phi_{J;k_1}(x)\,\phi_{J;k_2}(y)\,, \tag{30}$$

where $\phi_{J;k_1}(x)$ and $\phi_{J;k_2}(y)$ are two 1-D base functions depending on the coordinates x and y. In wavelet theory these functions are called level-J scaling functions and generate a 2-D tensor product MSR; see e.g. Schmidt (2001). As two examples the Daubechies scaling functions generate an orthogonal MSR, the endpoint-interpolating B-spline functions, however, provide a semi-orthogonal MSR; for the latter see e.g. Schmidt (2007). With the number K_J of scaling functions $\phi_{J;k}(z)$ with $z \in \{x, y\}$ and $k \in \{k_1, k_2\}$ we define the $(K_J \cdot K_J) \times 1$ vector $\mathbf{d}_J = \text{vec}\ \mathbf{D}_J$, wherein 'vec' means the vec-operator (Koch 1999). The $K_J \times K_J$ scaling coefficient matrix $\mathbf{D}_J$ is defined as

$$\mathbf{D}_J = \begin{bmatrix} d_{J;0,0} & d_{J;0,1} & \ldots & d_{J;0,K_J-1} \\ d_{J;1,0} & d_{J;1,1} & \ldots & d_{J;1,K_J-1} \\ \ldots & \ldots & \ldots & \ldots \\ d_{J;K_J-1,0} & d_{J;K_J-1,1} & \ldots & d_{J;K_J-1,K_J-1} \end{bmatrix} . \quad (31)$$

With the $K_J \times 1$ level-J scaling vector $\boldsymbol{\phi}_J(z) = (\phi_{J;k}(z))$ and considering the computation rules for the Kronecker product '$\otimes$' (see e.g. Koch 1999) Eq. (30) can be rewritten as

$$\begin{aligned} F_J(x, y) &= (\boldsymbol{\phi}_J^T(y) \otimes \boldsymbol{\phi}_J^T(x))\ \text{vec}\ \mathbf{D}_J \\ &= \boldsymbol{\phi}_J^T(x)\ \mathbf{D}_J\ \boldsymbol{\phi}_J(y) . \end{aligned} \quad (32)$$

With $j' = J - i_0$ and $0 < i_0 \leq J$ the MSR is given as

$$F_J(x, y) = F_{j'}(x, y) + \sum_{i=1}^{i_0} \sum_{\lambda=1}^{3} G_{J-i}^{\lambda}(x, y) , \quad (33)$$

wherein the low-pass filtered signal $F_{j'}(x, y)$ and the band-pass filtered detail signals $G_{J-i}^{\lambda}(x, y)$ are computable via the relations

$$\begin{aligned} F_{j'}(x, y) &= \boldsymbol{\phi}_{j'}^T(x)\ \mathbf{D}_{j'}\ \boldsymbol{\phi}_{j'}(y) , \\ G_{J-i}^1(x, y) &= \boldsymbol{\phi}_{J-i}^T(x)\ \mathbf{C}_{J-i}^1\ \boldsymbol{\psi}_{J-i}(y) , \\ G_{J-i}^2(x, y) &= \boldsymbol{\psi}_{J-i}^T(x)\ \mathbf{C}_{J-i}^2\ \boldsymbol{\phi}_{J-i}(y) , \\ G_{J-i}^3(x, y) &= \boldsymbol{\psi}_{J-i}^T(x)\ \mathbf{C}_{J-i}^3\ \boldsymbol{\psi}_{J-i}(y) \end{aligned} \quad (34)$$

by means of the $L_j \times 1$ level-j wavelet vectors $\boldsymbol{\psi}_j(z) = (\psi_{j,l}(z))$ with $L_j = K_{j+1} - K_j$.

The level-j and level-$(j-1)$ scaling functions $\phi_{j;k}(z)$ with $k = 0, \ldots, K_j - 1$ and $\phi_{j-1;n}(z)$ with $n = 0, \ldots, K_{j-1} - 1$ as well as the level-$(j-1)$ wavelet functions $\psi_{j-1;l}(z)$ with $l = 0, \ldots, L_{j-1} - 1$ are related to each other by means of the *two-scale relations*

$$\phi_{j-1;n}(z) = \sum_{k=2n-(K_j-1)}^{2n} p_{j;k}\ \phi_{j;2n-k}(z) , \quad (35)$$

$$\psi_{j-1;l}(z) = \sum_{k=2l-(K_j-1)}^{2l} q_{j;k}\ \phi_{j;2l-k}(z) \quad (36)$$

with given coefficients $p_{j;k}$ and $q_{j;k}$; for more details see e.g. (Schmidt 2012). The two Eqs. (35) and (36) can be reformulated as matrix equations

$$\begin{aligned} \boldsymbol{\phi}_{j-1}^T(x) &= \boldsymbol{\phi}_j^T(x)\ \mathbf{P}_j , \\ \boldsymbol{\psi}_{j-1}^T(x) &= \boldsymbol{\phi}_j^T(x)\ \mathbf{Q}_j \end{aligned} \quad (37)$$

with the $K_j \times K_{j-1}$ matrix $\mathbf{P}_j = (p_{j;k})$ and the $K_j \times L_{j-1}$ matrix $\mathbf{Q}_j = (q_{j;k})$. In Eqs. (34) we introduced the $K_{J-i} \times L_{J-i}$ matrix $\mathbf{C}_{J-i}^1$, the $L_{J-i} \times K_{J-i}$ matrix $\mathbf{C}_{J-i}^2$ and the $L_{J-i} \times L_{J-i}$ matrix $\mathbf{C}_{J-i}^3$. The corresponding 2-D downsampling equations read

$$\begin{bmatrix} \mathbf{D}_{j-1} & \mathbf{C}_{j-1}^1 \\ \mathbf{C}_{j-1}^2 & \mathbf{C}_{j-1}^3 \end{bmatrix} = \begin{bmatrix} \overline{\mathbf{P}}_j \\ \overline{\mathbf{Q}}_j \end{bmatrix} \mathbf{D}_j \begin{bmatrix} \overline{\mathbf{P}}_j^T & \overline{\mathbf{Q}}_j^T \end{bmatrix} \quad (38)$$

for $j = j' + 1, \ldots, J$ and mean the *pyramid algorithm*. The $K_{j-1} \times K_j$ matrix $\overline{\mathbf{P}}_j$ and the $L_{j-1} \times K_j$ matrix $\overline{\mathbf{Q}}$ are defined via the relation

$$\begin{bmatrix} \overline{\mathbf{P}}_j \\ \overline{\mathbf{Q}}_j \end{bmatrix} = \begin{bmatrix} \mathbf{P}_j & \mathbf{Q}_j \end{bmatrix}^{-1} . \quad (39)$$

Viewing on the rigorous combination explained before from the point of the MSR we (1) have to choose an appropriate number for the highest resolution level J defined in Eq. (30), (2) we perform the parameter estimation by means of an appropriate CM as defined in Sect. 3 and calculate the series coefficients collected in the matrix $\mathbf{D}_J$ (Eq. (31)) and (3) we calculate all detail signals by applying the pyramid algorithm according to Eqs. (34) and (38). Since all calculations within the pyramid algorithm are based on linear equation systems the covariance matrices of all sets of coefficients and the detail signals can be computed by means of the law of error propagation.

As an alternative to the procedure described before the spectral behavior of the different observation techniques can be used directly in the estimation procedure. Since, for instance, GOCE data cover a higher frequency part than GRACE data, the MSR comes directly into play. The procedure is visualized by the flowchart in Fig. 3. To be more specific, the high-resolution observation vector $\mathbf{y}_1$ with cofactor matrix $\mathbf{Q}_{1,1}$ of the first observation technique determines the coefficient matrix $\mathbf{D}_J$ (Eq. (31)) via a CM. The detail signals $G_{J-1} \in \{G_{J-1}^1, G_{J-1}^2, G_{J-1}^3\}$ to $G_{j_2+1} \in \{G_{j_2+1}^1, G_{j_2+1}^2, G_{j_2+1}^3\}$ of the highest levels $j_2 + 1, j_2 + 2, \ldots, J - 1$ are calculated via Eqs. (34) and (38). Applying in the next step again the pyramid algorithm the scaling coefficient matrix $\mathbf{D}_{j_2}$ of level j_2 is predicted as

$$\mathbf{D}_{j_2} = \overline{\mathbf{P}}_{j_2+1}\ \mathbf{D}_{j_2+1}\ \overline{\mathbf{P}}_{j_2+1}^T \quad (40)$$

with covariance matrix

$$\begin{aligned} D(\text{vec}\mathbf{D}_{j_2}) = (&\overline{\mathbf{P}}_{j_2+1} \otimes \overline{\mathbf{P}}_{j_2+1})\ D(\text{vec}\mathbf{D}_{j_2+1}) \\ &\cdot (\overline{\mathbf{P}}_{j_2+1}^T \otimes \overline{\mathbf{P}}_{j_2+1}^T) \end{aligned} \quad (41)$$

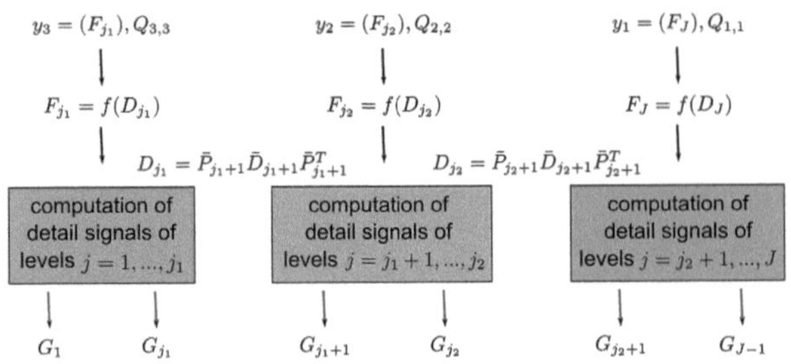

Fig. 3 Data combination within a MSR. The input data of different observation techniques are introduced into the evaluation model at different resolution levels. The coefficient matrix $\mathbf{D}_{j_2+1}$, estimated at the level $j_2 + 1$ is transferred to the lower level j_2 by means of the pyramid algorithm according to Eq. (40)

and improved by introducing the observation vector $\mathbf{y}_2$ with cofactor matrix $\mathbf{Q}_{2,2}$ of the second observation technique providing data of medium resolution. Continuing this process until the lowest level the MSR of the gravity field according to Eq. (30) is obtained.

Conclusion

In this paper we presented selected combination strategies for parameter estimation from input data of different geodetic observation techniques. Starting with the Gauss-Markov model we outline parameter estimation within different combination models including the estimation of unknown variance and covariance components. If furthermore prior information is available it can be used as additional information, e.g. by introducing pseudo observations from a given gravity field model. As discussed in various papers the variance components of the real observations and the prior information express the relative weighting and may be interpreted as regularization parameters; see e.g. Koch and Kusche (2001).

As an alternative approach to the rigorous combination we discussed the MSR for combining input data from observation techniques which are sensitive to different parts of the frequency spectrum. A further advantage of the MSR based on wavelet expansion is that usually a lot of elements of the matrices $\mathbf{C}_j^\lambda$ introduced in Eq. (34) are close to zero. Thus, as already standard in digital image processing efficient data compression techniques can be applied. This way, just the significant information of the signal is stored. Finally, it is worth to mention that all methods mentioned before can be formulated in the framework of Bayesian Inference (Koch 2000).

We discussed briefly the problem of the OSI deviation. This example demonstrates the importance of an realistic stochastic model. In our opinion it cannot be the main goal of data evaluation to achieve the smallest values for the estimated standard deviations. Instead of this we aim at *realistic accuracies*!

Finally we want to mention that future data combination goes far beyond the combination of geodetic techniques, since especially data from remote sensing missions have to be connected with geodetic data. In such a procedure, e.g. point measurements from GNSS have to be combined with superficial measurements from InSAR. These kinds of combination mean a further challenge for geodesy in the very near future.

References

Dettmering D, Schmidt M, Heinkelmann R, Seitz M (2011) Combination of different satellite observation data for regional ionosphere modeling. J Geod 85(12):989–998. doi:10.1007/s00190-010-0423-1

Fang X (2007) Statistische Analyse von Ausgleichungsergebnissen auf der Basis teilweise identischer Messwerte. Diploma thesis, Geodetic Institute, Leibniz University of Hannover

Gitlein O, Denker H, Müller J (2007) Local geoid computation by the spectral combination method. In: Jekeli C, Bastos L, Fernandes J (eds) Gravity, geoid and space missions. IAG symposia, vol 129. Springer, Berlin, pp 179–184

Gross R (2007) Earth rotation variations - long period. In: Schubert G (ed) Treatise on geophysics, vol 3 (Geodesy), Sect. 3.11. Elsevier, Amsterdam

Haagmans R, Prijatna K, Dahl-Omang O (2002) An alternative concept for validation of GOCE gradiometry results based on regional gravity. In: Proceedings of GG2002 meeting in Thessaloniki, pp 281–286

Heinkelmann R et al (2011) VLBI-derived troposphere parameter during CONT08. J Geod. doi:10.1007/s00190-011-0459-x

Koch KR (1999) Parameter estimation and hypothesis testing in linear models. Springer, Berlin

Koch KR (2000) Einführung in die Bayes-Statistik. Springer, Berlin

Koch KR, Kusche J (2001) Regularization of geopotential determination from satellite data by variance components. J Geod 76:259–268

Kühtreiber N, Abd-Elmotaal HA (2007) Ideal combination of deflection components and gravity anomalies for precise geoid computation. In: Tregonin P, Rizos C (eds) Dynamic planet, monitoring and understanding a dynamic planet with geodetic and oceanographic tools. IAG symposia, vol 130. Springer, Berlin, pp 259–265

Kutterer H, Krügel M, Tesmer V (2009). Towards an improved assessment of the quality of terrestrial reference frames. In: Drewes H (ed) Geodetic reference frames. IAG symposia, 134. Springer, New York, pp 67–72

Rummel R, Rothacher M, Beutler G (2005) Global geodetic observing system (GGOS): science rationale. J Geodyn 40:357–362

Schmidt M (2001) Grundprinzipien der Wavelet-Analyse und Anwendungen in der Geodäsie. Post doctoral thesis, Shaker, Aachen

Schmidt M (2007) Wavelet modeling in support of IRI. J Adv Sp Res 39:932–940

Schmidt M (2012) Towards a multi-scale representation of multi-dimensional signals. In: Sneeuw N et al (eds) VII Hotine-Marussi symposium on mathematical geodesy. IAG symposia, vol 137. Springer, Berlin, pp 119–127

Seitz M (2015) Combination of different space geodetic data types in order to compute terrestrial reference frames and time series of geophysical parameters. In: Kutterer H, Seitz F, Schmidt M (eds) Proceedings of the 1st international workshop on the quality of geodetic observation and monitoring system (QuGoMS), Munich,

2011. International association of geodesy symposia. Springer, Heidelberg, this volume

van Loon J (2008) Functional and stochastic modelling of satellite gravity data. In: Publication on geodesy, vol 67. Netherlands Geodetic Commission, Delft University of Technology

van Loon J, Kusche J (2007) Towards an optimal combination of satellite data and prior information. In: Tregonin P, Rizos C (eds) Dynamic planet, monitoring and understanding a dynamic planet with geodetic and oceanographic tools. IAG symposia, vol 130. Springer, Berlin, pp 345–353

On the Weighted Total Least Squares Solutions

X. Fang and H. Kutterer

Abstract

Nowadays the terminology Total Least Squares (TLS) is frequently used as a standard name of the estimation method for the errors-in-variables (EIV) model. Although a significant number of contribution have been published to adjust the EIV model, the computational advantages of the TLS problem are still largely unknown. In this contribution various approaches are applied for solving the weighted TLS problem, where the covariance matrix of the observation vector can be fully populated: 1. The auxiliary Lagrange multipliers are applied to give some implementations for solving the problem. 2. In contrast to the nonlinear Gauss–Helmert model (GHM) proposed by other authors, the model matrices and the inconsistency vector are analytically formulated within the GHM. 3. The gradient of the objective function is given when the weighted TLS problem is expressed as an unconstrained optimization problem. If the gradient equals to zero, the necessary conditions for the optimality are identical with the normal equation which is derived by Lagrange multipliers. Furthermore, a numerical example demonstrates the identical solution by the proposed algorithms.

Keywords

Errors-in-variables • Gauss–Helmert model • Gradient • Lagrange multipliers • Total least squares

1 Introduction

It is well known that the method of the Least Squares (LS), which has been developed by C.F. Gauss and A.M. Legendre in the nineteenth century, is applied to approximate solutions of these overdeterminedsystems $\mathbf{y} + \mathbf{v}_\mathbf{y} = \mathbf{A}\boldsymbol{\xi}$, where the traditional observation vector $\mathbf{y}$ affected by random errors has the corresponding residual vector $\mathbf{v}_\mathbf{y}$, and $\boldsymbol{\xi}$ is an unknown parameter vector. However, this is not always the case in the reality because the elements of the coefficient matrix can be the quantities observed in some geodetic problems. If the coefficient matrix is affected by random errors, the model is called errors-in-variables (EIV) model. The terminology, Total Least Squares (TLS), which is used for adjusting the EIV model to estimate the parameters, was introduced by Golub and Van Loan (1980) in numeric analysis, also seen in Van Huffel and Vandewalle (1991).

Recently, the investigation about the TLS estimation has been shown in quite a number of publications in geodesy. From the methodological point of view, the most frequently used approach, which rigorously adjusts the EIV model, is the closed form solution in terms of the singular value decomposition (SVD) of the data matrix (e.g., Teunissen 1988; Felus 2004; Akyilmaz 2007; Schaffrin and Felus 2008). Here, it should be mentioned that the generalized

X. Fang (✉)
School of Geodesy & Geomatics, Wuhan University, Wuhan, China
Geodetic institute, Leibniz University of Hanover, Hanover, Germany
e-mail: wflsfang1981@hotmail.com

H. Kutterer
Federal Agency for Cartography and Geodesy, Frankfurt, Germany
e-mail: hansjoerg.kutterer@bkg.bund.de

H. Kutterer et al. (eds.), *The 1st International Workshop on the Quality of Geodetic Observation and Monitoring Systems (QuGOMS'11)*, International Association of Geodesy Symposia 140,
DOI 10.1007/978-3-319-10828-5_7,

TLS solution as applied by Akyilmaz (2007) does not directly yield the solution for the structured TLS problem, see Schaffrin (2008). Another kind of methods uses auxiliary Lagrange multipliers to rearrange the TLS problem as a constrained minimization optimization, where the covariance matrix of the observations including the conventional observations and the observation in the model matrix is fairly general (e.g., Felus and Burtch 2009; Schaffrin and Wieser 2008). Furthermore, avoiding pitfalls the non-linear Gauss Helmert model (GHM) method proposed by Pope (1972) can solve the weighted TLS problem without any limitation of the variance covariance matrix (vcm). This also proves the TLS adjustment not referring to a new adjustment method but the adjustment for the model containing the uncertain model matrix (see, Neitzel and Petrovic 2008; Neitzel 2010; Schaffrin and Snow 2010). Note that in order to hold the consistency we will still use the term TLS throughout the report as a synonym for the LS estimation of the parameters in the EIV model.

Although the GHM method can solve the weighted TLS problem, the method is regarded as a particular one method to solve the nonlinear normal equation (Schaffrin 2007). The other algorithms cannot adjust the EIV model without the limitation of the weight information. In this contribution, the solutions for the TLS adjustment are established, which have the following properties: 1. The solutions do not have any limitation of the weight information. 2. They are not only a particular method to solve the nonlinear normal equations (i.e. the solutions are based on the different principles, see Chap. 3 for detail). 3. They have no limitations concerning the structure of the coefficient matrix $\mathbf{A}$. The presented solutions can be applied for solving weighted structured TLS problems. Furthermore, a numerical example demonstrates the identical solution by the proposed algorithms. The discussion about the relationship of the solutions, the sufficient conditions of the optimization problem is organized in the last part.

2 Mathematical Models of the TLS Problem

The LS estimation is the best linear unbiased estimation when the design matrix $\mathbf{A}$ is free of noise and the observation vector $\mathbf{y}$ is affected by random errors. This kind of estimation has frequently been applied in the Gauss–Markov model for the error adjustment. In contrast, an EIV model is a model similar to Gauss Markov model (GMM) except that the elements of the design matrix are observed with random errors. The LS adjustment is statistically motivated as a maximum likelihood estimator in a linear GMM, and the TLS as maximum likelihood estimator in the EIV model. The definition of the weighted TLS can be expressed as an optimization problem as follows

$$\begin{aligned} &\mathbf{v}^T\mathbf{P}\mathbf{v} = \min \\ &\textit{subject to } \mathbf{y} + \mathbf{v_y} = (\mathbf{A} + \mathbf{V_A})\,\boldsymbol{\xi} \end{aligned} \qquad (1)$$

with $\mathbf{v} = vec\left(\left[\mathbf{V_A}\ \mathbf{v_y}\right]\right)$ ('vec' denotes the operator that stacks one column of a matrix underneath the previous one). The matrix $\mathbf{A}$ affected by random errors should be corrected by the residual matrix $\mathbf{V_A}$. Note that there is an alternative formulation, which uses the error vector and matrix instead of the residual vector and matrix. i.e. error matrix $\mathbf{E_A} = -\mathbf{V_A}$, and error vector $\mathbf{e_y} = -\mathbf{v_y}$ (see, Schaffrin and Wieser 2008).

If one wants to take the stochastic property of all errors into account, the observations may be written in an extended vector $\mathbf{l} = vec\left(\left[\mathbf{A}\ \mathbf{y}\right]\right)$. Thus, the stochastic properties of the uncertainty can be characterized by the extended cofactor matrix

$$\mathbf{Q_{ll}} = \mathbf{P}^{-1} = \begin{bmatrix} \mathbf{Q_{AA}} & \mathbf{Q_{Ay}} \\ \mathbf{Q_{yA}} & \mathbf{Q_{y_y}} \end{bmatrix} = \begin{bmatrix} \mathbf{Q}_1 \\ \cdots \\ \mathbf{Q}_k \\ \cdots \\ \mathbf{Q}_{u+1} \end{bmatrix} \qquad (2)$$

$\mathbf{Q}_k$ is a $n \times n(u+1)$ matrix representing the variances and covariances between the elements of the kth column of the augmented matrix $\left[\mathbf{A}\ \mathbf{y}\right]$ and all elements. $\mathbf{Q_{ll}}$ and $\mathbf{P}$ are the symmetric and positive definite cofactor matrix and the weighted matrix of $\mathbf{l}$, respectively. $\mathbf{Q_{AA}}$ and $\mathbf{Q_{yy}}$ is the cofactor matrix for $\mathbf{A}$ and $\mathbf{y}$, and the cofactor matrices $\mathbf{Q_{Ay}}$ and $\mathbf{Q_{yA}}$ refer to the correlations between $\mathbf{A}$ and $\mathbf{y}$.

3 The Weighted TLS Solutions

3.1 Solutions Using Lagrange Multipliers

In this section we will show how to solve the fully weighted TLS with 'Lagrange multipliers'. According to the traditional Lagrange approach we form the target function as follows

$$\Phi\left(\mathbf{v}, \boldsymbol{\lambda}, \boldsymbol{\xi}\right) = \mathbf{v}^T\mathbf{P}\mathbf{v} + 2\boldsymbol{\lambda}^T\left(\mathbf{y} - \mathbf{A}\boldsymbol{\xi} - \mathbf{V_A}\boldsymbol{\xi} + \mathbf{v_y}\right) \qquad (3)$$

where $\boldsymbol{\lambda}$ is the Lagrange multipliers vector.

Setting the partial derivatives of the target function w.r.t. $\boldsymbol{\xi}, \mathbf{v}, \boldsymbol{\lambda}$ equal to $\mathbf{0}$, gives the necessary conditions as

$$\frac{1}{2}\frac{\partial\Phi}{\partial\boldsymbol{\xi}}\bigg|_{\hat{\boldsymbol{\xi}},\,\hat{\mathbf{v}},\,\hat{\boldsymbol{\lambda}}} = -\mathbf{A}^T\hat{\boldsymbol{\lambda}} - \hat{\mathbf{V}}_{\mathbf{A}}^T\hat{\boldsymbol{\lambda}} = \mathbf{0} \qquad (4)$$

$$\frac{1}{2}\frac{\partial \Phi}{\partial \mathbf{v}}\Big|_{\hat{\xi},\hat{\mathbf{v}},\hat{\lambda}} = \mathbf{P}\hat{\mathbf{v}} - \left[\hat{\xi}^T \otimes \mathbf{I}_n, -\mathbf{I}_n\right]^T \hat{\lambda} = \mathbf{P}\hat{\mathbf{v}} - \hat{\mathbf{B}}^T\hat{\lambda} = \mathbf{0} \tag{5}$$

$$\frac{1}{2}\frac{\partial \Phi}{\partial \lambda}\Big|_{\hat{\xi},\hat{\mathbf{v}},\hat{\lambda}} = \mathbf{y} - \mathbf{A}\hat{\xi} - \hat{\mathbf{B}}\hat{\mathbf{v}} = \mathbf{0} \tag{6}$$

where $\hat{\mathbf{B}}_{n\times n(u+1)} = \left[\hat{\xi}^T \otimes \mathbf{I}_n, -\mathbf{I}_n\right]$ with the full row rank. The symbol 'hat' of the matrix $\hat{\mathbf{B}}$ is because of the use of the estimated parameter vector $\hat{\xi}$. The operator $\otimes$ is the 'Kronecker–Zehfuss product' (e.g., Grafarend and Schaffrin 1993, p. 409), which is defined by $\mathbf{B}\otimes\mathbf{A} := [b_{ij}\cdot\mathbf{A}]$ if $\mathbf{B} = [b_{ij}]$.

From Eq. 5 we obtain the estimated residual vector as follows

$$\hat{\mathbf{v}} = \begin{bmatrix} vec\left(\hat{\mathbf{V}}_{\mathbf{A}}\right) \\ \hat{\mathbf{v}}_{\mathbf{y}} \end{bmatrix} = \begin{bmatrix} \hat{\mathbf{v}}_{\mathbf{A}} \\ \hat{\mathbf{v}}_{\mathbf{y}} \end{bmatrix} = \mathbf{Q}_{\mathrm{ll}}\hat{\mathbf{B}}^T\hat{\lambda}. \tag{7}$$

Inserting Eq. (7) to (6) $\hat{\lambda}$ can be expressed as

$$\hat{\lambda} = \left(\hat{\mathbf{B}}\mathbf{Q}_{\mathrm{ll}}\hat{\mathbf{B}}^T\right)^{-1}\left(\mathbf{y} - \mathbf{A}\hat{\xi}\right) = \hat{\mathbf{Q}}^{-1}\left(\mathbf{y} - \mathbf{A}\hat{\xi}\right) \tag{8}$$

with $\mathbf{Q} = (\mathbf{B}\mathbf{Q}_{\mathrm{ll}}\mathbf{B}^T)$.

Then by reinserting Eq. (8) to (7) we have

$$\hat{\mathbf{v}} = \mathbf{Q}_{\mathrm{ll}}\hat{\mathbf{B}}^T\hat{\mathbf{Q}}^{-1}\left(\mathbf{y} - \mathbf{A}\hat{\xi}\right). \tag{9}$$

Now, we use the vectorization of a transposed vector to be the same vector and $vec(\mathbf{ABC}) = (\mathbf{C}^T \otimes \mathbf{A})vec(\mathbf{B})$ (Koch 1999, p. 41) in order to derive

$$\hat{\mathbf{V}}_{\mathbf{A}}^T\hat{\lambda} = vec\left(\hat{\lambda}^T\hat{\mathbf{V}}_{\mathbf{A}}\mathbf{I}_{\mathbf{u}}\right) = \left(\mathbf{I}_{\mathbf{u}} \otimes \hat{\lambda}^T\right)\hat{\mathbf{v}}_{\mathbf{A}}. \tag{10}$$

With the help of Eqs. (10), (8), and (4) we obtain

$$\left(\mathbf{I}_u \otimes \hat{\lambda}^T\right)\hat{\mathbf{v}}_{\mathbf{A}} = \hat{\mathbf{V}}_{\mathbf{A}}^T\hat{\lambda} = \mathbf{A}^T\hat{\mathbf{Q}}^{-1}\left(\mathbf{A}\hat{\xi} - \mathbf{y}\right) \tag{11}$$

leading to

$$\left(\mathbf{I}_u \otimes \hat{\lambda}^T\right)\hat{\mathbf{v}}_{\mathbf{A}} + \mathbf{A}^T\hat{\mathbf{Q}}^{-1}\mathbf{y} = \mathbf{A}^T\hat{\mathbf{Q}}^{-1}\mathbf{A}\hat{\xi}. \tag{12}$$

Now, the parameter vector can be estimated by

$$\begin{aligned}\hat{\xi} &= \left(\mathbf{A}^T\hat{\mathbf{Q}}^{-1}\mathbf{A}\right)^{-1}\left(\hat{\mathbf{V}}_{\mathbf{A}}^T\hat{\lambda} + \mathbf{A}^T\hat{\mathbf{Q}}^{-1}\mathbf{y}\right) \\ &= \left(\mathbf{A}^T\hat{\mathbf{Q}}^{-1}\mathbf{A}\right)^{-1}\left(\left(\mathbf{I}_{\mathbf{u}} \otimes \hat{\lambda}^T\right)\hat{\mathbf{v}}_{\mathbf{A}} + \mathbf{A}^T\hat{\mathbf{Q}}^{-1}\mathbf{y}\right).\end{aligned} \tag{13}$$

If $\hat{\mathbf{V}}_{\mathbf{A}}$ is obtained through $\hat{\mathbf{V}}_{\mathbf{A}} = I\,vec_{n\times u}\left(\hat{\mathbf{v}}_{\mathbf{A}}\right)$, where the operator $I\,vec_{n\times u}$ is the opposite of the 'vec' operator and reshapes the vector as the assigned matrix form (Matlab's reshape), one can compute this solution with other two closed-form expressions. From Eqs. (4) and (8) we present the non-linear normal equation

$$\left(\mathbf{A}^T + \hat{\mathbf{V}}_{\mathbf{A}}^T\right)\hat{\mathbf{Q}}^{-1}\left(\mathbf{A}\hat{\xi} - \mathbf{y}\right) = \mathbf{0} \tag{14}$$

leading to

$$\left(\mathbf{A}^T + \hat{\mathbf{V}}_{\mathbf{A}}^T\right)\hat{\mathbf{Q}}^{-1}\mathbf{A}\hat{\xi} = \left(\mathbf{A}^T + \hat{\mathbf{V}}_{\mathbf{A}}^T\right)\hat{\mathbf{Q}}^{-1}\mathbf{y}. \tag{15}$$

The second solution can be expressed as

$$\hat{\xi} = \left(\left(\mathbf{A}^T + \hat{\mathbf{V}}_{\mathbf{A}}^T\right)\hat{\mathbf{Q}}^{-1}\mathbf{A}\right)^{-1}\left(\mathbf{A}^T + \hat{\mathbf{V}}_{\mathbf{A}}^T\right)\hat{\mathbf{Q}}^{-1}\mathbf{y}. \tag{16}$$

If we add $\left(\mathbf{A}^T + \hat{\mathbf{V}}_{\mathbf{A}}^T\right)\hat{\mathbf{Q}}^{-1}\hat{\mathbf{V}}_{\mathbf{A}}\hat{\xi}$ to both sides of Eq (15), the other solution can be expressed as follows

$$\hat{\xi} = \hat{\mathbf{N}}^{-1}\left(\mathbf{A} + \hat{\mathbf{V}}_{\mathbf{A}}\right)^T\hat{\mathbf{Q}}^{-1}\left(\mathbf{y} + \hat{\mathbf{V}}_{\mathbf{A}}\hat{\xi}\right) \tag{17}$$

with $\hat{\mathbf{N}} = \left(\mathbf{A} + \hat{\mathbf{V}}_{\mathbf{A}}\right)^T\hat{\mathbf{Q}}^{-1}\left(\mathbf{A} + \hat{\mathbf{V}}_{\mathbf{A}}\right)$.

In the non-linear problem, the estimated parameter vector cannot be separated from the predicted residual matrix $\hat{\mathbf{V}}_{\mathbf{A}}$ and even the estimated parameter vector per se. Based on the various closed-form expressions of the estimated parameter vector the algorithms can be correspondingly designed. However, the convergence behavior of these solutions will be discussed in another report.

3.2 Solutions Using the GHM Method

The EIV model can be adjusted through another class of the adjustment algorithm with linearization, namely the iterative non-linear GHM. The first successful investigation for solving the non-linear GHM was addressed by Pope (1972). Recently, based on the non-linear GHM Schaffrin and Snow (2010) put forward the regularized TLS solution in Tykhonov's sense to solve the circle fitting problem. Neitzel (2010) applied it to solve the 2D similarity transformation. However, the solution using the LS estimator is not explicitly algebraically formulated. (i.e. the detailed structures of the model (Jacobian) matrices and the inconsistency vector, which are represented by the parameters and the observations as well as residuals, are not presented and built into the final solution of the non-linear GHM method.)

Here, the nonlinear model $\mathbf{f}\left(\mathbf{l}+\mathbf{v},\ \boldsymbol{\xi}\right)=\mathbf{0}$ is linearized as (e.g., Schaffrin and Snow 2010)

$$\begin{aligned}&\tfrac{\partial \mathbf{f}}{\partial \boldsymbol{\xi}^T}\big|_{\mathbf{l}^0+\mathbf{v}^i,\boldsymbol{\xi}^i}\left(\boldsymbol{\xi}-\boldsymbol{\xi}^i\right)+\tfrac{\partial \mathbf{f}}{\partial \mathbf{l}^T}\big|_{\mathbf{l}^0+\mathbf{v}^i,\boldsymbol{\xi}^i}\left(\mathbf{v}-\mathbf{v}^i\right)\\&+\mathbf{f}\left(\mathbf{l}+\mathbf{v}^i,\boldsymbol{\xi}^i\right)=\mathbf{0}\end{aligned} \tag{18}$$

neglecting the terms of the higher order. Note that approximate values used for the position of linearization are not random. Thus, $\mathbf{l}^0=\mathbf{l}-\underset{\sim}{\mathbf{0}}$ (where the $\underset{\sim}{\mathbf{0}}$ denotes a random zero vector (or vector of 'pseudo-observation') of suitable size, in accordance with notion in Schaffrin and Snow (2010).

For the EIV model (1) we give the model matrices

$$\begin{aligned}\tfrac{\partial \mathbf{f}}{\partial \boldsymbol{\xi}^T}\big|_{\mathbf{l}^0+\mathbf{v}^i,\boldsymbol{\xi}^i}&=\mathbf{A}+\mathbf{V}_{\mathbf{A}}^i\\ \tfrac{\partial \mathbf{f}}{\partial \mathbf{l}^T}\big|_{\mathbf{l}^0+\mathbf{v}^i,\boldsymbol{\xi}^i}&=\left[\left(\boldsymbol{\xi}^i\right)^T\otimes\mathbf{I}_n,-\mathbf{I}_n\right]=\mathbf{B}^i\end{aligned} \tag{19}$$

and the inconsistency vector

$$\begin{aligned}&\mathbf{f}\left(\mathbf{l}+\mathbf{v}^i,\boldsymbol{\xi}^i\right)-\tfrac{\partial \mathbf{f}}{\partial \mathbf{l}^T}\big|_{\mathbf{l}^0+\mathbf{v}^i,\boldsymbol{\xi}^i}\mathbf{v}^i\\&=\left(\mathbf{A}+\mathbf{V}_{\mathbf{A}}^i\right)\boldsymbol{\xi}^i-\mathbf{y}-\mathbf{v}_{\mathbf{y}}^i-\left[\left(\boldsymbol{\xi}^i\right)^T\otimes\mathbf{I}_n,-\mathbf{I}_n\right]\mathbf{v}^i\\&=\mathbf{A}\boldsymbol{\xi}^i-\mathbf{y}=\mathbf{f}\left(\mathbf{l},\boldsymbol{\xi}^i\right)=\mathbf{w}^i.\end{aligned} \tag{20}$$

By using the model matrices and inconsistency vector, the alternative representation of the EIV model as linearized GHM reads

$$\left(\mathbf{A}+\mathbf{V}_{\mathbf{A}}^i\right)d\boldsymbol{\xi}+\mathbf{B}^i\mathbf{v}+\mathbf{w}^i=\mathbf{0}. \tag{21}$$

Based on the linearized model and Eq. 20 the solution can be obtained as follows

$$\begin{aligned}\hat{\boldsymbol{\xi}}^{i+1}&=\boldsymbol{\xi}^i+d\hat{\boldsymbol{\xi}}^{i+1}\\&=\boldsymbol{\xi}^i+\left(\mathbf{N}^i\right)^{-1}\left(\mathbf{A}+\mathbf{V}_{\mathbf{A}}^i\right)^T\left(\mathbf{Q}^i\right)^{-1}\left(-\mathbf{w}^i\right)\\&=\left(\mathbf{N}^i\right)^{-1}\left(\mathbf{A}+\mathbf{V}_{\mathbf{A}}^i\right)^T\left(\mathbf{Q}^i\right)^{-1}\left(\mathbf{y}-\mathbf{A}\boldsymbol{\xi}^i\right)+\boldsymbol{\xi}^i\\&=\left(\mathbf{N}^i\right)^{-1}\left(\mathbf{A}+\mathbf{V}_{\mathbf{A}}^i\right)^T\left(\mathbf{Q}^i\right)^{-1}\left(\mathbf{y}+\mathbf{V}_{\mathbf{A}}^i\boldsymbol{\xi}^i\right)\end{aligned} \tag{22}$$

with $\mathbf{N}^i=((\mathbf{A}+\mathbf{V}_{\mathbf{A}}^i)^T(\mathbf{Q}^i)^{-1}(\mathbf{A}+\mathbf{V}_{\mathbf{A}}^i))$and $\mathbf{Q}^i=\mathbf{B}^i\mathbf{Q}_{\mathbf{ll}}(\mathbf{B}^i)^T$.

Note that the residuals and parameters are updated with $\mathbf{v}^i=\hat{\mathbf{v}}^i-\underset{\sim}{\mathbf{0}}$, $\boldsymbol{\xi}^{i+1}=\boldsymbol{\xi}^i+d\hat{\boldsymbol{\xi}}^{i+1}-\underset{\sim}{\mathbf{0}}$. i.e. the residuals are not added on the previous residuals predicted. In contrast to the update of residuals, parameters are accumulated at the last stage. The process stops if the parameter or the extended residual vector does not change in the order of magnitude. The numerically obtained result is defined as the solution for the WTLS problem: $\hat{\boldsymbol{\xi}}_{WTLS}:=\boldsymbol{\xi}^{i+1}$.

3.3 Solutions Using the Gradient

In this part the TLS problem for the weighted case is alternatively defined as the minimum of a sum of weighted squared LS residuals (Fang 2011):

$$\min_{\boldsymbol{\xi}}(\mathbf{y}-\mathbf{A}\boldsymbol{\xi})^T\left(\mathbf{B}\mathbf{Q}_{\mathbf{ll}}\mathbf{B}^T\right)^{-1}(\mathbf{y}-\mathbf{A}\boldsymbol{\xi}). \tag{23}$$

The target function can be also classified as quasi indirect error adjustment (Wolf 1968, Eq. (2153, 10)). The objective function is similar to the objective function of the weighted LS case since only the weighted matrices are different. In this case the objective function is not subject to any constraints, and the variables such as the residual matrix, vector as well as the auxiliary vector of the Lagrange multipliers disappear.

The necessary condition for the local minimizer is given by

$$\left.\frac{\partial(\mathbf{y}-\mathbf{A}\boldsymbol{\xi})^T\left(\mathbf{B}\mathbf{Q}_{\mathbf{ll}}\mathbf{B}^T\right)^{-1}(\mathbf{y}-\mathbf{A}\boldsymbol{\xi})}{\partial\boldsymbol{\xi}}\right|_{\hat{\boldsymbol{\xi}}}=\mathbf{0}. \tag{24}$$

$f(\boldsymbol{\xi})=(\mathbf{y}-\mathbf{A}\boldsymbol{\xi})^T(\mathbf{B}\mathbf{Q}_{\mathbf{ll}}\mathbf{B}^T)^{-1}(\mathbf{y}-\mathbf{A}\boldsymbol{\xi})$ is scalar. The matrix analysis property (differentiation of a scalar function w.r.t. a vector) in Grafarend and Schaffrin (1993) is applied as follows

$$\frac{\partial f\left(\boldsymbol{\xi}\right)}{\partial\boldsymbol{\xi}}=\left[\tfrac{\partial f(\boldsymbol{\xi})}{\partial\xi_1}\cdots\tfrac{\partial f(\boldsymbol{\xi})}{\partial\xi_k}\cdots\tfrac{\partial f(\boldsymbol{\xi})}{\partial\xi_u}\right]^T, \tag{25}$$

where $\boldsymbol{\xi}=[\xi_1,\ldots,\xi_u]^T$ and $1\leq k\leq u$.

The first derivative w.r.t. the parameter vector can be extended in three parts according to the product rule (the well-known Leibniz's Law) as follows

$$\begin{aligned}&\tfrac{\partial(\mathbf{y}-\mathbf{A}\boldsymbol{\xi})^T\left(\mathbf{B}\mathbf{Q}_{\mathbf{ll}}\mathbf{B}^T\right)^{-1}(\mathbf{y}-\mathbf{A}\boldsymbol{\xi})}{\partial\xi_k}\\&=\tfrac{\partial(\mathbf{y}-\mathbf{A}\boldsymbol{\xi})^T}{\partial\xi_k}\left(\mathbf{B}\mathbf{Q}_{\mathbf{ll}}\mathbf{B}^T\right)^{-1}(\mathbf{y}-\mathbf{A}\boldsymbol{\xi})\\&+(\mathbf{y}-\mathbf{A}\boldsymbol{\xi})^T\tfrac{\partial\left(\mathbf{B}\mathbf{Q}_{\mathbf{ll}}\mathbf{B}^T\right)^{-1}}{\partial\xi_k}(\mathbf{y}-\mathbf{A}\boldsymbol{\xi})\\&+(\mathbf{y}-\mathbf{A}\boldsymbol{\xi})^T\left(\mathbf{B}\mathbf{Q}_{\mathbf{ll}}\mathbf{B}^T\right)^{-1}\tfrac{\partial(\mathbf{y}-\mathbf{A}\boldsymbol{\xi})}{\partial\xi_k}.\end{aligned} \tag{26}$$

Based on Eq. (26) the gradient can be analytically obtained (see Appendix)

$$\mathbf{g}=-2\left(\mathbf{A}+\mathbf{A}^*\right)^T\left(\mathbf{B}\mathbf{Q}_{\mathbf{ll}}\mathbf{B}^T\right)^{-1}(\mathbf{y}-\mathbf{A}\boldsymbol{\xi}). \tag{27}$$

Table 1 Observation data vector (x, y) and corresponding weights, taken from Neri et al. (1989) and also presented in Schaffrin and Wieser (2008) and Shen et al (2010)

Point number	x	Weights of x	y	Weights of y
1	0.0	1,000	5.9	1.0
2	0.9	1,000	5.4	1.8
3	1.8	500	4.4	4.0
4	2.6	800	4.6	8.0
5	3.3	200	3.5	20
6	4.4	80	3.7	20
7	5.2	60	2.8	70
8	6.1	20	2.8	70
9	6.5	1.8	2.4	100
10	7.4	1.0	1.5	500

Obtaining the stationary point should be fulfilled:

$$\hat{\mathbf{g}} = -2\left(\mathbf{A} + \hat{\mathbf{A}}^*\right)^T \left(\hat{\mathbf{B}}\mathbf{Q}_{\mathrm{ll}}\hat{\mathbf{B}}^T\right)^{-1} \left(\mathbf{y} - \mathbf{A}\hat{\boldsymbol{\xi}}\right) = \mathbf{0}, \quad (28)$$

which is identical to the normal Eq. (14), since the matrix $\hat{\mathbf{A}}^*$ is actually identical with $Ivec_{n\times u}\left(\left[\mathbf{Q}_{\mathrm{AA}}\mathbf{Q}_{\mathrm{Ay}}\right]\hat{\mathbf{B}}^T\mathbf{Q}^{-1}\left(\mathbf{y} - \mathbf{A}\hat{\boldsymbol{\xi}}\right)\right)$, namely $\hat{\mathbf{V}}_{\mathrm{A}}$. The normal equation can be identically gained from a different starting point. The gradient per se is also an essential quantity to design the algorithms for non-linear adjustment problems. Based on it, the descent direction can be calculated, and one of the oldest iterative descent methods is established (Teunissen 1990).

4 Test Results

In order to demonstrate the solution of the TLS problem, the slope ξ_1 and the intercept ξ_2 of a line are estimated and hence the model can be expressed as follows:

$$\mathbf{y} + \mathbf{v}_{\mathrm{y}} = \left[\left(\mathbf{x} + \mathbf{v}_{\mathrm{x}}\right), \mathbf{e}\right]\begin{bmatrix}\xi_1 \\ \xi_2\end{bmatrix}. \quad (29)$$

The weight matrix is omitted to be presented here, which can be found in Schaffrin and Wieser (2008).

The data of this example can be seen in Table 1. A comparison of our solutions to the solution of Schaffrin and Wieser (2008) is shown in Table 2. In fact, all of our solutions correspond with the solution of Schaffrin and Wieser (2008). However, our solutions do not have any limitation of the weight matrix.

Table 2 The parameter estimates using the data of Table 1

Parameter	Unweighted TLS solution	Solution of Schaffrin and Wieser	Solutions designed in the study
Parameter 1	−0.545561197	−0.480533407	−0.480533407
Parameter 2	5.784043775	5.479910224	5.479910224

Conclusions and Further Work

For the weighted TLS solutions presented it is obvious that the algorithm (17) has the same analytical formulation as the non-linear GHM method. The normal Eq. (28) which represents the necessary conditions of the optimality (23) is also identical with the normal Eq. (14) derived by the Lagrange multipliers. From this normal equation the solutions can be also differently designed. The different forms could have the beneficial convergence behavior, which can be investigated in the future.

It is broadly acknowledged that the method of Lagrange multipliers yields only necessary conditions for optimality in the constrained problems. Although the sufficient condition is fulfilled for the residual vector (Schaffrin and Wieser 2008), little importance has been attached to the study about the sufficient condition of weighted TLS problem for the parameter vector until now. The second derivatives of the objective function represent the sufficiency conditions of the weighted TLS solution. On the basis of the obtained gradient and the Hessian matrix (if we derive), the standard Newton algorithm is designed for the weighted TLS problem to bring some advantages versus the Gauss–Newton algorithm.

The proposed algorithm can be applied in geodetic transformation. By means of the proposed algorithms and properly designed covariance matrix, the structured TLS solution can be established.

Appendix

According to Eq. (26) the partial derivative can be arranged as follows

$$\begin{aligned}
&\frac{\partial(\mathbf{y}-\mathbf{A}\boldsymbol{\xi})^T\left(\mathbf{B}\mathbf{Q}_{\mathrm{ll}}\mathbf{B}^T\right)^{-1}(\mathbf{y}-\mathbf{A}\boldsymbol{\xi})}{\partial\xi_k} \\
&= 2\frac{\partial(\mathbf{y}-\mathbf{A}\boldsymbol{\xi})^T}{\partial\xi_k}\left(\mathbf{B}\mathbf{Q}_{\mathrm{ll}}\mathbf{B}^T\right)^{-1}(\mathbf{y}-\mathbf{A}\boldsymbol{\xi})\ldots \\
&+ (\mathbf{y}-\mathbf{A}\boldsymbol{\xi})^T\frac{\partial\left(\mathbf{B}\mathbf{Q}_{\mathrm{ll}}\mathbf{B}^T\right)^{-1}}{\partial\xi_k}(\mathbf{y}-\mathbf{A}\boldsymbol{\xi}) \\
&= -2\mathbf{a}_k^T\left(\mathbf{B}\mathbf{Q}_{\mathrm{ll}}\mathbf{B}^T\right)^{-1}(\mathbf{y}-\mathbf{A}\boldsymbol{\xi})\ldots \\
&+ (\mathbf{y}-\mathbf{A}\boldsymbol{\xi})^T\frac{\partial\left(\mathbf{B}\mathbf{Q}_{\mathrm{ll}}\mathbf{B}^T\right)^{-1}}{\partial\xi_k}(\mathbf{y}-\mathbf{A}\boldsymbol{\xi})
\end{aligned}$$

where $\mathbf{A} = [\mathbf{a}_{1,\ldots}\mathbf{a}_u]$ (i.e. $\mathbf{a}_k$ is the kth column of the matrix $\mathbf{A}$).

Then, the second part can be solved with

$$\frac{\partial \mathbf{A}^{-1}}{\partial \xi_k} = -\mathbf{A}^{-1}\frac{\partial \mathbf{A}}{\partial \xi_k}\mathbf{A}^{-1}$$

(Grafarend and Schaffrin 1993) as follows

$$\begin{aligned}
&(\mathbf{y}-\mathbf{A}\boldsymbol{\xi})^T \tfrac{\partial \mathbf{Q}^{-1}}{\partial \xi_k}(\mathbf{y}-\mathbf{A}\boldsymbol{\xi}) \\
&= -(\mathbf{y}-\mathbf{A}\boldsymbol{\xi})^T \mathbf{Q}^{-1} \tfrac{\partial \mathbf{Q}}{\partial \xi_k}\mathbf{Q}^{-1}(\mathbf{y}-\mathbf{A}\boldsymbol{\xi}) \\
&= -(\mathbf{y}-\mathbf{A}\boldsymbol{\xi})^T \mathbf{Q}^{-1}\mathbf{B}\mathbf{Q}_k^T\mathbf{Q}^{-1}(\mathbf{y}-\mathbf{A}\boldsymbol{\xi})\ldots \\
&\quad -(\mathbf{y}-\mathbf{A}\boldsymbol{\xi})^T \mathbf{Q}^{-1}\mathbf{Q}_k\mathbf{B}^T\mathbf{Q}^{-1}(\mathbf{y}-\mathbf{A}\boldsymbol{\xi}) \\
&= -2(\mathbf{y}-\mathbf{A}\boldsymbol{\xi})^T \mathbf{Q}^{-1}\mathbf{B}\mathbf{Q}_k^T\mathbf{Q}^{-1}(\mathbf{y}-\mathbf{A}\boldsymbol{\xi})
\end{aligned}$$

with $\mathbf{Q} = (\mathbf{B}\mathbf{Q}_{\mathrm{ll}}\mathbf{B}^T)$. The last step of the equation can use the following property: A scalar is given as $s = \mathbf{a}^T\mathbf{C}\mathbf{a}$ (exemplarily square matrix and vector). $\mathbf{a}^T\mathbf{C}\mathbf{a} = s = s^T = \mathbf{a}^T\mathbf{C}^T\mathbf{a}$. Combining the above equations, we finally obtain the gradient (the first derivative w.r.t. the parameter vector)

$$\mathbf{g} = \frac{\partial f(\boldsymbol{\xi})}{\partial \boldsymbol{\xi}} = -2\left(\mathbf{A}+\mathbf{A}^*\right)^T\mathbf{Q}^{-1}(\mathbf{y}-\mathbf{A}\boldsymbol{\xi})$$

with $\mathbf{A}^* = \left[\mathbf{Q}_1\mathbf{B}^T\mathbf{Q}^{-1}(\mathbf{y}-\mathbf{A}\boldsymbol{\xi}),\ldots,\mathbf{Q}_u\mathbf{B}^T\mathbf{Q}^{-1}(\mathbf{y}-\mathbf{A}\boldsymbol{\xi})\right]$ through

$$\begin{aligned}
&\tfrac{\partial f(\boldsymbol{\xi})}{\partial \xi_k} = -2\mathbf{a}_k^T\mathbf{Q}^{-1}(\mathbf{y}-\mathbf{A}\boldsymbol{\xi}) \\
&\quad -2(\mathbf{y}-\mathbf{A}\boldsymbol{\xi})^T\mathbf{Q}^{-1}\mathbf{B}\mathbf{Q}_k^T\mathbf{Q}^{-1}(\mathbf{y}-\mathbf{A}\boldsymbol{\xi})
\end{aligned}$$

References

Akyilmaz O (2007) Total least squares solution of coordinate transformation. Surv Rev 39(303):68–80

Fang X (2011) Weighted total least squares solutions for applications in Geodesy. Dissertation, Leibniz Universität Hannover

Felus F (2004) Application of total least squares for spatial point process analysis. J Surv Eng 130(3):126–133

Felus F, Burtch R (2009) On symmetrical three-dimensional datum conversion. GPS Solutions 13(1):65–74

Golub G, Van Loan C (1980) An analysis of the total least-squares problem. SIAM J Numer Anal 17(6):883–893

Grafarend E, Schaffrin B (1993) Ausgleichungsrechnung in linearen modellen. BI-Wissenschaftsverlag, Mannheim

Koch K (1999) Parameter estimation and hypothesis testing in linear models, 2nd edn. Springer, Berlin

Neitzel F (2010) Generalization of total least-squares on example of unweighted and weighted 2D similarity transformation. J Geod 84:751–762

Neitzel F, Petrovic S (2008) Total least squares (tls) im kontext der ausgleichung nach kleinsten quadraten am beispiel der ausgleichenden geraden. ZfV 133:141–148 (in German)

Neri F, Saitta G, Chiofalo S (1989) An accurate and straightforward approach to line regression analysis of error-affected experimental data. J Phys Ser E: Sci Instr 22:215–217

Pope A (1972) Some pitfalls to be avoided in the iterative adjustment of nonlinear problems. In: Proc 38th Ann Meet Am Soc Phot, Washington, pp 449–473

Schaffrin B (2007) Connecting the dots: the straight-line case revisited. ZfV 132:385–394 (in German)

Schaffrin B (2008) Correspondence, coordinate transformation. Surv Rev 40(307):102

Schaffrin B, Felus Y (2008) On the multivariate total least-squares approach to empirical coordinate transformations three algorithms. J Geod 82:373–383

Schaffrin B, Snow K (2010) Total least-squares regularization of tykhonov type and an ancient racetrack in Corinth. Linear Algebra Appl 432:2061–2076

Schaffrin B, Wieser A (2008) On weighted total least-squares adjustment for linear regression. J Geod 82:415–421

Shen Y, Li BF, Chen Y (2010) An iterative solution of weighted total least-squares adjustment. J Geod 85:229–238

Teunissen PJG (1988) The non-linear 2D symmetric Helmert transformation: an exact non-linear least-squares solution. Manuscr Geodet 62(1):1–16

Teunissen PJG (1990) Nonlinear least squares. Manuscr Geod 15:137–150

Van Huffel S, Vandewalle J (1991) The total least-squares problem. computational aspects and analysis, Society for Industrial and Applied Mathematics, Philadelphia

Wolf H (1968) Ausgleichungsrechnung nach der Methode der kleinsten Quadrate. Dümmler (in German)

Integration of Observations and Models in a Consistent Least Squares Adjustment Model

A. Heiker and H. Kutterer

Abstract

Models are often treated as deterministic in geodetic practice. Hence, inaccurate models directly affect the results of geodetic measurements. This paper proposes a method for the mutual validation of models and observed data. To consider the inaccuracy of models, data resulting from models are treated as stochastic parameter in a linear least squares adjustment. The required stochastic information is obtained by empirical auto and cross correlation functions. This approach is applied to the problem of the mutual validation of Earth orientation parameters, second degree gravity field coefficients and geophysical excitation functions. The results and the limitations of this approach are discussed.

Keywords

Auto correlation function • Data integration • Model quality • Time series • Validation of observations and models

1 Introduction

Most geodetic applications consider observed data as well as physical models. Hence, the results depend directly on the quality of measurements and models. Models describe the environment by mathematical relations, e.g., the changes of the refraction index due to variations of temperature and air pressure are modelled in electro-optical distance measurements. The quality of a model can be assessed by comparing the model with observed data, measured with sufficient accuracy. Vice versa, if the quality of a model is known measured data can be validated by comparing them with the model. This paper proposes a method for the mutual validation of models and observations.

The simplest method to judge the coincidence of models and observations is the calculation of empirical correlation coefficients. The three-cornered hat method (Premoli and Tavella 1993; Galindo et al. 2001) assesses the quality of time series by variance-covariance matrices (VCM). The VCMs are calculated by differences of time series. As differences instead of absolute values are used, the problem is under-determined (Galindo et al. 2001; Koot et al. 2006). Therefore, assumptions (zero correlations or the assumption of minimum correlations) are necessary to determine the VCMs. Galindo et al. (2001) have shown, that the results of the three-cornered hat method are insufficient, if the assumption of weak correlations between the reference time series and other time series are not fulfilled. We propose in Sect. 2 a further approach which obtains stochastic information and avoids the disadvantage of additional assumptions.

The mutual validation of observed data and models is performed within an universal, constrained, linear Gauß–Helmert model. The least squares adjustment includes a variance-covariance component estimation. Yu (1992)

A. Heiker (✉)
Geodätisches Institut, Leibniz-Universität Hannover,
Nienburger Straße 1, 30163 Hannover, Germany
e-mail: a.heiker@freenet.de; andrea.heiker@lgln.niedersachsen.de

H. Kutterer
Bundesamt für Kartographie und Geodäsie,
Richard-Strauss-Allee 11, 60598 Frankfurt, Germany

H. Kutterer et al. (eds.), *The 1st International Workshop on the Quality of Geodetic Observation and Monitoring Systems (QuGOMS'11)*, International Association of Geodesy Symposia 140,
DOI 10.1007/978-3-319-10828-5_8,

and Heiker (2013) have developed the necessary formulas. Based on the work of Kusche and Klees (2002), Kusche (2003), and Heiker (2013) propose an enhanced Monte Carlo solver for linear constrained adjustment models. Observations as well as model data are considered as stochastic (pseudo)observations. The interpretation of model data as stochastic parameters allows the determination of model residuals containing model inconsistencies. Missing stochastic information is approximated by auto and cross correlation functions. The complete adjustment model is presented in Sect. 2.

This approach is exemplary applied to the mutual validation of measured Earth orientation parameters (EOP), measured second degree gravity field coefficients (GFC2) and modelled geophysical excitation functions. Changes in the Earth's mass distribution impact the Earth's rotation on two ways. First, the tensor of inertia varies and induces variations of the Earth's gravity field and rotation. Second, the mass displacements cause currents, which chance change the angular momentum of the Earth and therefore the rotation of the Earth. The excitation functions model the mass displacements of oceans and atmosphere. The so called mass terms describe the changes in the Earth's rotation due to variations of the tensor of inertia and the motion terms the variations due to changes of the angular momentum (Gross 2007). The tensor of inertia is the fundamental link between the EOP, GFC2 and the excitation functions.

2 Integration in a Least Squares Adjustment Model

2.1 Functional Model

An universal, constrained, linear Gauß–Helmert functional model is two folded with

$$\begin{aligned} \mathbf{A}_H \Delta\boldsymbol{\beta} + \mathbf{B}\boldsymbol{\epsilon} - \mathbf{w}_H &= \mathbf{0} \\ \mathbf{A}_C \Delta\boldsymbol{\beta} \qquad - \mathbf{w}_C &= \mathbf{0} \end{aligned} \tag{1}$$

where $\boldsymbol{\epsilon} = \mathrm{E}(\mathbf{l}) - \mathbf{l} = \mathbf{0}$ and $\Delta\boldsymbol{\beta} = \boldsymbol{\beta} - \boldsymbol{\beta}_0$. The $n \times 1$ residual vector $\boldsymbol{\epsilon}$ contains the inconsistencies between the $n \times 1$ observation vector $\mathbf{l}$ and the expectation value $E(\mathbf{l})$ of the observations. The $u \times 1$ vector $\boldsymbol{\beta}_0$ denotes an approximation of the unknown, deterministic parameter vector $\boldsymbol{\beta}$. The $s \times 1$ misclosure vector $\mathbf{w}_H$ depends on the observation vector $\mathbf{l}$, whereas the $t \times 1$ vector $\mathbf{w}_C$ belongs to the deterministic constraint. The $t \times u$ matrix $\mathbf{A}_C$, the $s \times u$ matrix $\mathbf{A}_H$ and the $s \times n$ matrix $\mathbf{B}$ are full-ranked design matrices.

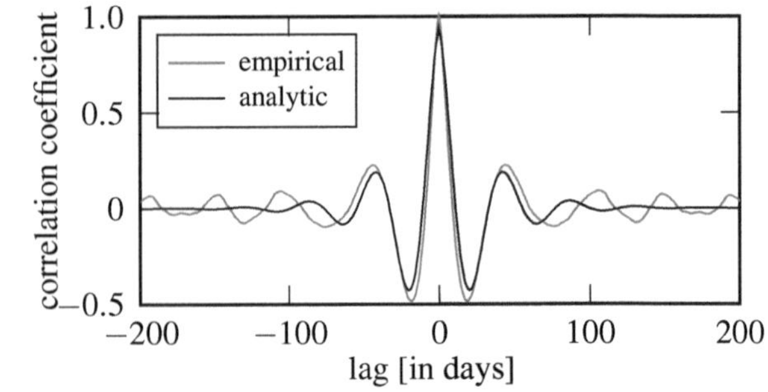

Fig. 1 Empirical function of LOD is displayed in *red*. The empirical function is approximated by the damped oscillation function (in *blue*)

2.2 Stochastic Model

Missing stochastic information are obtained by the empirical auto and cross correlation functions. As the VCMs should only contain noises, deterministic parts of the signal have to be removed before the calculation of empirical auto- and cross correlation functions. This leads to stationary time series. For example, the polar motion contains two main frequencies, the annual signal and the Chandler period. These two frequencies are caused by the Earth's rotation around the sun and by the inelastic deformation of the Earth's body. As these frequencies are caused by systematic effects they are removed. The VCMs are built as Toeplitz-matrices with identical values on the diagonals. If a VCM is directly built from empirical auto and cross correlation functions the matrix in some cases tends to be semi-definite due to numeric reasons. The approximation of the empirical auto and cross correlation functions by analytic functions leads usually to considerably better conditioned VCMs. Here, all auto and a few cross correlation functions are approximated by damped oscillation functions with

$$\sigma_{ij}(t) = \sigma_{max} e^{\delta|t - t_{max}|} \cos\omega\,(t - t_{max})\,. \tag{2}$$

where t denotes the lag, t_{max} the lag with the largest empirical correlation σ_{max}. δ denotes the damping factor and ω the angular frequency. Figure 1 shows the empirical auto correlation function of the length-of-day parameter (LOD) in red. The blue line denotes the approximated auto correlation function (see Eq. (2)). Cross correlation functions only are considered, if the largest empirical value of the function is above a specific threshold level.

The obtained VCM is divided in submatrices $\mathbf{V}_i$. The final VCM $\boldsymbol{\Sigma}_{ll}$, which describes the complete stochastic model, is then the sum of the matrices $\mathbf{V}_i$ multiplied with unknown variance-covariance components σ_i

$$\boldsymbol{\Sigma}_{ll} = \sum_i \sigma_i \mathbf{V}_i\,. \tag{3}$$

Note, that the index i of σ_i denotes a counter for the sum sign. The σ_i denote as well variance as covariance components with dedicated variance and covariance matrices $\boldsymbol{V}_i$.

2.3 Linear Least Squares Adjustment

The least squares minimization

$$\boldsymbol{\epsilon}' \boldsymbol{\Sigma}_{\text{ll}}^{-1} \boldsymbol{\epsilon} \rightarrow \min \tag{4}$$

leads to the system of normal equations

$$\begin{bmatrix} \mathbf{B}\boldsymbol{\Sigma}_{\text{ll}}\mathbf{B}' & \mathbf{A}_H & \mathbf{0} \\ \mathbf{A}'_H & \mathbf{0} & \mathbf{A}'_C \\ \mathbf{0} & \mathbf{A}_C & \mathbf{0} \end{bmatrix} \begin{bmatrix} \mathbf{k}_H \\ \Delta\boldsymbol{\beta} \\ \mathbf{k}_C \end{bmatrix} = \begin{bmatrix} \mathbf{w}_H \\ \mathbf{0} \\ \mathbf{w}_C \end{bmatrix} \tag{5}$$

with the vectors $\mathbf{k}_H$ and $\mathbf{k}_C$ denoting the Lagrange correlates due to the functional model in Eq. (1). The system of normal equations is solved by the (generalized) inversion of the normal equations matrix.

Then, the unknown variance-covariance components are obtained according to Heiker (2013) by

$$\sigma_i = \frac{\boldsymbol{\epsilon}'\mathbf{B}' \left(\mathbf{B}\boldsymbol{\Sigma}_{\text{ll}}\mathbf{B}'\right)^{-1} \mathbf{B}\mathbf{V}_i\mathbf{B}' \left(\mathbf{B}\boldsymbol{\Sigma}_{\text{ll}}\mathbf{B}'\right)^{-1} \mathbf{B}\boldsymbol{\epsilon}}{\operatorname{tr}\left(\mathbf{Q}_{11}\mathbf{B}\mathbf{V}_i\mathbf{B}'\right)} \tag{6}$$

where $\mathbf{Q}_{11}$ is the left, upper submatrix of the inversion of the normal equations matrix. The unknown parameters, the residuals and the variance-covariance components are determined by using the efficient iterative Monte-Carlo solver given in Heiker (2013).

3 Example

The link between the EOP, the GFC2 and the excitation functions is the tensor of inertia, here considered as unknown parameters. The mean and the trend are removed from all observed and modeled data to avoid secular effects from postglacial rebound. Thus, the unknown parameter vector $\Delta\boldsymbol{\beta}$ contains the deviations of the mean tensor of inertia.

The five second degree gravity field coefficients per each epoch are summarized in the stochastic observation vector $\mathbf{l}_{GFC2}$ and the six elements of the tensor of inertia are according to Hofmann-Wellenhof and Moritz (2006) linearly related by

$$\mathbf{A}_1\Delta\boldsymbol{\beta} - \left(\mathbf{l}_{GFC2} + \boldsymbol{\epsilon}_{GFC2}\right) = \mathbf{0}. \tag{7}$$

The linear approximation of the Euler-Liouville equation describes the EOP variations by the sum of the mass and the motion terms (Gross 2007). Thus, the tensor of inertia can be calculated from the difference of EOP and motion term. However, the linear approximation has the disadvantage that three of the six elements of the tensor of inertia vanish. The observed EOP are summarized in the vector $\mathbf{l}_{EOP}$ and $\mathbf{l}_{MO}$ contain the motion terms, then it holds

$$\mathbf{A}_2\Delta\boldsymbol{\beta} - \left(\mathbf{B}_1\left(\mathbf{l}_{EOP} + \boldsymbol{\epsilon}_{EOP}\right) - \mathbf{B}_2\left(\mathbf{l}_{MO} + \boldsymbol{\epsilon}_{MO}\right)\right) = \mathbf{0}. \tag{8}$$

The mass terms of the excitation functions in vector $\boldsymbol{l}_{MA}$ are equal to three elements of the tensor of inertia multiplied by a constant factor

$$\mathbf{A}_3\Delta\boldsymbol{\beta} - \left(\mathbf{l}_{MA} + \boldsymbol{\epsilon}_{MA}\right) = \mathbf{0}. \tag{9}$$

Equations (7)–(9) are merged to

$$\begin{bmatrix} \mathbf{A}_1 \\ \mathbf{A}_2 \\ \mathbf{A}_3 \end{bmatrix} \Delta\boldsymbol{\beta} + \begin{bmatrix} \mathbf{0} & -\mathbf{I} & \mathbf{0} & \mathbf{0} \\ -\mathbf{B}_1 & \mathbf{0} & \mathbf{0} & \mathbf{B}_2 \\ \mathbf{0} & \mathbf{0} & -\mathbf{I} & \mathbf{0} \end{bmatrix} \begin{bmatrix} \boldsymbol{\epsilon}_{EOP} \\ \boldsymbol{\epsilon}_{GFC2} \\ \boldsymbol{\epsilon}_{MA} \\ \boldsymbol{\epsilon}_{MO} \end{bmatrix} \cdots - \begin{bmatrix} \mathbf{l}_{GFC} \\ \mathbf{B}_1\mathbf{l}_{EOP} - \mathbf{B}_2\mathbf{l}_{MO} \\ \mathbf{l}_{MA} \end{bmatrix} = \mathbf{0}. \tag{10}$$

The trace of the tensor on inertia is not uniquely determinable by Eqs. (7), (8) or (9). Hence, the equation system has to be regularized by an additional constraint. The trace of the tensor of inertia remains constant (Rochester and Smylie 1974) which leads to the linear equation

$$\mathbf{A}_C\Delta\boldsymbol{\beta} = \mathbf{0}. \tag{11}$$

Equations (10) and (11) describe a functional model according to Eq. (1).

Following time series are available for a mutual validation of EOP, GFC2 and excitation functions:

- EOP 08 C04 time series from the International Earth Rotation Service (see www.iers.org),
- GRACE and SLR gravity field coefficients from the university of Texas at Austin, Center for Space Research,
- two different atmospheric and oceanic excitation functions from the geophysical fluid center: NCEP/ECCO vs. ECMWF/OMCT (further information: http://geophy.uni.lu/).

The five given time series are validated in four independent combinations:

- EOP + GRACE + NCEP/ECCO,
- EOP + SLR + NCEP/ECCO,
- EOP + GRACE + ECMWF/OMCT,
- EOP + SLR + ECMWF/OMCT.

The full priori VCMs for the EOP and the excitation functions are determined according to Sect. 2. As the available GFC2 time series are too short to determine meaningful correlations, the VCM for the GFC2 is a diagonal matrix built from the standard deviations provided by the Center of

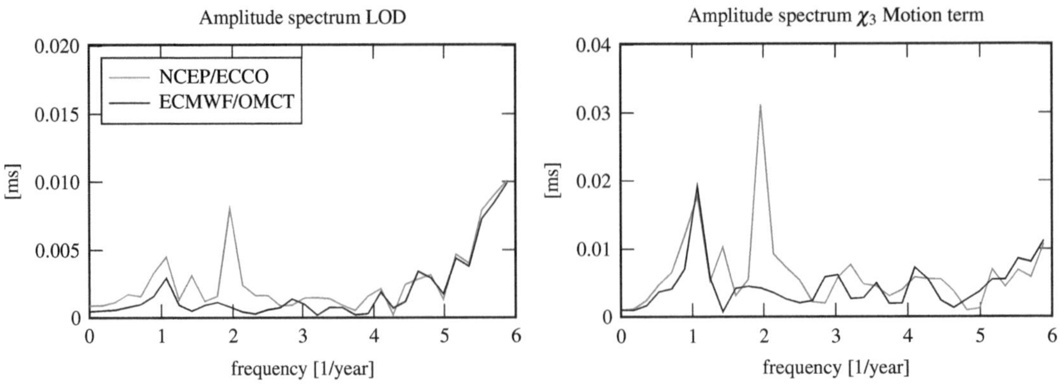

Fig. 2 Amplitude spectra of LOD residuals (*left*) and the χ_3 motion term residuals (*right*). Amplitudes are transformed in milliseconds per day

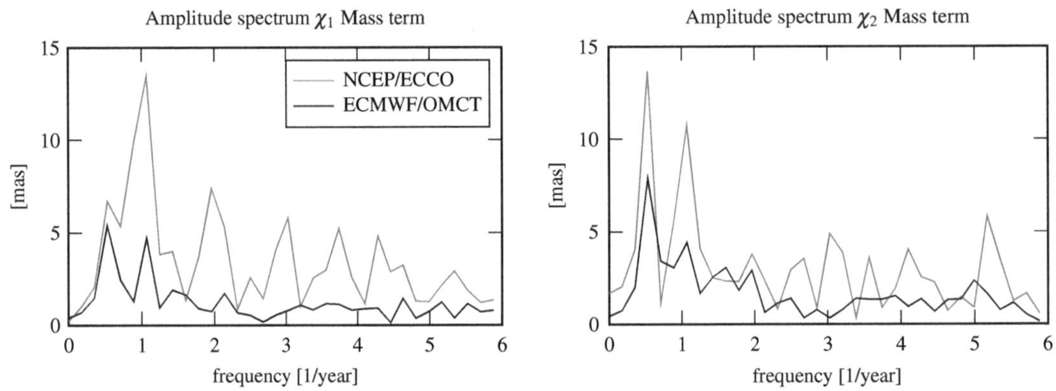

Fig. 3 Amplitude spectra of χ_1 and χ_2 mass terms residuals. Amplitudes are transformed in milliarcseconds

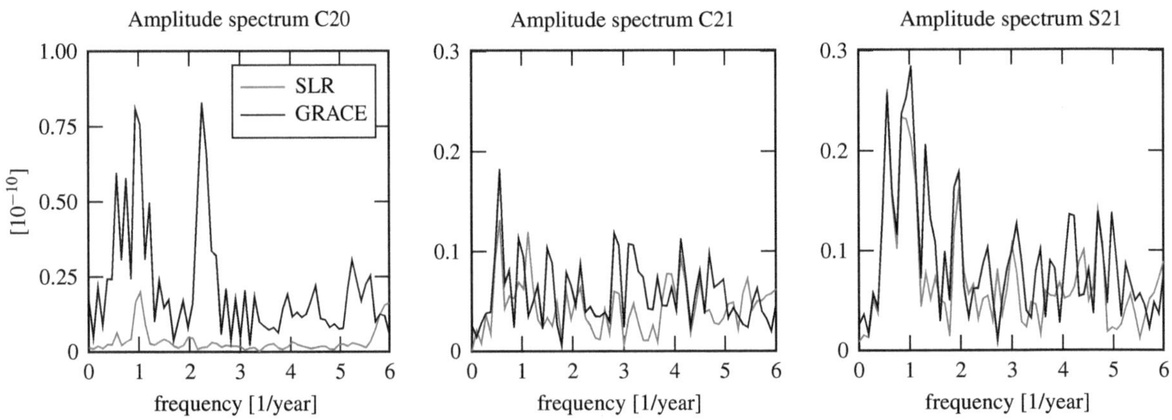

Fig. 4 Amplitude spectra of the residuals of the second degree gravity field coefficients

Space Research. In each combination five variance components for

- equatorial EOP: polar motion,
- axial EOP: length of day,
- equatorial excitation functions: χ_1 / χ_2 mass and motion terms,
- axial excitation functions: χ_3 mass and motion terms,
- second degree gravity field coefficients.

are estimated. The results of the least squares adjustments are residuals for each time series, the adjusted tensor of inertia and adjusted variance covariance components (Heiker 2013). If the residuals show systematic effects, inconsistencies between observed data and models are existing. All residuals are analyzed in the frequency domain to relate inconsistencies to periods. Figures 2, 3 and 4 show the amplitude spectra of some residuals.

Table 1 Variance components after 25 iterations

	Full VCM	Diagonal VCM
Polar motion	16.8462	0.4527
Length of day	22.1600	0.8307
Equatorial excitation functions	6.9196	0.7988
Axial excitation functions	94.8439	4.9342
GFC2	0.4459	0.2801

The graphs in Figs. 2 and 3 are obtained from the combination of two different excitation functions with the EOP and SLR GFC2. The NCEP/ECCO model contains in Fig. 2 a semiannual period which is not present in the ECWMWF/OMCT model. According to Eq. (8) solely the difference between the EOP and the motions terms can be validated, but not their absolute value. Hence, the semiannual peak in LOD is due to inconsistencies in the NCEP/ECCO χ_3 motion term. The amplitudes increase near the frequency of six times a year. These are artefacts as daily EOP are compared with monthly GFC2 (cf. Nyquist-Shannon sampling theorem). According to Fig. 3, the combination with the NCEP/ECCO model shows larger variations in the mass terms than the combination with the ECMWF/OMCT model. Thus, the ECMWF/OMCT model has obviously a higher degree of consistency. However, the reason for the higher degree of consistency might be due to the fact that the gravity field coefficients are obtained from background models using the ocean model OMCT driven by ECMWF atmospheric data.

The residuals in Fig. 4 are obtained by combining SLR and GRACE gravity field coefficients with the EOP and the ECMWF/OMCT excitation functions. The GRACE C20 residuals show a large annual period and a large 161-day-period which is due to the wrong ocean tide model used in the GRACE-processing (Chen at al. 2008). This known fact has led to the recommendation of the GRACE community to replace the GRACE C20 coefficients by values obtained by SLR (GRACE Technical Note no. 5 2011). The C21 and S21 gravity field coefficients show merely small differences between GRACE and SLR. The S21 coefficient residuals show a dominant annual frequency in both graphs. The gravity field coefficients C22 and S22 can not be validated at all, as they are not redundantly determined by the functional model.

The influence of the priori stochastic model on the least squares adjustment is analyzed by using identical time series but different stochastic models. The first run is performed with full VCMs for the EOP and the excitation functions and the second run only considers the diagonal elements of the VCMs. Table 1 contains the adjusted variance components. The variance components are clearly smaller if uncorrelated stochastic models are used. Figure 5 shows the amplitude spectra of the residuals of the χ_3 mass term. The residuals

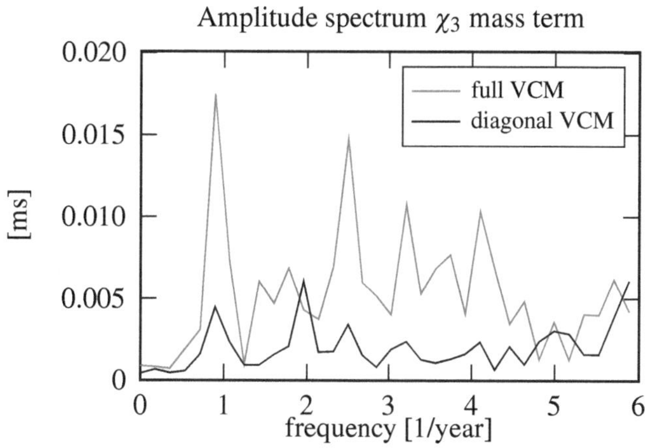

Fig. 5 Amplitude spectra of the χ_3 mass term residuals from different VCM

are considerably larger, if correlations are considered. The two largest peaks of the amplitude spectra are even assigned to different frequencies.

Conclusions and Outlook

From a Bayesian point of view, our approach implicitly assumes a normal distribution of the misclosure vector $\mathbf{w}_H$ with the expectation value $E(\mathbf{w}_H) = \mathbf{A}_H \Delta\boldsymbol{\beta}$ and the variance covariance matrix $D(\mathbf{w}_H) = \mathbf{B}\boldsymbol{\Sigma}_{ll}\mathbf{B}'$. If the misclosure vector is equal to $\mathbf{w}_H = \mathbf{Bl}$ and the (pseudo)observation vector $\mathbf{l}$ is normally distributed, this assumption is true. Otherwise, if the functional model result by a Taylor expansion of an arbitrary nonlinear function or the observation vector lacks normal distribution, the results of our approach would be compromised by linearizion errors or by a distribution errors. If the implicit assumptions are strongly violated, one might consider to use a Bayesian approach. The Bayes statistic is able to deal with nonlinear functional relations and to consider every kind of priori distributions.

This paper presents a method for the mutual validation of models and data by means of a linear, constrained least squares adjustment. As our approach has confirmed some known facts, this approach is adequate for the mutual validation of EOP, GFC2 and geophysical excitation functions. The findings are summarized as follows.

- Once more, this example has led to the trivial conclusion, that the functional and the stochastic model have to be sensitive to the given problem. The difference of the EOP and the motion term can be validated but not their absolute values. The validation of C22 and S22 is impossible.
- The ECMWF/OMCT excitation functions show a higher degree of consistency with the EOP and GFC2 than the NCEP/ECCO model.

- The S2 tide aliasing in GRACE C20 coefficients published in Chen at al. (2008) is confirmed with our approach.
- The comparison of two different stochastic models has led to different results. Neglected correlations might lead to too optimistic results.

Further time series are analyzed. There are some hints that correlations might lead to frequency shifts in the residuals (compare Fig. 5). Hence, the role of correlations has to be investigated more closely.

Acknowledgements These results have been derived within the work on the project *Mutual validation of EOP and gravity field coefficients*. This project is part of the research unit *Earth Rotation and global Geodynamic processes* funded by the German Research Foundation (FOR 584; see: www.erdrotation.de). This is gratefully acknowledged.

We like to thank Minkang Chen who provided SLR gravity field coefficients and gave valuable advices.

References

Chen M, Ries J (2011) GRACE Technical Note No. 5. http://isdc.gfz-potsdam.de/index.php?name=UpDownload&req=viewsdownload&sid=5

Chen JL, Wilson CR, Seo KW (2008) S2 tide aliasing in GRACE time-variable gravity solutions. J Geod 83:679–687

Galindo FJ, Ruiz JJ, Giachino E, Premoli A, Tavella P (2001) Estimation of the covariance matrix of individual standards by means of comparison measurements. In: Advanced mathematical and computational tools in metrology V. Series on advances in mathematics for applied sciences, vol 57. World Scientific, Singapore, pp 179–183

Gross R (2007) Earth rotation variations - long period. In: Herring T (ed) Treatise on geophysics, vol 3. Elsevier, Amsterdam

Heiker A (2013) Mutual validation of Earth orientation parameters, geophysical excitation functions and second degree gravity field coefficients. PhD thesis, Universität Hannover, http://dgk.badw.de/fileadmin/docs/c-697.pdf

Hofmann-Wellenhof B, Moritz H (2006) Physical geodesy, 2nd edn. Springer, Berlin

Koot L, Viron O, Dehant V (2006) Atmospheric angular momentum time-series: characterization of their internal noise and creation of a combined series. J Geod 79:663–674

Kusche J (2003) A Monte-Carlo technique for weight estimation in satellite geodesy. J Geod 76:641–652

Kusche J, Klees R (2002) Regularization of gravity field estimation from satellite gravity gradients. J Geod 76:359–368

Premoli A, Tavella P (1993) A revisited three-cornered hat method for estimating frequency standard instability. IEEE Trans Instrum Meas 42(1):7–13

Rochester MG, Smylie DE (1974) On changes in the trace of the earth's inertia tensor. J Geophys Res 79:4948–4951

Yu Z (1992) A generalization theory of estimation of variance-covariance components. Manuscripta Geodaetica 17:295–301

Comparison of Different Combination Strategies Applied for the Computation of Terrestrial Reference Frames and Geodetic Parameter Series

Manuela Seitz

Abstract

The combination of space geodetic techniques is today and becomes in future more and more important for the computation of Earth system parameters as well as for the realization of reference systems. Precision, accuracy, long-term stability and reliability of the products can be improved by the combination of different observation techniques, which provide an individual sensitivity with respect to several parameters. The estimation of geodetic parameters from observations is mostly done by least squares adjustment within a Gauß-Markov model. The combination of different techniques can be done on three different levels: on the level of observations, on the level of normal equations and on the level of parameters. The paper discusses the differences between the approaches from a theoretical point of view. The combination on observation level is the most rigorous approach since all observations are processed together ab initio, including all pre-processing steps, like e.g. outlier detection. The combination on normal equation level is an approximation of the combination on observation level. The only difference is that pre-processing steps including an editing of the observations are done technique-wise. The combination on the parameter level is more different: Technique-individual solutions are computed and the solved parameters are combined within a second least squares adjustment process. Reliable pseudo-observations (constraints) have to be applied to generate the input solutions. In order to realize the geodetic datum of the combined solution independently from the datum of the input solutions, parameters of a similarity transformation have to be set up for each input solution within the combination. Due to imperfect network geometries, the transformation parameters can absorb also non-datum effects. The multiple parameter solution of the combination process leads to a stronger dependency of the combined solution on operator decisions and on numerical aspects.

Keywords

Combination • DORIS • GPS • ITRF • Normal equation level • Observation level • SLR • VLBI

1 Introduction

The combination of different space geodetic techniques is a common procedure in order to compute precise geodetic products today. Combining different observation types, the individual potentials of the different techniques with respect

M. Seitz (✉)
Deutsches Geodätisches Forschungsinstitut,
Alfons-Goppel-Str. 11, 80539 München, Germany
e-mail: seitz@dgfi.badw.de

H. Kutterer et al. (eds.), *The 1st International Workshop on the Quality of Geodetic Observation and Monitoring Systems (QuGOMS'11)*, International Association of Geodesy Symposia 140,
DOI 10.1007/978-3-319-10828-5_9,

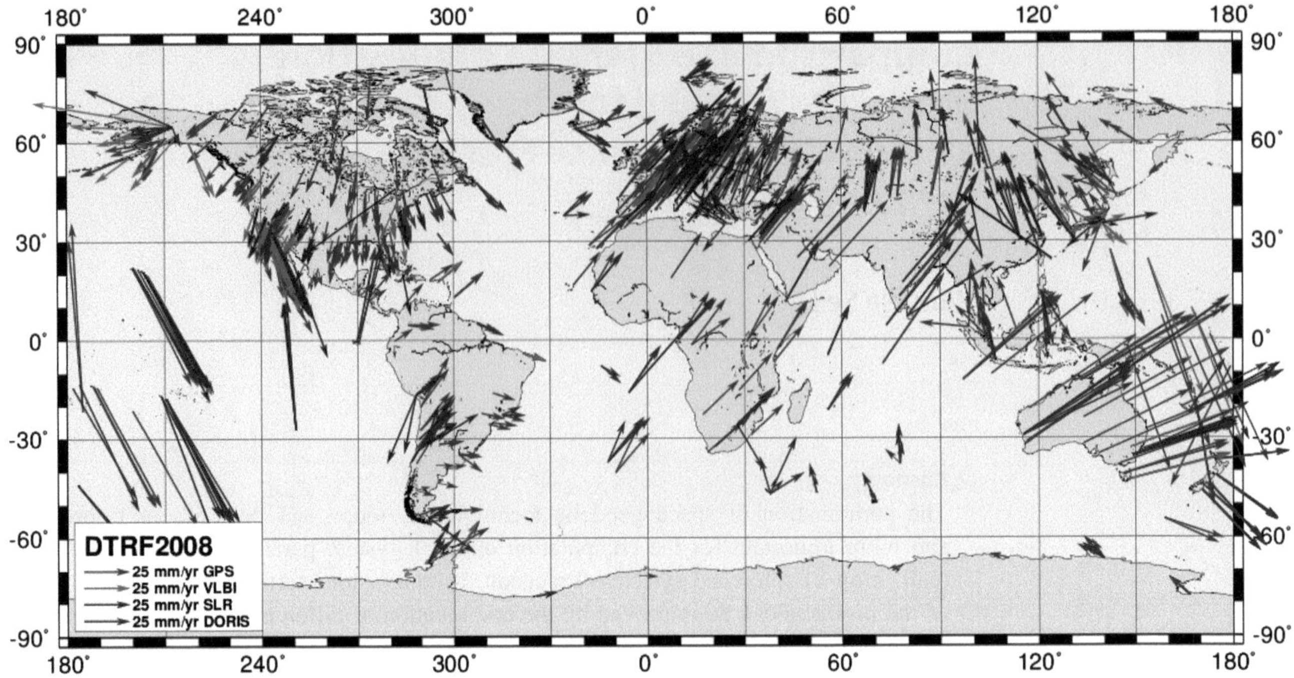

Fig. 1 Horizontal station velocity field of DTRF2008

to the determination of certain geodetic parameters can be exploited for the combined product. But also the increased redundancy due to the larger number of observations leads to a higher precision of the parameters. Examples for combinations are the (a) realization of the International Terrestrial Reference System (ITRS) which is computed by combining data of Very Long Baseline Interferometry (VLBI), Global Navigation Satellite Systems (GNSS), Satellite Laser Ranging (SLR) and Doppler Orbitography and Radiopositioning Integrated by Satellite (DORIS) (e.g. Seitz et al. 2012), (b) the computation of gravity field products by combining satellite and terrestrial data (e.g. Kern et al. 2003) and (c) ionosphere models and troposphere parameters derived from multi-technique data (e.g. Dettmering et al. 2011; Krügel et al. 2007). The integration of different space geodetic techniques in order to ensure a long-term and precise monitoring of the geodetic parameters is also the goal of the Global Geodetic Observing System (GGOS) a component of the International Association of Geodesy (IAG) (Plag and Pearlman 2009). The derivation of geodetic parameters from the observations is usually done by using the linear Gauß-Markov model (see also Schmidt et al. 2013). The combination of different techniques can be performed on the three levels within a least squares adjustment: the observation, the normal equation and the parameter level. This paper describes the mathematical fundamentals of the different approaches and discusses the pros and cons with regard to the ITRS realization from a theoretical point of view.

The ITRS is a global Earth related cartesian coordinate system which is well defined (IERS 2010). It is realized by the International Terrestrial Reference Frame (ITRF) which consists of station positions and velocities of global distributed GNSS, VLBI, SLR and DORIS observing stations. It is computed from long-term observation time series of the four techniques, which span between 15 and 25 years. Figure 1 shows the horizontal velocity field of the ITRS realization DTRF2008 (Seitz et al. 2012). Consistently to the station coordinate time series of Earth Orientation Parameters (EOP) are estimated. DTRF2008 is computed by combining the four space geodetic techniques which contribute to the determination of station coordinates and EOP according to their individual potentials.

2 Least Squares Adjustment by Gauß-Markov Model

In general the linearized observation equation describes the expectation values of the observation vector **b** as a linear combination of known coefficients and the unknown parameters **p** (Koch 1999)

$$E(\mathbf{b}) = \mathbf{Ax} \tag{1}$$

with

$\mathbf{b}$	$n \times 1$ vector of observations (measurement minus "computed with a priori values")
$E(\mathbf{b})$	$n \times 1$ vector of the expectation values of observations
$\mathbf{A}$	$n \times u$ coefficient matrix
$\mathbf{x} = \mathbf{p} - \mathbf{p_0}$	$u \times 1$ vector of unknowns (vector of corrections of $\mathbf{p_0}$)
$\mathbf{p}$	$u \times 1$ parameter vector
$\mathbf{p_0}$	$u \times 1$ vector of a priori values of $\mathbf{p}$

Equation (1) is the deterministic part of the Gauß-Markov model. The relation between observations and parameters are given by physical or mathematical principles. Linearity is, if necessary, achieved by a linearization of the original observation equations. Therefore, the observation equations are expanded into Taylor series. Usually, the number of observations n is larger than the number of unknowns u in order to reduce the impact of one observation on the estimates (Koch 1999). For $n > u$ the equation system $\mathbf{Ax} = \mathbf{b}$ is generally not consistent. From the addition of the vector of observation errors $\mathbf{v}$, the consistent observation equation is obtained

$$\mathbf{Ax} = \mathbf{b} + \mathbf{v} \tag{2}$$

with $E(\mathbf{b} + \mathbf{v}) = E(\mathbf{b})$ because of $E(\mathbf{v}) = \mathbf{0}$.

It is expected that the observations are random values. The variance-covariance matrix of the observations is assumed to be known, except of the variance factor σ_0^2

$$\mathbf{C}_{bb} = \sigma_0^2 \mathbf{P}^{-1} \tag{3}$$

with

$\mathbf{C}_{bb}$	$n \times n$ variance-covariance matrix of observations
$\mathbf{P}$	$n \times n$ positive definite weight matrix of observations.

Equation (3) is the stochastic part of the Gauß-Markov model.

For $n > u$ the solution of Eq. (2) is not unique. A solution can be get, if the squared sum of observation residuals $\mathbf{v} = \mathbf{Ax} - \mathbf{b}$ is minimized (least squares adjustment). The solution means the best linear unbiased estimation. The corresponding normal equation reads (e.g. Schmidt et al. 2013)

$$\mathbf{A}^{\mathrm{T}}\mathbf{PA}\hat{\mathbf{x}} = \mathbf{A}^{\mathrm{T}}\mathbf{Pb}. \tag{4}$$

The solution of the normal equation is given by

$$\hat{\mathbf{x}} = (\mathbf{A}^{\mathrm{T}}\mathbf{PA})^{-1}\mathbf{A}^{\mathrm{T}}\mathbf{Pb}. \tag{5}$$

With the normal equation matrix $\mathbf{N} = \mathbf{A}^{\mathrm{T}}\mathbf{PA}$ (which has to be regular, see below) and the right hand side $\mathbf{y} = \mathbf{A}^{\mathrm{T}}\mathbf{Pb}$ the equation can be written in a compact form

$$\hat{\mathbf{x}} = \mathbf{N}^{-1}\mathbf{y}. \tag{6}$$

For the variance also an unbiased estimation is required. That means, the equation $E(\hat{\sigma}_0^2) = \sigma_0^2$ must be fulfilled, which can be reached by the estimation

$$\hat{\sigma}_0^2 = \frac{\hat{\mathbf{v}}^{\mathrm{T}}\mathbf{P}\hat{\mathbf{v}}}{n-u}. \tag{7}$$

The relation between the squared sum of observations and the squared sum of residuals is expressed as

$$\mathbf{b}^{\mathrm{T}}\mathbf{Pb} - \hat{\mathbf{v}}^{\mathrm{T}}\mathbf{P}\hat{\mathbf{v}} = \mathbf{y}^{\mathrm{T}}\hat{\mathbf{x}}. \tag{8}$$

The variance-covariance matrix of the estimated parameters is obtained from the variance-covariance matrix of observations $\hat{\mathbf{C}}_{\mathbf{bb}} = \hat{\sigma}_0^2\mathbf{P}^{-1}$, with the estimated variance factor $\hat{\sigma}_0^2$ by error propagation

$$\begin{aligned}\hat{\mathbf{C}}_{\hat{\mathbf{x}}\hat{\mathbf{x}}} &= (\mathbf{N}^{-1}\mathbf{A}^{\mathrm{T}}\mathbf{P})\hat{\sigma}_0^2\mathbf{P}^{-1}(\mathbf{N}^{-1}\mathbf{A}^{\mathrm{T}}\mathbf{P})^{\mathrm{T}} \\ &= \hat{\sigma}_0^2(\mathbf{A}^{\mathrm{T}}\mathbf{PA})^{-1}.\end{aligned} \tag{9}$$

Solving the normal equation with Eq. (6) it is required that the matrix $\mathbf{N}$ is of full rank. This is usually not the case. For example in ITRS realization, the matrix $\mathbf{N}$ has a rank deficiency with respect to datum parameters, namely with respect to the orientation of the frame. In order to achieve a regular normal equation matrix, pseudo-observations are added to the normal equation system

$$\hat{\mathbf{x}} = (\mathbf{N} + \mathbf{D})^{-1}(\mathbf{y} + \mathbf{d}) \tag{10}$$

wherein, $\mathbf{D}$ is the normal equation matrix of pseudo-observations and $\mathbf{d}$ is the right hand side of the normal equation of pseudo-observations [according to Eqs. (4)–(6)]. In order to remove the rank deficiency with respect to the orientation of the frame, no-net-rotation conditions in form of pseudo-observations are added (Angermann et al. 2004). In the following, it is assumed, that the initial normal equation system is free of pseudo-observations and all constraints are added on the normal equation level.

3 Combination Strategies

Different space geodetic techniques are sensitive to common geodetic parameters (e.g. station coordinates and EOP). Performing technique-individual least squares adjustments,

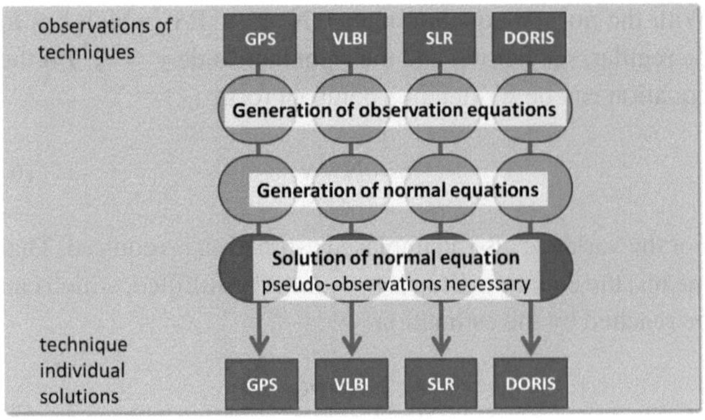

Fig. 2 Individual estimation of geodetic parameters from the observations of different techniques

independent solutions of the same parameters are obtained as it is graphically shown in Fig. 2.

The combination can be performed on the different levels of least squares adjustment as it is shown in Fig. 3. If the processing could consistently be done for the techniques (including also the data editing) and the stations would be distributed homogeneously all three approaches would lead to the same results. But this is usually, and in particular for ITRF computation, not the case. Thus, the three combination approaches provide solutions which differ slightly, illustrated by three individual solutions in Fig. 3. In the following paragraphs, the combination procedures are discussed in more detail.

3.1 Combination on Observation Level

According to Eqs. (2) and (3) the observation equation system for each of the m techniques k reads

$$\mathbf{A}_k \mathbf{x}_k = \mathbf{b}_k + \mathbf{v}_k \tag{11}$$

$$\mathbf{C}_{\mathbf{b}_k \mathbf{b}_k} = \sigma^2 \mathbf{P}_k^{-1} \tag{12}$$

with k=1,..,m. Preassumptions for a combination are (1) that the equations are related to the same parameters and (2) that the same a priori reduction models are used for modelling the same phenomena (e.g. solid Earth tides, ocean loading, troposphere refraction, ...). Then the individual observation equations can be composed to one system

$$\begin{bmatrix} \mathbf{A}_1 \\ \cdots \\ \mathbf{A}_m \end{bmatrix} \mathbf{x} = \begin{bmatrix} \mathbf{b}_1 \\ \cdots \\ \mathbf{b}_m \end{bmatrix} + \begin{bmatrix} \mathbf{v}_1 \\ \cdots \\ \mathbf{v}_m \end{bmatrix} \tag{13}$$

$$\mathbf{C}_{bb} = \sigma^2 \begin{bmatrix} \mathbf{P}_1^{-1} & \mathbf{0} & \mathbf{0} \\ \mathbf{0} & \ddots & \mathbf{0} \\ \mathbf{0} & \mathbf{0} & \mathbf{P}_m^{-1} \end{bmatrix}. \tag{14}$$

Pre-processing procedures, which includes also editing of the observation data, are performed. Afterwards the least squares adjustment is done according to Eqs. (5)–(9) considering Eq. (10) if necessary.

3.2 Combination on Normal Equation Level

The basic observation equations which are written for each of the techniques are identical to those of the combination on observation level (see Eq. (11)), considering the requirement of using the same standards for parametrization and a priori models. The data editing is done separately for each technique. Before the combination is done, the observation equations are transformed to normal equations applying the condition that the squared sum of residuals is minimized.

The normal equation for the technique k reads

$$\mathbf{N}_k \hat{\mathbf{x}}_k = \mathbf{y}_k. \tag{15}$$

The normal equations of all techniques are combined by adding those elements of the normal equations that are related to the same parameters. Thus, if the normal equations

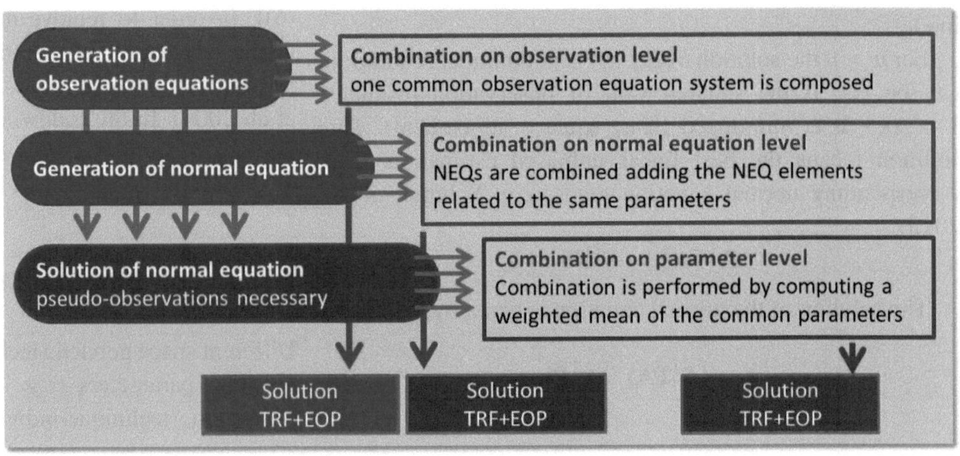

Fig. 3 Combination methods based on different levels of least squares adjustment process shown for the example of the consistent computation of the Terrestrial Reference Frame (TRF) and the EOP

do not have the same size $u \times u$, the individual equations must be expanded to the full amount of all parameters which shall be solved. This is done by adding zero lines and columns to the normal equation matrix $\mathbf{N}_\mathrm{k}$ and zero elements to the right hand side of normal equation $\mathbf{y}_\mathrm{k}$. Then all the normal equation systems are sorted by the same order of parameters. The combined normal equation system is written as (Gerstl et al. 2001)

$$\mathbf{N} = \frac{1}{\sigma_1^2}\mathbf{N}_1 + \ldots + \frac{1}{\sigma_m^2}\mathbf{N}_\mathrm{m}, \quad \mathbf{N} \in \mathrm{u} \times \mathrm{u} \tag{16}$$

$$\mathbf{y} = \frac{1}{\sigma_1^2}\mathbf{y}_1 + \ldots + \frac{1}{\sigma_m^2}\mathbf{y}_\mathrm{m}, \quad \mathbf{y} \in \mathrm{u} \times 1 \tag{17}$$

$$\mathbf{b}^\mathrm{T}\mathbf{P}\mathbf{b} = \frac{1}{\sigma_1^2}\mathbf{b}_1^T\mathbf{P}\mathbf{b}_1 + \ldots + \frac{1}{\sigma_m^2}\mathbf{b}_\mathrm{m}^T\mathbf{P}\mathbf{b}_\mathrm{m}, \tag{18}$$

considering the estimated variance factors σ_k^2. The combined system $\mathbf{N}\hat{\mathbf{x}} = \mathbf{y}$ is solved by using Eqs. (5)–(9) considering Eq. (10) if necessary.

3.3 Combination on Parameter Level

Combining techniques on the parameter level individual technique solutions are performed initially, like it is illustrated in Fig. 2. The adjusted parameters are then combined in a second least squares adjustment process by considering the full variance-covariance matrices (see Fig. 3). Consequently, the observation equation reads different from the combination on observation and normal equation level

$$\mathbf{I}\mathbf{x}_\mathrm{k} = \hat{\mathbf{x}} + \bar{\mathbf{v}}_\mathrm{k} \tag{19}$$

$$\mathbf{C}_{\hat{\mathbf{x}}_\mathrm{k}\hat{\mathbf{x}}_\mathrm{k}} = \sigma^2(\mathbf{N}_\mathrm{k} + \mathbf{D}_\mathrm{k})^{-1} = \sigma^2\mathbf{P}_\mathrm{k}^{-1}, \tag{20}$$

wherein $\mathbf{D}_\mathrm{k}$ is the normal equation matrix of pseudo-observations (see also Eq. (10)). That means, in case of a rank deficiency of the matrix $\mathbf{N}_\mathrm{k}$ pseudo-observations are added in order to generate the individual technique solutions. The combined system is obtained by composing the observation equations to one equation system

$$\mathbf{I}\begin{bmatrix}\mathbf{x}_1\\ \ldots \\ \mathbf{x}_\mathrm{m}\end{bmatrix} = \begin{bmatrix}\hat{\mathbf{x}}_1\\ \ldots \\ \hat{\mathbf{x}}_\mathrm{m}\end{bmatrix} + \begin{bmatrix}\bar{\mathbf{v}}_1\\ \ldots \\ \bar{\mathbf{v}}_\mathrm{m}\end{bmatrix} \tag{21}$$

$$\mathbf{C}_{\hat{x}\hat{x}} = \sigma^2\begin{bmatrix}\mathbf{P}_1^{-1} & \mathbf{0} & \mathbf{0}\\ \mathbf{0} & \ddots & \mathbf{0}\\ \mathbf{0} & \mathbf{0} & \mathbf{P}_\mathrm{m}^{-1}\end{bmatrix}. \tag{22}$$

The equation system is solved by a least squares adjustment [see Eqs. (5)–(9) and Eq. (10) if necessary].

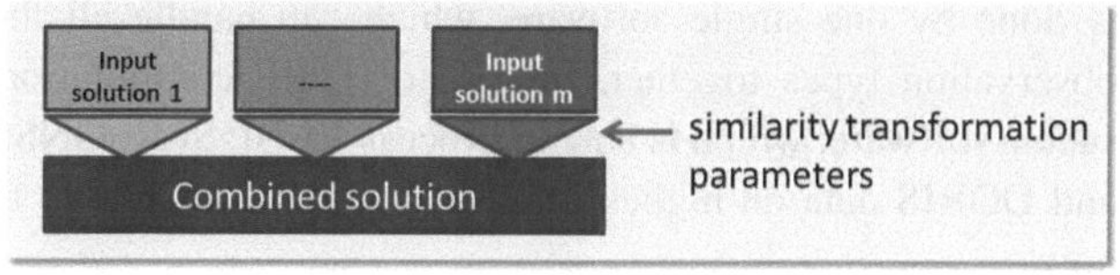

Fig. 4 Combination on parameter level: parameters of similarity transformation are set up per input solution

In ITRF computation, the input solutions are solved by applying pseudo-observations in form of minimum conditions in order to reduce the rank deficiency of the individual techniques with respect to parameters of the geodetic datum. Thus, in case of GPS, SLR and DORIS no-net-rotation conditions (see e.g. Seitz 2009) are used. In case of VLBI, which is not related to the orbit dynamics of a satellite and thus not sensitive to the Earth's centre of mass, additionally no-net-translation conditions are needed to realize the origin of the frame.

Two different aspects, concerning the application of pseudo-observations in case of combination on parameter level, have to be discussed. (1) Applying pseudo-observations it must be considered, that the combined solution depends on the variances and co-variances obtained from the individual technique solutions [see Eq. (20)]. Thus, reliable pseudo-observations must be applied in order to provide variances that reflect only the uncertainty of the deterministic part of the Gauß-Markov model and not an uncertainty with respect to the geodetic datum. That means, loose constraint solutions (see e.g. Angermann et al. 2004) cannot be used for a combination on parameter level, but they have to be preliminary resolved applying minimum datum conditions with suitable standard deviations (in ITRF computation a few millimetres or less). (2) To be able to realize the datum of the combined solution independently from the input solutions, parameters of a similarity transformation have to be set up for each individual input solution in order to restore the rank deficiency of the combined equation system with respect to geodetic datum parameters (see Fig. 4). The singularity can then be removed by arbitrary conditions.

4 Comparison of the Different Combination Strategies

The combination on observation level is the most rigorous combination model. All observation types are processed together starting with the generation of observation equations using the same parameterizations and reduction models. Preprocessing steps as the detection of outliers and the editing of the observation data can be done using the full amount of available observations. In an optimal case, the data analysis is done by one single software, which can handle all the observation types together. In case of ITRF computation, such a software, which is able to process VLBI, SLR, GNSS and DORIS data on highest standard, is not available up to now.

The combination on normal equation level is – under certain conditions – a good approximation of the combination on observation level. If the observation equations are generated using the same parametrization and reduction models and no constraints are added before the normal equations are combined, the combination on normal equation level is comparable to the combination on observation level. The only difference is that the detection of outliers and the data editing is done technique-wise. However, the effect on the solution is, particularly in the case of ITRF computation, assumed to be unverifiable.

A further aspect has to be discussed for the example of ITRS realization. In order to be able to handle the large normal equation matrices, parameters which are not of very direct interest are reduced before the combination. Parameters which cannot be estimated very stable, as e.g. clock parameters, are slightly constrained before reduction. These constraints cannot be removed anymore in the combination, even if the parameter would indirectly benefit from the combination to such an extent that the constraint would be unnecessary, e.g. if correlations to other parameters are reduced. Thus, in order to avoid deformations of the solution, the a priori constraints on the reduced parameters must be introduced very carefully. A more rigorous way is not to constrain and reduce parameters before combination.

The combination on parameter level shows clear differences with respect to the combination on observation and normal equation level. Individual technique solutions are performed adding the necessary datum conditions in form of pseudo-observations. Only a minimum number of pseudo-observations is allowed to avoid any over-constraining and hence the deformation of the solution. Furthermore, it must be considered that the combined solution depends on the variances and co-variances obtained from the technique solutions, so that reliable pseudo-observations must be applied while generating the input solutions.

In order to ensure that the datum of the combined solution can be realized independently, it is necessary to set up parameters of a similarity transformation for each of the input solutions. The estimated similarity transformation parameters and consequently the combined solution depend on the set of stations used for the parameter set up. Due to the inhomogeneous global distribution of stations the transformation parameters are correlated and can absorb also non-datum effects, which becomes particularly critical in case of a poor network geometry.

The combination on solution level is not a straightforward approach due to the multiple application of pseudo-observations and the subsequent removement of datum information by the set up of transformation parameters. Added to that is the fact, that these steps are not independent from operator decisions (Which stations are used for the set up of transformation parameters?).

Table 1 Comparison of the different combination methods

	Observation level	Normal equation level	Parameter level
Common pre-analysis of observations	Yes	No	No
Pseudo-observations needed for combination	No	No	Yes
Additional transformation parameters needed	No	No	Yes
Rigorous method	Yes	Widely	No

Table 1 summarizes the most important characteristics of the combination methods.

5 Combination in the Geodetic Practice: The Realization of ITRS

The ITRS realization computed at DGFI is one example for the combination on normal equation level (Seitz et al. 2012). Figure 5 shows the flowchart of the combination procedure. The input normal equations are extracted from files in SINEX (Solution INdependent EXchange) format. The format description is available under http://www.iers.org (2014-09-29). SINEX allows for the storage of normal equations as well as of solutions. For the latest ITRS realization (DTRF2008) GNSS, SLR and DORIS input data are provided as solutions and the normal equation has to be reconstructed using the information about the applied constraints given in the SINEX files. The formulas used for the reconstruction of the normal equation from SINEX can be inter alia found in Seitz et al. (2012). The VLBI contribution was provided in form of normal equations and could be used directly from the SINEX files.

A combination of normal equations is done in the process at two different stages (1) the combination of the normal equation time series of one technique to one normal equation per technique and (2) the combination of the technique normal equations. The ITRF2005 was the first ITRF based on time series of weekly or session-wise input data (Altamimi et al. 2007; Angermann et al. 2007). The advantage of input time series compared to multi-year input solutions is that the analysis of the parameter time series (station positions, datum parameters and EOP) can be done consistently for all the techniques within the combination process.

The parameters, which are relevant in the ITRS realization DTRF2008 are specified in Fig. 6. The figure gives

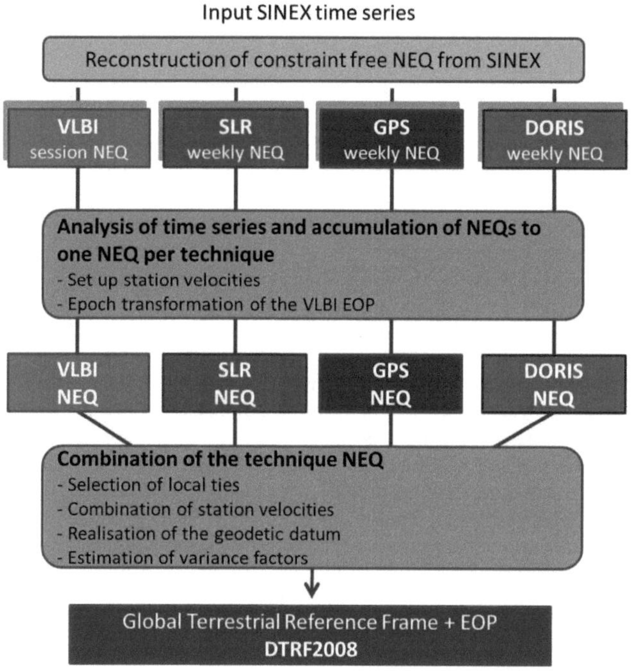

Fig. 5 Flowchart of ITRS realization at DGFI

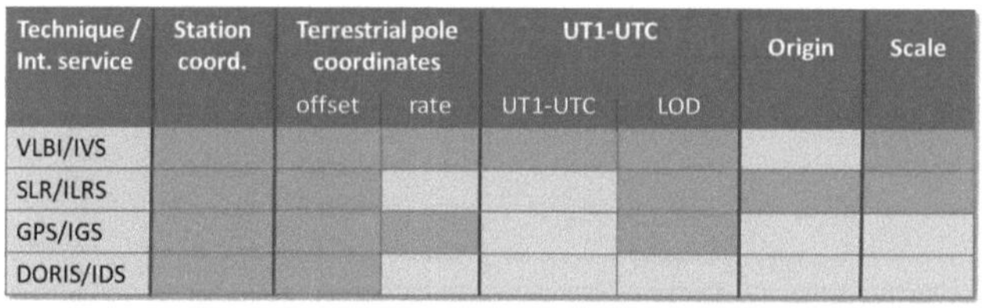

Fig. 6 Parameters of the ITRS realization DTRF2008. *Dark colour* means that the technique contributes to the parameter

an overview about the contribution of the techniques to the determination of the several parameters. While station coordinates and the coordinates of the terrestrial pole are provided by all the techniques, UT1-UTC can only be derived from VLBI in an absolute sense. GNSS and SLR contribute to the UT1-UTC estimates by the first derivative in time. The actually provided parameter is LOD, which is related to the UT1-UTC by LOD $= - d/dt$(UT1-UTC). Concerning the datum parameters, the origin is derived from SLR observations only, while for the realization of the scale SLR and VLBI observations are considered. Since also GPS and DORIS observations provide information on origin and scale (which should not be used for ITRF datum realization as they are affected by systematics) the normal equations are extended by scale as well as geocenter parameters, which correspond to the degree 0 and degree 1 gravity field coefficients, respectively.

Examples for the combination on parameter level are e.g. the computation of the ITRF2008 (Altamimi et al. 2011) and that of EOP series IERS C04 (ftp://hpiers.obspm.fr/iers/eop/eopc04/C04.guide.pdf, 2011-0-02). As the combination on observation level is the most intricate approach, ITRS related products are not yet computed by combination on observation level. Therefore, the IERS installed the Working Group on the Combination at the Observation Level (COL), which deals with the development of methods, strategies and software necessary for the combination of the different techniques on the observation level (http://www.iers.org, 2014-09-29).

Summary and Conclusions

The combination of different observation types can be performed on the three levels of least squares adjustment within a Gauß-Markov model: the observation, the normal equation and the parameter level. For applications in ITRF computation the approaches show differences with respect to the data processing and/or to the mathematical concepts and do not lead to identical results. The combination on observation level is the most rigorous combination method as the whole processing line starting from the preprocessing steps is run using the full amount of available observation data.

The combination on normal equation level is a good approximation of the combination on observation level, if the observation equations are written by using the same standards for models and parametrization and if no constraints are added to the observation or normal equations before the combination. Thus, both, the combination on observation and on normal equation level can be recommended for the computation of geodetic products.

The combination on solution level differs clearly from the two other approaches. Technique-wise solutions are performed and the resulting parameters are combined afterwards by a second least squares adjustment process. The multiple addition of pseudo-observations and the subsequent reconstruction of rank deficiencies by set ups of transformation parameters is problematic as the latter step induce a stronger dependency of the final solution on operator impacts (impact of selected stations on the results) and numerical aspects (correlations between the parameters of similarity transformation).

While combination on normal equation level is a standard, combination on the observation level is not performed for ITRS realization today since an analysis software, which allows the processing of all the different data types in a contemporary way, is not available yet. In order to overcome this problem, different groups working on the development of multi-technique softwares. The IERS initiated the Working Group on the Combination at the Observation Level which deals with the development of combination strategies for the IERS products of the future by the use of the expertise of the different groups.

References

Altamimi Z, Collilieux X, Legrand J, Garayt B, Boucher C (2007) ITRF2005: a new release of the international terrestrial reference frame based on time series of station positions and Earth orientation parameters. J Geophys Res 112(B09401). doi:10.1029/2007JB004949

Altamimi Z, Collilieux X, Métivier L (2011) ITRF2008: an improved solution of the international terrestrial reference frame. J Geod. doi:10.1007/s00190-011-0444-4

Angermann D, Drewes H, Krügel M, Meisel B, Gerstl M, Kelm R, Müller H, Seemüller W, Tesmer V (2004) ITRS combination center at DGFI - a terrestrial reference frame realization 2003. Deutsche Geodätische Kommission, Reihe B, München

Angermann D, Drewes H, Krügel M, Meisel B (2007) Advances in terrestrial reference frame computation. In: Tregoning P, Rizos C (eds) Dynamic planet. International Association of Geodesy Symposia, 130:595–602. doi:10.1007/978-3-540-49350-1_86

Dettmering D, Schmidt M, Heinkelmann R, Seitz M (2011) Combination of different space-geodetic observations for regional ionosphere modeling. J Geod. doi:10.1007/s00190-010-0423-1

Gerstl M, Kelm R, Müller H, Ehrnsperger W (2001) DOGSCS Kombination und Lösung großer Gleichungssysteme. Interner Bericht, DGFI, München

IERS (2010) IERS Conventions (2010). IERS Tech. Note 36. Verlag des Bundesamtes für Geodäsie und Kartographie, Frankfurt am Main

Kern M, Schwarz K, Sneeuw N (2003) A study on the combination of satellite, airborne, and terrestrial gravity data. J Geod 77(3–4):217–225. doi:10.1007/s00190-003-0313-x

Koch K (1999) Parameter estimation and hypothesis testing in linear models. Springer, Berlin

Krügel M, Thaller D, Tesmer V, Rothacher M, Angermann D, Schmid R (2007) Tropospheric parameters: combination studies based on homogeneous VLBI and GPS data. J Geod 81(6–8):515–527. doi:10.1007/s00190-006-0127-8

Plag H-P, Pearlman M (eds) (2009) The global geodetic observing system: meeting the requirements of a global society on a changing planet in 2020. Springer, Berlin

Schmidt M, Göttl F, Heinkelmann R (2013) Towards the combination of data sets from various observation techniques. In: Kutterer H, Seitz F, Schmidt M (eds) The 1st International Workshop on the Quality of Geodetic Observation and Monitoring Systems (QuGOMS'11). International Association of Geodesy Symposia. Springer, Heidelberg, in this volume

Seitz M (2009) Combination of space geodetic techniques in order to realize the terrestrial reference system. Deutsche Geodätische Kommission, München (in German)

Seitz M, Angermann D, Bloßfeld M, Drewes H, Gerstl M (2012) The 2008 DGFI realization of ITRS: DTRF2008. J Geod 86:1097–1123. doi:10.1007/s00190-012-0567-2

W-Ratio Test as an Integer Aperture Estimator: Pull-in Regions and Ambiguity Validation Performance

T. Li and J. Wang

Abstract

Global Navigation Satellite Systems (GNSS) carrier phase integer ambiguity resolution is an indispensible step in generating highly accurate positioning results. As a quality control step, ambiguity validation, which is an essential procedure in ambiguity resolution, allows the user to make sure the resolved ambiguities are reliable and correct. Many ambiguity validation methods have been proposed in the past decades, such as R-ratio, F-ratio, W-ratio tests, and recently a new theory named integer aperture estimator. This integer aperture estimator provides a framework to compare the other validation methods with the same predefined fail-rate, even though its application in practice can only be based on simulations.

As shown in literature, the pull-in regions of different validation methods may have a variety of shapes which may dictate the closeness of such validation methods to the optimal integer least-squares method. In this contribution, the W-ratio is shown to be connected with the integer aperture theory with an exact formula. The integer least-squares pull-in region for W-ratio is presented and analysed. The results show that the W-ratio's pull-in region is close to the integer least-squares pull-in region. We have performed numerical experiments which show that the W-ratio is a robust way of validating the resolved ambiguities.

Keywords

Ambiguity validation test • GNSS • Integer aperture estimator • W-ratio

1 Introduction

Nowadays, Global Navigation Satellite Systems (GNSS) are widely used in navigation, surveying, mapping, etc. The demand for high precision GNSS positioning is still increasing. Generally, there are two types of measurements: code and carrier phase. The positioning accuracy based on code measurements and carrier phase measurements are in meter level and centimeter to millimeter level, respectively. Hence, carrier phase measurements are essential for precise positioning.

As a drawback, each carrier phase contains an integer ambiguity in the number of wavelengths that needs to be resolved. Traditionally used estimation methods, e.g. Least-squares, or Kalman filtering, can only provide us with float (or the real-valued) solutions and fixing the float ambiguities (ambiguity resolution) to integers is not an easy task. An amount of literature (e.g. Teunissen 1995; Han 1997; Wang et al. 1998) can be found to study the integer ambiguity resolution problem, and one of the most popular approaches is the so-called LAMBDA (Least-squares AMBiguity Decorrelation Adjustment) proposed by Teunissen (1995). With the float solution and variance–covariance matrix of the ambiguities from Least-squares, a search is carried out to find out integer ambiguity candidates inside the hyper-ellipsoid which is defined based on the float solution and the variance–covariance matrix. Instead of one integer candidate, usually the first best and the second best ambiguity combinations are

T. Li (✉) • J. Wang
School of Surveying and Spatial Information Systems, The University of New South Wales, Sydney, NSW 2052, Australia
e-mail: tao.li@unsw.edu.au

H. Kutterer et al. (eds.), *The 1st International Workshop on the Quality of Geodetic Observation and Monitoring Systems (QuGOMS'11)*, International Association of Geodesy Symposia 140,
DOI 10.1007/978-3-319-10828-5_10,

compared to make sure there is strong confidence in using the best combination in positioning. Consequently another problem called integer ambiguity validation emerged.

Various ambiguity validation methods have been proposed, such as F-ratio test, R-ratio, difference test, project test (Verhagen 2005), and W-ratio (Wang et al. 1998, 2000). From a statistical point of view, the critical values to validate the resolved ambiguity of these methods can be generated according to their distributions or empirical values. In another approach to ambiguity validation, Integer Aperture (IA) estimator (Teunissen 2003a, 2003b) has been developed. In Verhagen (2005), the IA estimator was considered as a framework for all the other classical validation methods, and the geometries of different validation methods are then reflected through various aperture pull-in regions.

In this contribution, the W-ratio test is first presented, and then the deduction of the W-ratio test as an integer aperture estimator is given, as well as its aperture pull-in region, finally the performance of the W-ratio is analysed with real data.

2 Parameter Estimation

The raw GNSS measurements are affected by many error sources, such as troposphere delay, ionosphere delay, clock error, etc. Through a double-differencing procedure, such systematic errors can be reduced over a short baseline, and the baseline components and integer ambiguities are the remaining parameters to be estimated. Without loss of generality, the double differenced functional models are given as follows:

$$\Delta\nabla\varnothing = \frac{1}{\lambda}\Delta\nabla\rho + \Delta\nabla N + \varepsilon_{\varnothing} \qquad (1)$$

$$\Delta\nabla P = \Delta\nabla\rho + \varepsilon_P \qquad (2)$$

where $\Delta\nabla$ is the double differencing operator between satellites and receivers, $\varnothing$ and P are carrier phase measurements and code measurements respectively, λ is the carrier phase wavelength, ρ is the geometric distance between satellites and receivers, N are the integer ambiguities in cycles, and ε represents the noise of the two types of measurements.

With an approximate rover position given, the above models can be linearized as

$$l = Ax + v \qquad (3)$$

where l is the vector of the observations, v is the vector of observation errors, x is the vector of unknowns $x = (x_r, \Delta\nabla N)^T$, A is the design matrix of both coordinates A_{x_r} and ambiguities A_a, and x_r are the coordinates.

It is interesting to note that Eq. (3) and the double differenced stochastic model in the following are the so-called Gauss–Markov model:

$$D(l) = \sigma_0^2 Q = \sigma_0^2 P^{-1} \qquad (4)$$

where D is the covariance matrix, σ_0^2 is an a priori variance factor, Q and P are the cofactor matrix and weight matrix of the measurements.

By applying the classical least-squares approach, which minimizes $v^T P v$, the unknown parameters and their covariance are uniquely estimated as:

$$\widehat{x} = (\widehat{x}_r, \widehat{a})^T = (A^T P A)^{-1} A^T P l,$$

$$\widehat{Q}_{\hat{x}} = (A^T P A)^{-1} = \begin{bmatrix} Q_{\hat{x}_r} & Q_{\hat{x}_r \hat{a}} \\ Q_{\hat{a}\hat{x}_r} & Q_{\hat{a}} \end{bmatrix} \qquad (5)$$

where $\widehat{a}$ represents the float solution of the integer ambiguities. Then the posteriori variance is

$$\widehat{s}_0^2 = \frac{\widehat{v}^T P \widehat{v}}{f} = \frac{\Omega_0}{f} \qquad (6)$$

with f is the degree of freedom.

3 W-Ratio Statistical Test

The W-ratio has been proposed by Wang et al. (1998), with the purpose of discriminating two sets of best integer candidates—the most likely candidate and the second most likely candidate. The likelihood ratio method (Koch 1988) and the artificial nesting method (Hoel 1947) may be applied to construct the discrimination tests, which both yield the following test statistic:

$$W = \frac{d}{\sqrt{Var(d)}} \geq c \qquad (7)$$

where

$$d = \Omega_2 - \Omega_1, \ Var(d) = \rho^2 Q_d \qquad (8)$$

$$\Omega_i = \Omega_0 + (\check{a}_i - \widehat{a})^T Q_{\hat{a}}^{-1} (\check{a}_i - \widehat{a}) \qquad (9)$$

c is the critical value, and ρ^2 could be decided by users either from an a priori variance σ_0^2 or from an a posteriori variance $\widehat{s}_0^2$; $\check{a}_i$ represents the integer ambiguities. By applying the variance–covariance propagation law, the variance for d can be derived as $Q_d = 4 \times (\check{a}_2 - \check{a}_1)^T Q_{\hat{a}}^{-1} (\check{a}_2 - \check{a}_1)$. Assuming

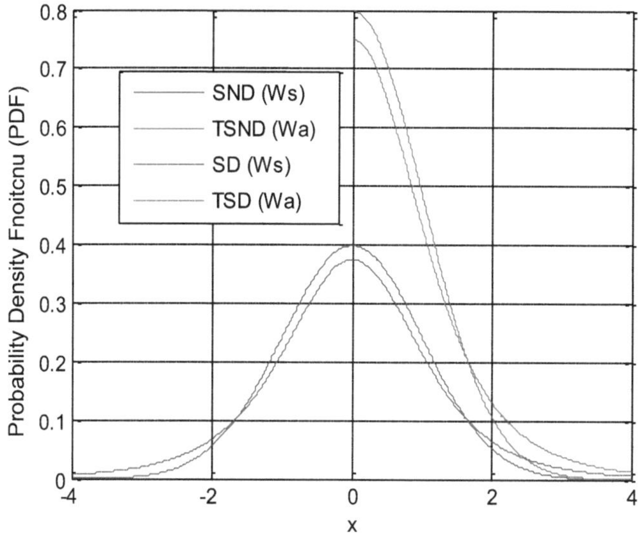

Fig. 1 Probability density function of SND, TSND, SD, TSD (degree of freedom 4)

that W_a and W_s are the two ratios corresponding to a priori variance σ_0^2 and a posteriori variance $\widehat{s}_0^2$, and under the assumption that the fixed ambiguities are deterministic quantities, they are supposed to have a truncated standard normal distribution (TSND) and a truncated student t distribution (TSD) respectively, from which the critical values can be easily obtained (Wang et al. 1998).

From the definition of the TSND, TSD and one constraint of the W-ratio ($d \geq 0$), we can obtain the probability density function (PDF) and cumulative distribution function (CDF) for the W-ratio. Figure 1 shows the PDF of both Wa-ratio and Ws-ratio.

Apparent from Fig. 1, due to the constraint of $d \geq 0$, the PDF of TSND and standard normal distribution (SND) are different, as well as the TSD and student t distribution (SD). The critical value for TSND is in fact a special case of SND, and it is slightly larger than the SND critical value, which implies that the accepted number of epochs for TSND will be less than SND, however, more reliable.

4 W-Ratio as an Integer Aperture Estimator

On the basis of the integer estimator, the IA theory was first introduced by Teunissen (2003a), and the IA estimator $\bar{a}$ is developed as:

$$\bar{a} = \sum_{z \in Z^n} z\omega_z(\widehat{a}) + \widehat{a}\left(1 - \sum_{z \in Z^n} \omega_z(\widehat{a})\right) \tag{10}$$

with the indicator function $\omega_z(x)$ defined as:

$$\omega_z(x) = \begin{cases} 1 & \text{if } x \in \Omega_z \\ 0 & \text{otherwise} \end{cases} \tag{11}$$

where z is an integer vector and the centre of Ω_z. x represents the float ambiguity vector. The Ω_z are the aperture pull-in regions, and their union $\Omega \subset R^n$ is the aperture space, which is also translational invariant.

With the above definition, three outcomes can be distinguished as: (1) $\widehat{a} \in \Omega_a$ Success: correct integer estimation; (2) $\widehat{a} \in \Omega \backslash \Omega_a$ Failure: incorrect integer estimation; (3) $\widehat{a} \notin \Omega$ Undecided: ambiguity not fixed to an integer. Then the probabilities of success (P_s), failure (P_f) and undecided (P_u) can be derived accordingly.

In the case of a GNSS model, $f_{\widehat{a}}(x)$ represents the probability density function of the float ambiguities, and is usually assumed to be normally distributed. The IA estimator allows the user to choose a pre-defined fail-rate, and then determine the critical value accordingly.

In Verhagen (2005), the pull-in regions of integer aperture bootstrapping, integer aperture least-squares have been explored, as well as other IA estimators. Since integer least-squares is optimal, the IA least-squares estimator can be considered to be a better solution than the others, and the pull-in region has been shown to be a hexagon. According to simulations, the success-rate, fail-rate and undecided-rate can be obtained respectively.

In a similar way, assume $\sigma_0^2 = 1$, the W-ratio can be considered as one of the integer aperture estimators and its pull-in region is derived as follows:

$$\frac{\Omega_2 - \Omega_1}{2\sqrt{(\check{a}_2 - \check{a}_1)^T Q_{\widehat{a}}^{-1} (\check{a}_2 - \check{a}_1)}} \geq c$$

$$\Leftrightarrow \frac{(\check{a}_2 - \widehat{a})^T Q_{\widehat{a}}^{-1} (\check{a}_2 - \widehat{a}) - (\check{a}_1 - \widehat{a})^T Q_{\widehat{a}}^{-1} (\check{a}_1 - \widehat{a})}{2\sqrt{(\check{a}_2 - \check{a}_1)^T Q_{\widehat{a}}^{-1} (\check{a}_2 - \check{a}_1)}} \geq c$$

replace a with a generic form of x, we have:

$$\Leftrightarrow \frac{(\check{x}_2 - \widehat{x})^T Q_{\widehat{a}}^{-1} (\check{x}_2 - \widehat{x}) - (\check{x}_1 - \widehat{x})^T Q_{\widehat{a}}^{-1} (\check{x}_1 - \widehat{x})}{2\sqrt{(\check{x}_2 - \check{x}_1)^T Q_{\widehat{a}}^{-1} (\check{x}_2 - \check{x}_1)}} \geq c$$

with the property of integer translation, $\check{x}_1$ has been moved to the zero vector. By defining $z = \check{x}_2 - \check{x}_1 = \check{x}_2$, the deduction goes as follows:

$$\Leftrightarrow \frac{(z - \widehat{x})^T Q_{\widehat{a}}^{-1} (z - \widehat{x}) - \widehat{x}^T Q_{\widehat{a}}^{-1} \widehat{x}}{2\sqrt{z^T Q_{\widehat{a}}^{-1} z}} \geq c$$

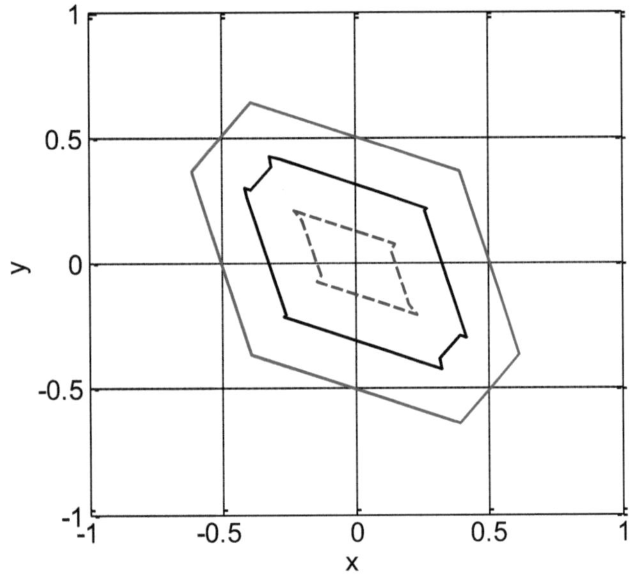

Fig. 2 W-ratio aperture pull-in region (*black dash* for critical value of 0.5, *blue dash dot* for critical value of 1) and integer least-squares pull-in region (*red, solid line*)

$$\Leftrightarrow z^T Q_{\hat{a}}^{-1} z - 2z^T Q_{\hat{a}}^{-1} \hat{x} \geq 2c\sqrt{z^T Q_{\hat{a}}^{-1} z}$$

$$\Leftrightarrow \frac{z^T Q_{\hat{a}}^{-1} \hat{x}}{\sqrt{z^T Q_{\hat{a}}^{-1} z}} \leq \frac{1}{2}\sqrt{z^T Q_{\hat{a}}^{-1} z} - c$$

where z is the closest integer other than zero vector. Note that the final formula for the pull-in region of W-ratio is different from the difference test and the projector test, whose pull-in regions as an integer aperture estimator are provided in Verhagen (2005). By comparing these three ambiguity validation methods, a noticeable difference is that the determination of critical values for the difference test and the projector test are either empirical or based on non-strict distribution, whereas W-ratio is based on the truncated normal distribution or truncated student *t* distribution. Consequently, the size of their pull-in regions varies with the critical values.

The shape of the W-ratio's pull-in region, however, is similar to those of the integer aperture least-squares, the integer aperture difference test and the projector's pull-in region; see Figures 5 and 10 in Verhagen (2005). For a two dimensional case, the W-ratio's pull-in region can be constructed with six intersecting half-spaces which are bounded by the planes orthogonal to z with the condition above. Figure 2 shows the pull-in region of the W-ratio test, with

$$Q = [0.1392 - 0.0486; -0.0486\ 0.1583];$$

In this case, the a priori variance for the W-ratio test is considered as 1.0. The pull-in region of the integer least-squares is shown as red solid line, with the other two pull-in regions of the W-ratio tests shown in black and blue. Obviously the W-ratio pull-in region is also close to the integer least-squares pull-in region.

5 Numerical Analysis

Under the framework of the integer aperture estimator, the W-ratio could be applied in another way instead of depending on its truncated distribution. As suggested in Teunissen (2003a), Verhagen (2005), the critical value should be determined from a given fail-rate. Based on the fail-rate, simulations are carried out to find the corresponding critical value. The exact procedures are: (1) given variance–covariance matrix of the ambiguities and the fail-rate; (2) determine the critical value according to the given fail-rate and then use the critical value to perform the ambiguity validation test. For the purpose of reliable results, the sample size should be as large as possible. A discussion about the influence of the sample size on the ambiguity validation results can be found in Li and Wang (2012).

With the purpose of analysing the W-ratio's performance in real applications, a 30 min static data set with six satellites was utilized, and the data was gathered on 6th, June, 2010, Sydney, Australia, with a sampling rate and an elevation angle as 1 s and 15 degrees, respectively. After least-squares estimation, we can obtain the float solution together with their variance–covariance matrix, and using the exact procedures above and a pre-defined fail-rate, simulations can be applied to determine the critical values.

Due to the heavy computational burden of simulation for each epoch, a certain epoch was selected to illustrate the application of the W-ratio instead. In Figs. 3 and 4, both the ADOP (ambiguity dilution of precision) values and Wa-ratio values are plotted. It is shown that there is just a minor change in the ADOP values, so the variance–covariance matrix of the nine hundredth epoch was chosen to simulate the critical value with a pre-defined fail rate as 0.001.

The results are listed in Table 1, with a critical value of 2.81 for TSND and 2.85 for WIA (W-ratio IA estimator). For both approaches, the correct acceptances are quite similar (1794, 1792), as well as the wrongly rejected ones (6, 8). These results are extremely close to the truth so that for all the resolved ambiguities it can be assumed to be correct.

Another 20 min kinematic data set, which was collected on 9th, June, 2010, Sydney, Australia, has been used to describe the W-ratio's performance. There were six GPS satellites tracked with dual-frequency observations available, and the sampling rate and the elevation angle are the same as in the previous data set.

After estimation on an epoch by epoch basis, the float solution and its variance and covariance were obtained.

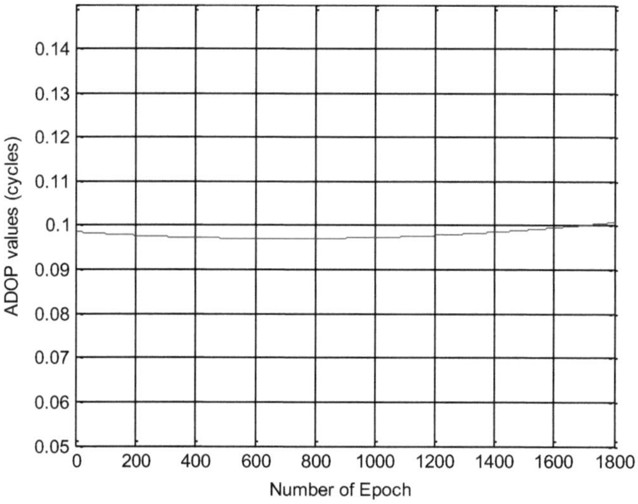

Fig. 3 ADOP values changing with epoch number in static case

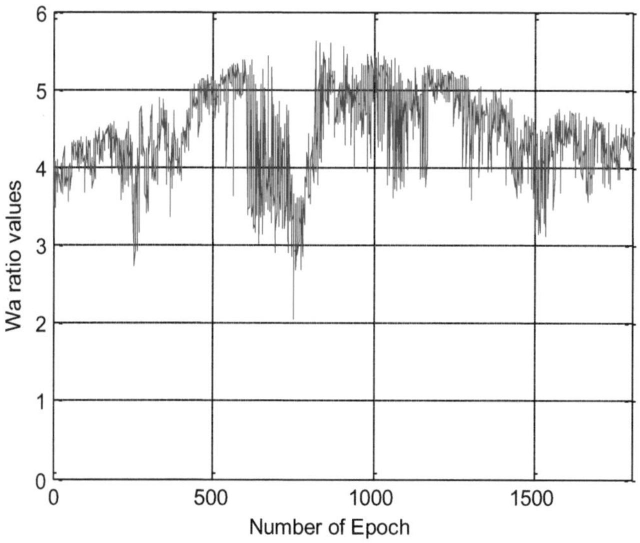

Fig. 4 Wa-ratio statistical values in static case

Table 1 Wa-ratio performance in static case, CA = correct acceptance; WA = wrong acceptance; CR = correct rejection; WR = wrong rejection

	CA	WA	CR	WR
TSND ($\alpha = 0.005$)	1,794	0	0	6
WIA (Pf = 0.001)	1,792	0	0	8

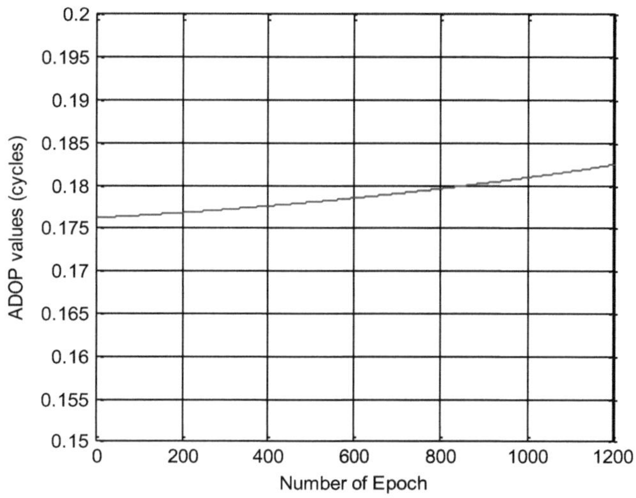

Fig. 5 ADOP values changing with epoch number in kinematic case

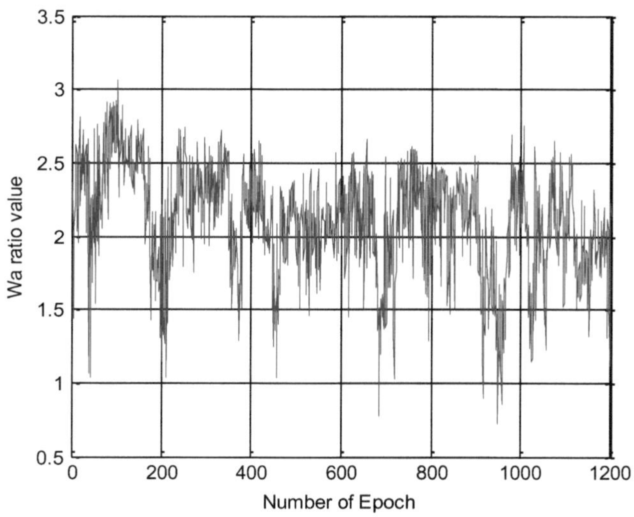

Fig. 6 Wa-ratio statistical values in kinematic case

Table 2 Wa-ratio performance in the kinematic case, CA = correct acceptance; WA = wrong acceptance; CR = correct rejection; WR = wrong rejection

	CA	WA	CR	WR
TSND ($\alpha = 0.05$)	823	6	160	211
WIA (Pf = 0.015)	905	21	145	129

The ADOP values plotted in Fig. 5 are roughly ranging from 0.176 to 0.183 cycles, and in order to effectively apply the IA for the W-ratio test, the geometry of the six hundredth epoch was chosen to determine the critical value. In Fig. 6, the statistical values of the Wa-ratio were plotted. Table 2 shows the validation results with respect to different way of determining the critical values. A 95 % confidence level yields a significance level of 1.96 for TSND, and the W-ratio as an integer aperture estimator (WIA), with a predefined fail-rate of 0.015 generates a critical value of 1.81. Comparing with the WIA, the number of correctly accepted (CA) epochs for the W-Ratio with TSND is smaller, which means that the critical value for TSND is conservative. However, the wrongly accepted (WA) number, correctly rejected (CR) number and the wrongly rejected number (WR) change significantly.

Concluding Remarks

In this contribution, various ways of utilizing the W-ratio test have been discussed. Instead of determining the critical value from the standard normal distribution, a preferable way is to consider the constraint and thus, to apply the truncated normal distribution, with a creation of the look-up table to specify the critical values. Besides, under the framework of the integer aperture estimator, the pull-in region of W-ratio is presented, and the critical value could be generated with a given fail-rate according to simulations, which is also another way to apply the W-ratio. In case of a short observation period, as the satellite geometry doesn't change too much, the user is capable of obtaining the critical values by simulating the critical values with one epoch data. However, more investigations should be carried out in the future to study the performance of the W-ratio test as an integer aperture estimator. The application of integer aperture theory mainly depends on simulation, which only requires geometry information regardless of the float solution (or the quality of the observations). This issue needs further analysis.

References

Han S (1997) Quality control issues relating to instantaneous ambiguity resolution for real-time GPS kinematic positioning. J Geod 71(6):351–361

Hoel PG (1947) On the choice of forecasting formulas. J Am Stat Assoc 42:605–611

Koch KR (1988) Parameter estimation and hypothesis testing in linear models. Springer, New York

Li T, Wang JL (2012) Some remarks on GNSS integer ambiguity validation methods. Surv Rev, 44:230–238

Teunissen PJG (1995) The least-squares ambiguity decorrelation adjustment: a method for fast GPS integer ambiguity estimation. J Geod 70:65–82

Teunissen PJG (2003a) Integer aperture GNSS ambiguity resolution. Artif Satell 38(3):79–88

Teunissen PJG (2003b) A carrier phase ambiguity estimator with easy-to-evaluate fail-rate. Artif Satell 38(3):89–96

Verhagen S (2005) The GNSS integer ambiguities: estimation and validation. PhD thesis, Publications on Geodesy, 58, Netherland Geodetic Commission, Delft

Wang J, Stewart MP, Tsakiri M (1998) A discrimination test procedure for ambiguity resolution on-the-fly. J Geod 72(11):644–653

Wang J, Stewart MP, Tsakiri M (2000) A comparative study of the integer ambiguity validation procedures. Earth, Planets Space 52(10):813–817

Performing 3D Similarity Transformation Using the Weighted Total Least-Squares Method

J. Lu, Y. Chen, X. Fang, and B. Zheng

Abstract

The 3D similarity transformation models, e.g. Bursa model is usually applied in geodesy and photogrammetry. In general, they are suitable in small angle 3D transformation. However, a lot of large 3D transformations need to be performed. This contribution describes a 3D similarity transformation model suitable for any angle rotation, where the nine elements in the rotation matrix are used to replace the three rotation angles as unknown parameters. In the coordinate transformation model, the Errors-In-Variables (EIV) model will be adjusted according to the theory of Least Squares (LS) method within the nonlinear Gauss–Helmert (GH) model. At the end of the contribution, case studies are investigated to demonstrate the coordinate transformation method proposed in this paper. The results show that using the linearized iterative GH model the correct solution can be obtained and this mixed model can be applied no matter whether the variance covariance matrices are full or diagonal.

Keywords

Errors-In-Variables (EIV) model • Gauss–Helmert model • Orthonormal matrix • Weighted Total Least-Squares (WTLS)

1 Introduction

3D similarity transformation methods for nearly aligned systems with small rotation angles are employed in many cases, because the mathematical model can be simplified into a linear system. Leick (2004) reviewed the similarity transformation methods for these nearly aligned systems. However, in many disciplines, e.g. photogrammetry, computer vision and surveying engineering etc., the rotation angles are generally not small. The simplified models are not suitable anymore in this large transformation.

During the computational process, if the number of observations is more than the number of unknown parameters, the Least-Squares (LS) adjustment within the Gauss–Markov (GM) model is usually used to calculate the transformation parameters from a redundant set of equations. LS estimation is the best linear unbiased estimation, when the errors in the

J. Lu (✉)
Shanghai Real Estate Science Research Institute, Shanghai, China

Institute of Photogrammetry and GeoInformation, Leibniz University Hanover, Hanover, Germany
e-mail: lujue1985@126.com

Y. Chen
Department of Surveying and Geo-informatics, Tongji University, Shanghai, China

Key Laboratory of Advanced Surveying Engineering of State Bureau of Surveying and Mapping, Shanghai, China

X. Fang
School of Geodesy and Geomatics, Wuhan University, Wuhan, China

B. Zheng
Shanghai Cadastre Management Center, Shanghai, China

H. Kutterer et al. (eds.), *The 1st International Workshop on the Quality of Geodetic Observation and Monitoring Systems (QuGOMS'11)*, International Association of Geodesy Symposia 140,
DOI 10.1007/978-3-319-10828-5_11,

observation vector are normally distributed, and the matrix of variables (the coefficient matrix) is error-free.

However, in coordinate transformation, various types of random errors exist in the observation vector and the matrix of variables. The Total Least-Squares (TLS) approach provides a solution, when all the data are affected by random errors and can solve estimation problems in the so-called Errors-In-Variables (EIV) model. Essentially, the EIV model is a kind of nonlinear model. The nonlinear LS adjustment has been investigated by Pope (1972). For estimating the parameters in the EIV model, the TLS approach was introduced (e.g. Golub and Van Loan 1980; Van Huffel and Vandewalle 1991). In recent years, this method has also been developed by Schaffrin (2006), Schaffrin and Wieser (2008), Schaffrin and Felus (2009) and Neitzel (2010) solved the TLS within the EIV model as a special case of the method of LS within the nonlinear Gauss–Helmert (GH) model.

For applications, the EIV model is widely employed in geodesy or engineering survey (e.g. Felus and Schaffrin 2005; Akyilmaz 2007; Schaffrin et al. 2009; Schaffrin and Felus 2008; Lu et al. 2008; Felus and Burtch 2009; Neitzel 2010).

The 3D similarity transformation method proposed in this paper can suit not only small but also large angle rotation. Based on this nonlinear model, the solution within the linearized iterative GH model will be applied to solve the EIV model.

In Felus and Burtch (2009) the EIV model is calculated by a closed form Procrustes approach, which limits the size of the matrix and the accuracy of the result. In Neitzel (2010), the EIV model is performed in the 2D transformation model. Compared with this previous work, the GH model solved in this paper does not impose any restrictions on the form of functional relationship between the quantities involved in the model, and so the transformation parameters can be calculated, no matter how complex the EIV model would be. Furthermore, in contrast to the linear 2D transformation model in Neitzel (2010), the 3D transformation model described here is nonlinear and more complicated. Finally, the calculation process presented in this paper permits to solve the coordinate transformation, no matter if the weight matrix is diagonal or not.

We present our transformation method, which can suit any angle in the next section. The linearized iterative solution within the GH model is used to solve the EIV model in Sect. 3. The performance of our method is demonstrated by the results of simulations and experiments with real data in Sect. 4. Finally, last section gives "concluding remarks".

2 Nonlinear 3D Similarity Transformation

3D similarity coordinate transformation is the process of converting spatial data from one coordinate system to another:

$$\begin{bmatrix} X \\ Y \\ Z \end{bmatrix} = \begin{bmatrix} X_0 \\ Y_0 \\ Z_0 \end{bmatrix} + \mu \mathbf{M}(\alpha_1, \alpha_2, \alpha_3) \begin{bmatrix} x \\ y \\ z \end{bmatrix} \tag{1}$$

where $\begin{bmatrix} X, Y, Z \end{bmatrix}$ denote the coordinates in the target system $O-XYZ$, $\begin{bmatrix} x, y, z \end{bmatrix}$ represent the coordinates of the corresponding point in the source system $o-xyz$, (X_0, Y_0, Z_0) are the three translation parameters, μ is the scale factor, and α_1, α_2, α_3 stand for the three rotation angles, respectively, which produce the rotation matrix $\mathbf{M}$

$$\mathbf{M}(\alpha_1, \alpha_2, \alpha_3) = \mathbf{M_3}(\alpha_3)\mathbf{M_2}(\alpha_2)\mathbf{M_1}(\alpha_1) = \begin{bmatrix} a_1 & a_2 & a_3 \\ b_1 & b_2 & b_3 \\ c_1 & c_2 & c_3 \end{bmatrix} \tag{2}$$

The nine elements $(a_1, a_2, a_3, b_1, b_2, b_3, c_1, c_2, c_3)$ form an orthonormal matrix $\mathbf{M}^T\mathbf{M} = \mathbf{M}\mathbf{M}^T = \mathbf{I}|_{3\times 3}$ ($\mathbf{I}|_{3\times 3}$ denotes the 3×3 identity matrix) with $\det(\mathbf{M}) = +1$.

The usual mathematical model of the transformation is the simplified model, such as the Bursa and the Molodensky model, which assume that the rotation parameters are small and a linear approximation is valid. However, if the rotation angles are large, the simplified models are not suitable anymore.

In this contribution, the nine elements in the rotation matrix will be used to replace the rotation angles as unknown parameters. So the transformation model with 13 unknown parameters can be obtained as:

$$\begin{bmatrix} X \\ Y \\ Z \end{bmatrix} = \begin{bmatrix} X_0 \\ Y_0 \\ Z_0 \end{bmatrix} + \mu \begin{bmatrix} a_1 & a_2 & a_3 \\ b_1 & b_2 & b_3 \\ c_1 & c_2 & c_3 \end{bmatrix} \begin{bmatrix} x \\ y \\ z \end{bmatrix} \tag{3}$$

However, in the rotation matrix $\mathbf{M}$, only three of the nine elements are independent, and the remaining six parameters can be represented by nonlinear functions of these three values. According to the property of the orthonormal matrix

$\mathbf{M}^T\mathbf{M} = \mathbf{M}\mathbf{M}^T = \mathbf{I}|_{3\times 3}$, six constraints can be organized in various kinds. One of them can be expanded as:

$$\begin{cases} a_1^2 + a_2^2 + a_3^2 - 1 = 0 \\ b_1^2 + b_2^2 + b_3^2 - 1 = 0 \\ c_1^2 + c_2^2 + c_3^2 - 1 = 0 \\ a_1a_2 + b_1b_2 + c_1c_2 = 0 \\ a_1a_3 + b_1b_3 + c_1c_3 = 0 \\ a_2a_3 + b_2b_3 + c_2c_3 = 0 \end{cases} \tag{4}$$

In this paper, the constraints (4) will be converted into pseudo-observation equations. Thus, the weights of these six pseudo-observation equations will be set much larger than others. So the unknown parameters will be estimated by the weighted adjustment. In general case, the initial values of the parameters can be set as $\mu^0 = 1$, $(X_0^0, Y_0^0, Z_0^0) = (0, 0, 0)$, $\mathbf{M}^0 = \mathbf{I}|_{3\times 3}$, which are equivalent to aligned systems at the beginning. The orthogonality has been tested and is endured up to numerical inaccuracies.

3 The Nonlinear 3D Similarity Transformation Solved by the Solution Within the Nonlinear GH Model

In the 3D similarity transformation model described in Eq. (3) the number of unknown parameters is 13. If the number of corresponding points is k, then the number of observation equations is n (here $n = 3k$). Combined with six constraints from Eq. (4), at least two three-dimensional points and one elevation point are required to determine the 13 parameters; in general, more corresponding points are measured, and an adjustment process is required for computing the best fitting parameters from the redundant data.

The LS adjustment is employed for estimation of the unknown parameters in many cases. But there is a basic assumption that only observations are affected by random errors. This assumption implies that just the data in the target coordinate system include errors, but coordinates in the source system are true and error-free. In this case a Gauss–Markov (GM) model is suitable.

However, the assumption that all the random errors are confined to the observation vector is often not true. In many cases, errors occur not only in the observation vector, but also in the coefficient data matrix. In this case, the TLS approach is the proper method for treating this EIV model.

The starting point for the TLS adjustment is the definition of a quasi-linear model. However, the 3D similarity transformation model described in the last section is nonlinear. To calculate the nonlinear Weighted TLS (WTLS) problem, the rigorous evaluation in a nonlinear GH model will be performed.

Because $\{X_i, Y_i, Z_i\}$ and $\{x_i, y_i, z_i\}$ are both observations, random errors $e_{X_i}, e_{Y_i}, e_{Z_i}$ and $e_{x_i}, e_{y_i}, e_{z_i}$ have to be considered:

$$\begin{aligned} X_i &- e_{X_i} - \mu a_1 (x_i - e_{x_i}) - \mu a_2 (y_i - e_{y_i}) \\ &- \mu a_3 (z_i - e_{z_i}) - X_0 = 0 \\ Y_i &- e_{Y_i} - \mu b_1 (x_i - e_{x_i}) - \mu b_2 (y_i - e_{y_i}) \\ &- \mu b_3 (z_i - e_{z_i}) - Y_0 = 0 \\ Z_i &- e_{Z_i} - \mu c_1 (x_i - e_{x_i}) - \mu c_2 (y_i - e_{y_i}) \\ &- \mu c_3 (z_i - e_{z_i}) - Z_0 = 0 \end{aligned} \tag{5}$$

Since the Eq. (4) is converted into pseudo-observation equations, errors are also included in:

$$\begin{cases} 1 - e_{31} - a_1^2 - a_2^2 - a_3^2 = 0 \\ 1 - e_{32} - b_1^2 - b_2^2 - b_3^2 = 0 \\ 1 - e_{33} - c_1^2 - c_2^2 - c_3^2 = 0 \\ -e_{34} - a_1a_2 - b_1b_2 - c_1c_2 = 0 \\ -e_{35} - a_1a_3 - b_1b_3 - c_1c_3 = 0 \\ -e_{36} - a_2a_3 - b_2b_3 - c_2c_3 = 0 \end{cases} \tag{6}$$

The error vector is:

$$\begin{aligned} \mathbf{e}\,|_{(2n+6)\times 1} &:= \left[\mathbf{e}_2^T, \mathbf{e}_1^T, \mathbf{e}_3^T\right]^T \\ \mathbf{e}_2^T &= \left[\cdots e_{X_i}\ e_{Y_i}\ e_{Z_i} \cdots\right] \\ \mathbf{e}_1^T &= \left[\cdots e_{x_i}\ e_{y_i}\ e_{z_i} \cdots\right] \\ \mathbf{e}_3^T &= \left[e_{31}\ e_{32}\ e_{33}\ e_{34}\ e_{35}\ e_{36}\right] \end{aligned} \tag{7}$$

where the subscript i indicates the corresponding point. As usual, variances and covariance of the observations have to be taken into account. Transferring the accuracies into the corresponding weight matrices $\mathbf{P}_1$, $\mathbf{P}_2$ and $\mathbf{P}_3$, the objective function to be minimized reads:

$$\mathbf{e}^T\mathbf{P}\mathbf{e} = \mathbf{e}_2{}^T\mathbf{P}_2\mathbf{e}_2 + \mathbf{e}_1{}^T\mathbf{P}_1\mathbf{e}_1 + \mathbf{e}_3{}^T\mathbf{P}_3\mathbf{e}_3 = \min \tag{8}$$

The implicit form of the functional relation is established by the Eqs. (5) and (6), which are nonlinear. The solution of this EIV model can be obtained through an evaluation within the GH model.

The nonlinear differentiable equations (5) and (6) can be combined and written as:

$$\mathrm{f}(\mathbf{e}, \boldsymbol{\xi})\,|_{(n+6)\times 1} = \left[\cdots \mathrm{f}_i(\mathbf{e}, \boldsymbol{\xi}) \cdots, \mathrm{f}_C(\mathbf{e}, \boldsymbol{\xi})\right]^T = 0 \tag{9}$$

With the parameter vector

$$\boldsymbol{\xi} = \left[X_0\ Y_0\ Z_0\ \mu\ a_1\ a_2\ a_3\ b_1\ b_2\ b_3\ c_1\ c_2\ c_3\right]^T \tag{10}$$

In nonlinear 3D similarity transformation $\mathbf{f}(\mathbf{e}, \boldsymbol{\xi})$ is:

$$\mathbf{f}(\mathbf{e}, \boldsymbol{\xi})\big|_{(n+6)\times 1} = \mathbf{0} = -\begin{bmatrix} \cdots \\ X_i - e_{X_i} - \mu a_1 (x_i - e_{x_i}) - \mu a_2 (y_i - e_{y_i}) - \mu a_3 (z_i - e_{z_i}) - X_0 \\ Y_i - e_{Y_i} - \mu b_1 (x_i - e_{x_i}) - \mu b_2 (y_i - e_{y_i}) - \mu b_3 (z_i - e_{z_i}) - Y_0 \\ Z_i - e_{Z_i} - \mu c_1 (x_i - e_{x_i}) - \mu c_2 (y_i - e_{y_i}) - \mu c_3 (z_i - e_{z_i}) - Z_0 \\ \cdots \\ \mathbf{f}_C(\mathbf{e}_3, \boldsymbol{\xi}) \end{bmatrix} \quad (11)$$

where $\mathrm{f_C}(\mathbf{e_3}, \boldsymbol{\xi}) = \mathbf{0}$ represents the six pseudo-observation equations (6).

The linearized condition equations can be written as:

$$\mathbf{f}(\mathbf{e}, \boldsymbol{\xi}) \approx \mathbf{A}^0 (\boldsymbol{\xi} - \boldsymbol{\xi}^0) + \mathbf{B}^0 (\mathbf{e} - \mathbf{e}^0) + \mathbf{f}(\mathbf{e}^0, \boldsymbol{\xi}^0) = \mathbf{0} \quad (12)$$

involving the matrices of partial derivatives:

$$\mathbf{A}^0 (\mathbf{e}, \boldsymbol{\xi}) = \frac{\partial \mathbf{f}(\mathbf{e}, \boldsymbol{\xi})}{\partial \boldsymbol{\xi}^T} \quad (13)$$

and

$$\mathbf{B}^0 (\mathbf{e}, \boldsymbol{\xi}) = -\frac{\partial \mathbf{f}(\mathbf{e}, \boldsymbol{\xi})}{\partial \mathbf{e}^T} \quad (14)$$

So in nonlinear 3D similarity transformation $\mathbf{A}^0(\mathbf{e}, \boldsymbol{\xi})$ can be built as:

$$\mathbf{A}^0 \big|_{(n+6)\times 13} = \begin{bmatrix} \mathbf{A}_1^0 \\ \mathbf{A}_2^0 \end{bmatrix} \quad (15)$$

where $\mathbf{A}_1^0$ stands for the coefficient matrix of the error equations, which is linearized from Eq. (5). So the factors in $\mathbf{A}_1^0$ except zeros are written as:

$$\begin{aligned}
&\mathbf{A}_1^0 (3i-2, 1) = 1 \\
&\mathbf{A}_1^0 (3i-2, 4) = a_1^0 (x_i - e_{x_i}) + a_2^0 (y_i - e_{y_i}) \\
&\quad + a_3^0 (z_i - e_{z_i}) \\
&\mathbf{A}_1^0 (3i-2, 5) = \mu^0 (x_i - e_{x_i}) \\
&\mathbf{A}_1^0 (3i-2, 6) = \mu^0 (y_i - e_{y_i}) \\
&\mathbf{A}_1^0 (3i-2, 7) = \mu^0 (z_i - e_{z_i}) \\
&\mathbf{A}_1^0 (3i-1, 2) = 1 \\
&\mathbf{A}_1^0 (3i-1, 4) = b_1^0 (x_i - e_{x_i}) + b_2^0 (y_i - e_{y_i}) \\
&\quad + b_3^0 (z_i - e_{z_i}) \\
&\mathbf{A}_1^0 (3i-1, 8) = \mu^0 (x_i - e_{x_i}) \qquad\qquad (i = 1, \ldots, k) \\
&\mathbf{A}_1^0 (3i-1, 9) = \mu^0 (y_i - e_{y_i}) \\
&\mathbf{A}_1^0 (3i-1, 10) = \mu^0 (z_i - e_{z_i}) \\
&\mathbf{A}_1^0 (3i, 3) = 1 \\
&\mathbf{A}_1^0 (3i, 4) = c_1^0 (x_i - e_{x_i}) + c_2^0 (y_i - e_{y_i}) \\
&\quad + c_3^0 (z_i - e_{z_i}) \\
&\mathbf{A}_1^0 (3i, 11) = \mu^0 (x_i - e_{x_i}) \\
&\mathbf{A}_1^0 (3i, 12) = \mu^0 (y_i - e_{y_i}) \\
&\mathbf{A}_1^0 (3i, 13) = \mu^0 (z_i - e_{z_i})
\end{aligned} \quad (16)$$

$\mathbf{A}_2^0$ in Eq. (15) presents the coefficient matrix of the additional error equations, which is linearized from Eq. (6):

$$\mathbf{A}_2^0 = -\begin{bmatrix}
0 & 0 & 0 & 0 & 2a_1^0 & 2a_2^0 & 2a_3^0 & 0 & 0 & 0 & 0 & 0 & 0 \\
0 & 0 & 0 & 0 & 0 & 0 & 0 & 2b_1^0 & 2b_2^0 & 2b_3^0 & 0 & 0 & 0 \\
0 & 0 & 0 & 0 & 0 & 0 & 0 & 0 & 0 & 0 & 2c_1^0 & 2c_2^0 & 2c_3^0 \\
0 & 0 & 0 & 0 & a_2^0 & a_1^0 & 0 & b_2^0 & b_1^0 & 0 & c_2^0 & c_1^0 & 0 \\
0 & 0 & 0 & 0 & a_3^0 & 0 & a_1^0 & b_3^0 & 0 & b_1^0 & c_3^0 & 0 & c_1^0 \\
0 & 0 & 0 & 0 & 0 & a_3^0 & a_2^0 & 0 & b_3^0 & b_2^0 & 0 & c_3^0 & c_2^0
\end{bmatrix} \quad (17)$$

$\mathbf{B}^0(\mathbf{e}, \boldsymbol{\xi})$ is decomposed as:

$$\mathbf{B}_1^0 \big|_{n\times n} = \frac{\partial \mathbf{f}(\mathbf{e}, \boldsymbol{\xi})}{\partial \mathbf{e}_1{}^T} = \begin{bmatrix}
-\mu a_1{}^0 & -\mu a_2{}^0 & -\mu a_3{}^0 & \cdots & 0 & 0 & 0 \\
-\mu b_1{}^0 & -\mu b_2{}^0 & -\mu b_3{}^0 & \cdots & 0 & 0 & 0 \\
-\mu c_1{}^0 & -\mu c_2{}^0 & -\mu c_3{}^0 & \cdots & 0 & 0 & 0 \\
\vdots & \vdots & \vdots & \ddots & \vdots & \vdots & \vdots \\
0 & 0 & 0 & \cdots & -\mu a_1{}^0 & -\mu a_2{}^0 & -\mu a_3{}^0 \\
0 & 0 & 0 & \cdots & -\mu b_1{}^0 & -\mu b_2{}^0 & -\mu b_3{}^0 \\
0 & 0 & 0 & \cdots & -\mu c_1{}^0 & -\mu c_2{}^0 & -\mu c_3{}^0
\end{bmatrix} \quad (18)$$

$$\mathbf{B}_2^0 \big|_{n\times n} = \frac{\partial \mathbf{f}(\mathbf{e}, \boldsymbol{\xi})}{\partial \mathbf{e}_2{}^T} = \mathbf{I}\big|_{n\times n} \quad (19)$$

$$\mathbf{B}_3^0 \big|_{6\times 6} = \frac{\partial \mathbf{f}(\mathbf{e}, \boldsymbol{\xi})}{\partial \mathbf{e}_3{}^T} = \mathbf{I}\big|_{6\times 6} \quad (20)$$

Here $\mathbf{B}_1^0$, $\mathbf{B}_2^0$ and $\mathbf{B}_3^0$, respectively, denote the matrices of partial derivatives w.r.t. $\mathbf{e}_1$, $\mathbf{e}_2$ and $\mathbf{e}_3$. According to Eq. (14), $\mathbf{B}^0(\mathbf{e}, \boldsymbol{\xi})$ is:

$$\mathbf{B}^0 \big|_{(n+6)\times(2n+6)} = \begin{bmatrix} \mathbf{B}_2^0 & \mathbf{B}_1^0 & \mathbf{0}\big|_{n\times 6} \\ \mathbf{0}\big|_{6\times n} & \mathbf{0}\big|_{6\times n} & \mathbf{B}_3^0 \end{bmatrix} \quad (21)$$

with the vector of misclosures:

$$\boldsymbol{\omega}^0 = -\mathbf{B}^0\mathbf{e}^0 + \mathbf{f}\left(\mathbf{e}^0, \boldsymbol{\xi}^0\right)\Big|_{(n+6)\times 1} = \begin{bmatrix} \cdots \\ X_i - \mu^0 {a_1}^0 x_i - \mu^0 {a_2}^0 y_i - \mu^0 {a_3}^0 z_i - {X_0}^0 \\ Y_i - \mu^0 {b_1}^0 x_i - \mu^0 {b_2}^0 y_i - \mu^0 {b_3}^0 z_i - {Y_0}^0 \\ Z_i - \mu^0 {c_1}^0 x_i - \mu^0 {c_2}^0 y_i - \mu^0 {c_3}^0 z_i - {Z_0}^0 \\ \cdots \\ 1 - \left(a_1^0\right)^2 - \left(a_2^0\right)^2 - \left(a_3^0\right)^2 \\ 1 - \left(b_1^0\right)^2 - \left(b_2^0\right)^2 - \left(b_3^0\right)^2 \\ 1 - \left(c_1^0\right)^2 - \left(c_2^0\right)^2 - \left(c_3^0\right)^2 \\ -a_1^0 a_2^0 - b_1^0 b_2^0 - c_1^0 c_2^0 \\ -a_1^0 a_3^0 - b_1^0 b_3^0 - c_1^0 c_3^0 \\ -a_2^0 a_3^0 - b_2^0 b_3^0 - c_2^0 c_3^0 \end{bmatrix} \tag{22}$$

and cofactor matrices of $\mathbf{e}_1$, $\mathbf{e}_2$ and $\mathbf{e}_3$:

$$\mathbf{Q}_1\big|_{n\times n} = \mathbf{P}_1^{-1} \tag{23}$$

$$\mathbf{Q}_2\big|_{n\times n} = \mathbf{P}_2^{-1} \tag{24}$$

$$\mathbf{Q}_3\big|_{6\times 6} = \mathbf{P}_3^{-1} = \left(\delta \cdot \mathbf{I}\big|_{6\times 6}\right)^{-1} \tag{25}$$

Here, δ is a sufficiently large constant that presents the weights of the six pseudo-observation equations.

If the data has been adjusted previously, the coordinates in source and target systems might be related. Considering this situation, the more general form of the cofactor matrix is:

$$\mathbf{Q}\big|_{(2n+6)\times(2n+6)} = \begin{pmatrix} \mathbf{Q}_2 & \mathbf{Q}_{21} & \mathbf{0}\big|_{n\times 6} \\ \mathbf{Q}_{12} & \mathbf{Q}_1 & \mathbf{0}\big|_{n\times 6} \\ \mathbf{0}\big|_{6\times n} & \mathbf{0}\big|_{6\times n} & \mathbf{Q}_3 \end{pmatrix} \tag{26}$$

where $\mathbf{Q}_{21}$ and $\mathbf{Q}_{12}$ denote the covariance matrix of $\mathbf{e}_2$ and $\mathbf{e}_1$. So compared with the calculation process in Neitzel (2010), in which the weighted matrix is diagonal, the observations here can be correlated.

The estimation for the unknown parameters from the solution of the linear equations system will be obtained as follows (Neitzel 2010):

$$\begin{bmatrix} \mathbf{B}^0\mathbf{Q}\left(\mathbf{B}^0\right)^T & \mathbf{A}_0 \\ \left(\mathbf{A}^0\right)^T & \mathbf{0} \end{bmatrix} \begin{bmatrix} \widehat{\boldsymbol{\lambda}}^1 \\ \widehat{\boldsymbol{\xi}}^1 - \widehat{\boldsymbol{\xi}}^0 \end{bmatrix} + \begin{bmatrix} \boldsymbol{\omega}^0 \\ \mathbf{0} \end{bmatrix} = \mathbf{0} \tag{27}$$

and the first error vector is:

$$\tilde{\mathbf{e}}^1 = \mathbf{Q}\left(\mathbf{B}^0\right)^T \widehat{\boldsymbol{\lambda}}^1 \tag{28}$$

This is an iterative calculation process. After stripping the randomness of the solution $\tilde{\mathbf{e}}^1$ and $\widehat{\boldsymbol{\xi}}^1$, they are used in the next iteration step as their approximations.

Table 1 Coordinates of corresponding points with errors ($\sigma_0 = 0.001$) in two coordinate systems

No.	1	2	3	4	5
X_1	−0.0011	100.0003	−99.9994	−99.9989	100.0015
Y_1	0.0001	99.9985	−100.0007	−100.0010	100.0023
Z_1	−0.0006	100.0007	−100.0001	100.0008	−100.0003
X_2	9.9986	−83.7289	103.7279	4.2617	15.7378
Y_2	10.0014	128.3343	−108.3338	−143.4093	163.4101
Z_2	10.0007	98.4045	−78.4039	93.8759	−73.8769

Table 2 The estimated transformation parameters resulting from different algorithms with the data of Table 1

	Exact values	LS method	WLS method	Iterative GH model
X_0(m)	10.0000	9.9992	9.9992	9.9992
Y_0(m)	10.0000	10.00031	10.00026	10.00026
Z_0(m)	10.0000	9.99901	9.99983	9.99983
μ	1.01	0.84101740	1.0099958	1.0099958
α_1	10°	4°58′16.677″	9°59′59.583″	9°59′59.583″
α_2	30°	25°12′26.447″	30°00′01.145″	30°00′01.145″
α_3	60°	58°04′48.891″	59°59′59.888″	59°59′59.888″
$\widehat{\sigma}_0^2$	0.00127	14.46322	0.00182	0.00148

4 Case Study

In the following section, a simulation and a real data example will be used to examine the nonlinear 3D transformation method and the computation method.

In the first example, we simulate a 3D similarity transformation with large rotation angles. In this simulation, all variables contain normally distributed errors with zero mean, $\sigma_0 = 0.001$, and all data have the same accuracy. The transformation parameters are $\mu = 1.01$, $(\alpha_1, \alpha_2, \alpha_3) = (10°, 30°, 60°)$, and $(X_0, Y_0, Z_0) = (10.00000, 10.00000, 10.00000)$ (m). Five corresponding points were given in the two coordinate systems with the coordinate values given in Table 1. The unit of the coordinates is meter.

The weight matrix of the six pseudo-observation equations is $\mathbf{P}_3 = 10^{10} \cdot \mathbf{I}|_{6\times 6}$. The initial values of the parameters are set as $\mu^0 = 1$, $(X_0^0, Y_0^0, Z_0^0) = (0, 0, 0)$, $\mathbf{M}^0 = \mathbf{I}|_{3\times 3}$.

During the calculation process, the iteration is stopped, if the results satisfy the condition $\left\| \widehat{\boldsymbol{\xi}}^h - \widehat{\boldsymbol{\xi}}^{h-1} \right\| < 10^{-8}$ (h is the number of iterations). Table 2 displays the estimated transformation parameters calculated by different transformation methods and computation models.

In Table 2, the second column presents the exact value and the third column displays the estimated parameters by the Bursa model using LS method. Column 4 and 5 show the results calculated by the nonlinear model proposed in this paper. The former presents the parameters using the

Table 3 Coordinates of control points in two coordinate systems from Felus and Burtch (2009)

No.	x	y	z	X	Y	Z	Weights
1	30	40	10	290	150	15	1
2	100	40	10	420	80	2	2
3	100	130	10	540	200	20	2.5
4	30	130	10	390	300	5	4

Table 4 Comparison of the transformation parameters resulting from different algorithms with the unequal data of Table 3

	Row-wise weighted GM model	Algorithm in Felus and Burtch (2009)	Solution within the linearized iterative GH model
X_0(m)	193.96132	188.97714	188.97713
Y_0(m)	104.42440	101.51720	101.51720
Z_0(m)	−32.49900	−33.38008	−33.38008
μ	2.1301997	2.1761269	2.1761270
α_1	$-0°30'51.4666''$	$-0°30'51.4666''$	$-0°30'51.4670''$
α_2	$4°31'21.1255''$	$4°31'21.1255''$	$4°31'21.1253''$
α_3	$33°32'19.5111''$	$33°32'19.5111''$	$33°32'19.5113''$
SSE	7659.90	319.004	319.004

Weighted Least Squares (WLS) method, while the latter presents the results calculated by the EIV model within the GH model.

Because the Bursa model is not suitable anymore in this large angle transformation, the correct results cannot be obtained. But with the nonlinear transformation model, no matter which calculation model is chosen, the results can converge to the correct parameters. The number of iterations by WLS method is 5, via the solution within the GH model just 4 besides, the results calculated by the EIV model are more reasonable, because all observation data which are affected by random errors are corrected.

The variance component ($\widehat{\sigma}_0^2$) shows that the solution of the WLS method and the iterative GH model method are all matching the parameters well, but the latter one is closer to the exact values.

The second numerical example originates from Felus and Burtch (2009). The accuracy of the data is varying, and the weights are represented in the last column of Table 3. The unit of the coordinates is meter.

In order to solve this EIV model, Felus and Burtch (2009) employ a closed form Procrustes method. In our solution, the weights for the six pseudo-observation equations are still 10^{10}. The initial values of the parameters and the stopping criteria of the iteration are set identical to the first example. The estimated transformation parameters are displayed in Table 4, where SSE denotes the sum of squared errors.

The number of iterations for the algorithm in Felus and Burtch (2009) and the linearized iterative weighted GH model are four in both cases. The residuals of the corresponding points obtained from the solution within the unequally weighted GH model are presented in Table 5.

Table 5 Residuals of the variables in the numerical example 2

	Point 1	Point 2	Point 3	Point 4
e_X(m)	0.9855	−1.3720	−3.8094	1.5944
e_Y(m)	−1.4834	−3.4764	3.2994	−0.8270
e_Z(m)	−3.3302	0.9734	−0.8199	1.6354
e_x(m)	−2.9929	−2.9385	10.4971	−3.1787
e_y(m)	1.7975	6.7221	−1.4685	−1.2974
e_z(m)	7.0490	−1.7622	1.3278	−0.9406

Comparing the results in Table 4 and the residuals in Table 5, differences and similarities in the results can be analyzed. First of all, the nonlinear 3D similarity transformation model described in this paper can estimate the correct transformation parameters assuming there are no systematic errors in the measurements. Secondly, since the errors are obviously distributed in both the source and the target coordinate systems, the EIV model is preferable for solving this problem. This can be detected from the SSE measure in Table 4. The SSE values calculated by the EIV model are much smaller than those calculated by the GM model. Finally, the error components presented in Table 5 demonstrate the key property of the EIV model, which treats the source and the target coordinate systems equally under the assumption that there are errors in all variables.

Concluding Remarks

This contribution investigates a method for 3D similarity coordinate transformations. During the adjustment, the solution of the WTLS method is demonstrated by means of the linearized iterative GH model. The conclusions are summarized as follows:

1. Unlike the Bursa model and the Molodensky model, which are mainly suitable in small angle 3D transformation, the method for 3D similarity coordinate transformations proposed here can suit any angle transformation, and this method is not sensitive to the initial values. In other words, the orthogonality of the rotation matrix has been tested and is endured up to numerical inaccuracies.
2. The solution within the linearized iterative GH model can be used as an alternative WTLS method for computing the exact solution, but it is more general with respect to the possible weight matrix.
3. Compared with other solutions, SSE or $\widehat{\sigma}_0^2$ of the iterative solution within the linearized GH model are smaller.
4. With the EIV model, the errors in the source and the target coordinate systems can be presented directly in

every point. In other words, statistical information on the transformed points can easily be obtained.

Acknowledgement Financial support: National Natural Science Foundation of China, Grant No. 41074017.

References

Akyilmaz O (2007) Total least squares solution of coordinate transformation. Surv Rev 39(303):68–80

Felus YA, Burtch RC (2009) On symmetrical three-dimensional datum conversion. GPS Solut 13(1):65–74

Felus YA, Schaffrin B (2005) Performing similarity transformations using the error-in-variables model. In: ASPRS 2005 annual conference, Baltimore, March, pp 7–11

Golub HG, Van Loan FC (1980) An analysis of the total least squares problem. SIAM J Numer Anal 17(6):883–893

Leick A (2004) GPS satellite surveying, 3rd edn. Wiley, Hoboken

Lu J, Chen Y, Zheng B (2008) Research study on three-dimensional datum transformation using total least squares. J Geod Geodyn 28(5):77–81

Neitzel F (2010) Generalization of total least-squares on example of unweighted and weighted 2D similarity transformation. J Geod 84(12):751–762

Pope AJ (1972) Some pitfalls to be avoided in the iterative adjustment of nonlinear problems. In: Proceedings of the 38th annual meeting of the American society of photogrammetry, Washington, pp 449–477

Schaffrin B (2006) A note on constrained total least-squares estimation. Linear Algebra Appl 417:245–258

Schaffrin B, Felus YA (2008) On the multivariate total least-squares approach to empirical coordinate transformation: three algorithms. J Geod 82(6):373–383

Schaffrin B, Felus YA (2009) An algorithmic approach to the total least-squares problem with linear and quadratic constraints. Stud Geophys Geod 53:1–16

Schaffrin B, Wieser A (2008) On weighted total least-squares adjustment for linear regression. J Geod 82(7):415–421

Schaffrin B, Neitzel F, Uzum S (2009) Empirical similarity transformation via TLS-adjustment: exact solution vs. Cadzow's approximation. In: International geomatics forum, Qingdao, pp 28–30

Van Huffel S, Vandewalle J (1991) The total least squares problem. Computational aspects and analysis. Front Appl Math 9:1–87 [SIAM, Philadelphia]

Comparison of SpatialAnalyzer and Different Adjustment Programs

C. Herrmann, M. Lösler, and H. Bähr

Abstract

Net adjustment is one of the basic tools for various surveying tasks. Among the transformation of coordinates or the analysis and comparison of geometries, the adjustment of geodetic networks is an important part of the surveyor's work. The market offers a number of software solutions, both commercial and freeware.

Seeing the range of software solutions, the question arises, whether the programs give equivalent results. Earlier evaluations of net adjustment programs, partly including New River Kinematics' SpatialAnalyzer (SA), revealed on the one hand almost identical adjustment results for the classic programs. On the other hand, the evaluations showed that SA, using a different mathematical model (bundle adjustment), yields clearly distinguishable deviations. Hence, in this paper the authors focused on SA with the classic programs as reference. The first part of the comparison deals with the results of evaluating a terrestrial network. As programs do not account for the earth's curvature in a standardized way, the chosen network is of small size to minimize the influence of the curvature to an insignificant level.

The second part of the paper compares the results of the evaluation of basic geometries (plane, circle, cylinder, sphere) using SA and other software packages with the least squares solution obtained in a rigorous Gauss–Helmert model (GHM).

Keywords

Form fitting • Net adjustment • Quality of geodetic software • Rigorous Gauss–Helmert model

1 Introduction

A study from Schwieger et al. (2010) took a brief look on commercially available software products for net adjustment. The authors discussed the user requirements for such software and the various quality parameters dedicated to assess reliability, efficiency and accuracy. The comparison of the numerical results focused on the estimated coordinates of the network points and a couple of quality parameters. Deviations up to several millimetres in the coordinates between the results of the different programs were observed.

Lösler and Bähr (2010) extended the list of compared programs a little, including open source software and freeware

C. Herrmann (✉)
Karlsruhe Institute of Technology, Geodetic Institute, Englerstr. 7, 76131 Karlsruhe, Germany
e-mail: christoph.herrmann@kit.edu

H. Bähr
Steenhouwerskade 18, 9718 DB Groningen, Netherlands

M. Lösler
Frankfurt University of Applied Sciences, Niebelungenplatz 1, 60318, Frankfurt am Main, Germany

H. Kutterer et al. (eds.), *The 1st International Workshop on the Quality of Geodetic Observation and Monitoring Systems (QuGOMS'11)*, International Association of Geodesy Symposia 140,
DOI 10.1007/978-3-319-10828-5_12,

as well. They focused on the estimated coordinates as a result solely. SpatialAnalyzer (SA), taking part in their comparison, revealed deviating results with respect to the other programs and some characteristics concerning the data processing. Consequently the present paper focuses on SA with some other programs as reference. The authors extended the study by a comparison of different form fitting algorithms to discuss the availability of quality parameters of the estimated geometries.

2 Net Adjustment

2.1 SpatialAnalyzer

This commercial software is developed and distributed by New River Kinematics. It is designed for an industrial environment and mainly used for quality control purposes. The software architecture of SA allows the user to connect and directly operate a large variety of measuring equipment (total stations, laser trackers, scanners etc.). SA presents the measurements on-line in a CAD environment.

Compared to the classic adjustment programs, SA uses a different mathematical model. Instead of the common approach of directly adjusting observations in one step, the software uses concatenated similarity transformations. In SA the tachymetric observations (distances, horizontal directions and vertical angles) cannot be used for the adjustment directly. Instead, SA calculates local coordinates of all target points per station. Thus, each station and the measurements taken there, form an independent (sub-) system with individual orientation. The adjustment is then performed by simultaneously concatenating the station subsystems of the network via similarity transformations (Calkins 2002). Up to seven transformation parameters (translations in x, y and z, rotations about the three axes and one scale factor) can be estimated individually for each station. The adjustment process is initially carried out in an arbitrary coordinate system. To finally acquire the coordinates in the target system, the adjusted network is transformed to the point group of the initial values of the network points, again via similarity transformation.

The other programs included in this study are GNU Gama, Java Graticule 3D (JAG3D), Leica Geo Office (LGO), Netz3D and NetzCG.

GNU Gama is developed by Aleš Čepek. The software is open source and capable of adjusting geodetic networks consisting of, e.g. observed distances, angles, height differences and/or observed coordinates (see Čepek 2011).

JAG3D is developed by Michael Lösler and is open source. The program offers adjustment of geodetic networks in 1D, 2D or 3D. Furthermore routines for coordinate transformation, form fitting and coordinate conversion are included.

LGO is distributed by Leica Geosystems. It is commercial software to evaluate geodetic measurements. The mathematical model of LGO's computation module MOVE3 is rigorously ellipsoidal (Grontmij 2011).

Netz3D is developed by the Geodetic Institute Karlsruhe. It is a program for the adjustment of three dimensional networks.

NetzCG is developed by the Geodetic Institute Karlsruhe and COS Systemhaus OHG. It is an integrated net adjustment tool for AutoCAD. NetzCG automatically separates horizontal position and height and adjusts them separately.

2.2 Network

The network for this comparison was kindly provided by COS Geoinformatik GbR. It consists of 72 sets of measurements (slope distances, horizontal directions and vertical angles) taken on 6 stations with 23 network points in total. The maximum distance between two points is approximately 31 m.

As mentioned above, the programs account differently for the earth's curvature. The influence of the deflection of the vertical increases with the network's size. Witte and Schmidt (2000) give a rule of thumb to assess the effect on the height between two network points with

$$k = \frac{s^2}{2R} \tag{1}$$

where s is the horizontal distance and R is the earth's mean radius. The effect is smaller than 0.1 mm for distances below 36 m. This motivates the choice of a small network, minimizing the influence of the curvature to an insignificant level.

All the programs offer to calculate the adjustment with a priori uncertainty values. Unfortunately the handling differs with each program. To produce comparable results, the authors chose a distance uncertainty of 0.3 mm and an angle uncertainty of 5.5 arcsec (1.7 mgon) for all the software packages.

The reader might wonder why the value for the angles is that large and why the authors chose absolute values rather than using a distance-depending stochastic model. The fact, that the programs cope differently with the a priori uncertainties, made it necessary to choose this approach. Especially the stochastic model of SA lacks the option to take centering or aiming uncertainties into account. The user is only able to define an absolute value (1 sigma level) for the angle uncertainty of the horizontal and vertical angles separately. The stochastic model of the direction uncertainty with a distance-dependent approach is as follows:

$$\sigma_{direction} = \sqrt{b_1{}^2 + \left(\frac{b_2}{s} \cdot \rho\right)^2} \tag{2}$$

where b_1 is the direction uncertainty of the instrument, b_2 is the distance-dependent part representing an aiming or centering uncertainty and s is the distance to the target point. ρ is for converting $b_2/\ s$ into an angle value, e.g. $180/\pi$. It is obvious that the influence of the aiming is largest at short distances. Hence, especially in a network of small size, the aiming uncertainty contributes significantly to the overall uncertainty budget of a point and cannot be neglected. Due to the rather small size of the network the authors chose the relatively high absolute value of 5.5 arcsec (1.7 mgon).

The programs differ in the stochastic model of the distance uncertainty as well. Equation (3) is implemented in LGO, NetzCG and Netz3D. JAG3D calculates the distance uncertainty according to the law of propagation of variances with (4). The model (5) is implemented in Gama. Similar to the model of the directions, a_1 is the absolute uncertainty of the distance measurement and a_2 is the distance-dependent part. With $a_3 = 1$, (5) is the same as (3). Because of the different models, the authors chose an absolute value for the distance uncertainty. Compared to the horizontal directions, this has a rather small effect, especially when measuring short distances.

$$\sigma_{distance} = a_1 + a_2 \cdot s \tag{3}$$

$$\sigma_{distance} = \sqrt{a_1{}^2 + (a_2 \cdot s)^2} \tag{4}$$

$$\sigma_{distance} = a_1 + a_2 \cdot s^{a_3} \tag{5}$$

2.3 Results

The adjustment was carried out with four constraints for the datum defect (three translations and one rotation parameter) as it is appropriate for tachymetric 3D networks (Illner 1983). Table 1 provides an overview of the differences in coordinates and standard deviations between the results of the compared programs. Gama represents the results of JAG3D and Netz3D, too, because the three of them provided identical values, in coordinates as well as standard deviations. This result shows the equally good quality of open source software compared to commercially available solutions.

Using an alternative mathematical model, the results of SA are similar to the other programs with a maximum deviation of 0.5 mm. Taking into account the introduced distance uncertainty of 0.3 mm and the tachymetric application, this result is satisfying. On the other hand, the standard deviations of SA's solution are up to four times larger than the ones of Gama (representing JAG3D and Netz3D, too, as stated above) (Table 1) and are only calculated for actually measured points. Point 3333, which was determined by setting up a station there, is not included in the covariance matrix of SA's net adjustment routine. The available covariance matrix is only of a 3×3 block diagonal structure. The reason for the differences of the standard deviations could not be distinguished clearly. The developers have been notified on this discrepancy.

Concerning classic geodetic measurements, SA lacks some basic features. There are no options for instrument and reflector heights. Consequently they have to be zero or the offset has to be adjusted manually. In contrast to the classic programs, single observations cannot be excluded from the adjustment process (e.g. because of a gross error). If one of the polar elements of a measured point is missing, the other two will be excluded as well, because SA does not use the observations directly, as described in Sect. 2.1.

The comparison also revealed minor deviations between LGO and NetzCG on the one hand, to the group of Gama, JAG3D and Netz3D on the other hand. The maximum value of the differences is 0.2 mm. These deviations are easily explained, again through the mathematical models of LGO and NetzCG. LGO works with a rigorously ellipsoidal coordinate system. NetzCG separates horizontal position and height automatically and adjusts the two "systems" separately.

3 Form Fitting

A common way for the evaluation of point clouds is the form fitting. Regular geometries, like planes, circles and cylinders, are fitted to the measured points. Through estimating the form parameters, it is possible to derivate the characteristics of the object. Those parameters can be the radius of a sphere or the normal vector of a plane, and by that its orientation, just to name a few. The parameters can later be used to assess the form in terms of quality control (e.g. dimensional accuracy).

As the reference for the comparison, the authors realized the approximate and the rigorous GHM with MATLAB. They compared this implementation to the form fitting tools of SA and the software packages mentioned below. By using this implementation, the authors could distinguish whether the software packages obtain the least-squares solution via the rigorous or the approximate GHM. In contrast to the rigorous model, the approximate model does not update the initial values of the adjusted observations with every iteration. For further information on the rigorous evaluation of the GHM see Lenzmann and Lenzmann (2004) and Neitzel (2010).

Table 1 Coordinate differences and standard deviations of SA and the other programs

Point	SA—Gama [mm]			SA—LGO [mm]			SA—NetzCG [mm]			Standard deviation SA [mm]			Standard deviation Gama [mm]		
	X	Y	Z	X	Y	Z	X	Y	Z	X	Y	Z	X	Y	Z
1007	−0.1	0.0	0.2	−0.1	0.0	0.1	−0.1	0.0	0.2	0.42	0.27	0.29	0.09	0.06	0.07
1008	0.1	0.0	0.3	0.0	0.0	0.1	0.1	0.0	0.2	0.43	0.24	0.25	0.10	0.06	0.07
1009	−0.1	0.0	0.2	−0.1	0.0	0.1	−0.1	0.0	0.2	0.44	0.32	0.34	0.12	0.10	0.09
1098	0.1	0.1	0.0	0.1	0.2	0.1	0.1	0.1	0.0	0.46	0.36	0.41	0.07	0.06	0.07
3333	0.2	0.1	−0.1	0.2	0.2	0.0	0.2	0.1	−0.1				0.14	0.12	0.12
101	0.0	0.0	−0.1	0.0	0.0	−0.2	0.0	0.0	−0.2	0.46	0.42	0.44	0.18	0.16	0.17
102	−0.3	0.1	0.1	−0.3	0.1	0.1	−0.3	0.1	0.1	0.53	0.53	0.54	0.22	0.21	0.23
103	0.2	−0.2	0.0	0.2	−0.2	0.0	0.2	−0.2	0.0	0.53	0.52	0.53	0.21	0.21	0.22
104	0.0	−0.1	0.1	0.0	−0.2	0.1	0.0	−0.1	0.2	0.37	0.38	0.40	0.12	0.14	0.15
105	0.0	0.0	0.0	−0.1	0.0	0.0	0.0	−0.1	0.2	0.37	0.46	0.45	0.15	0.19	0.20
106	−0.1	0.0	0.0	−0.1	0.0	0.0	−0.1	0.0	0.0	0.29	0.35	0.28	0.11	0.13	0.11
206	−0.1	−0.4	−0.2	−0.1	−0.3	0.1	0.0	−0.4	−0.2	0.67	0.45	0.67	0.13	0.08	0.11
401	0.0	0.0	0.0	0.0	0.0	−0.1	0.0	0.0	0.0	0.53	0.36	0.38	0.21	0.14	0.15
402	−0.3	0.2	0.0	−0.3	0.1	−0.1	−0.3	0.1	−0.1	0.72	0.53	0.71	0.34	0.22	0.35
501	0.1	0.1	0.0	0.2	0.1	−0.1	0.1	0.1	0.0	0.41	0.35	0.31	0.16	0.14	0.13
504	0.3	0.0	−0.4	0.3	−0.1	−0.5	0.2	−0.1	−0.4	0.67	0.53	0.66	0.28	0.22	0.30
505	0.2	0.1	0.0	0.2	0.1	−0.1	0.2	0.1	0.0	0.58	0.54	0.58	0.24	0.22	0.25
506	0.0	0.0	0.0	0.0	0.0	−0.1	0.0	0.0	0.0	0.21	0.26	0.19	0.10	0.10	0.09
602	−0.1	0.0	0.1	0.0	0.0	0.2	−0.1	0.0	0.1	0.50	0.50	0.48	0.19	0.20	0.20
603	−0.1	−0.1	−0.1	−0.1	0.0	0.1	−0.1	0.0	0.1	0.44	0.38	0.44	0.15	0.14	0.17
604	−0.1	0.0	−0.2	−0.1	0.0	0.0	−0.1	0.0	−0.2	0.42	0.38	0.44	0.13	0.14	0.17
605	0.2	0.1	−0.1	0.2	0.2	0.1	0.2	0.2	0.0	0.44	0.50	0.46	0.20	0.20	0.19
606	−0.1	0.0	−0.1	−0.1	0.1	0.1	−0.1	0.0	−0.1	0.35	0.39	0.35	0.13	0.15	0.14

Gama represents JAG3D and Netz3D

The Least Squares Geometric Elements (LSGE) is a MATLAB toolbox freely offered on www.eurometros.org. The toolbox provides estimation of parameters for standard geometries like lines, planes, spheres and cylinders etc.

The Form Fitting Toolbox is part of the program JAG3D by Michael Lösler. It offers the estimation of form parameters through a GHM for two- and three-dimensional functions (e.g. lines, n-degree polynomials, ellipsoids).

For the comparison of the software packages four basic geometries were chosen. The sample data was taken from the following studies: plane, Drixler (1994); sphere, Jäger et al. (2005); cylinder, Späth (2000a) and circle, Späth (2000b). The following equations depict the functional model for each geometry.

The Hessian normal form (6) is one way to describe a plane. $n_0 = [n_x\ n_y\ n_z]^T$ represents the normalized normal vector. d is the shortest distance of the plane to the point of origin. $P_i = [x_i\ y_i\ z_i]^T$ is a point on the plane.

$$n_0{}^T P_i = d \tag{6}$$

The only form parameter of the sphere is its radius r. The radius is defined as the distance between the center point $P_0 = [x_0\ y_0\ z_0]^T$ and the sphere's surface. The center point defines the sphere's position. All points $P_i = [x_i\ y_i\ z_i]^T$ with the distance r to P_0 lie on the sphere. The functional model can be written in vector form as follows, where the double bars denote the length of the vector:

$$\| P_i - P_0 \| = r \tag{7}$$

Reducing the dimension from 3D to 2D enables to describe a circle with (7). However, the conversion of the 2D geometry into the three dimensional space succeeds only with the use of auxiliary quantities (Späth 2000b). Usually, a circle is derived from intersecting two geometries, for instance a plane and a sphere. The combination of two rather simple functional models like (6) and (7) leads easily to the estimation of the form (Eschelbach and Haas 2003). Hereby the normal vector of the plane determines the orientation

Table 2 Estimated parameters of the forms

	Sphere/m	Plane/m	Circle/m	Cylinder/m
x_0	9.99972450		21,303.5851708	0.23012344
y_0	7.99980653		22,913.70679085	−0.29012746
z_0	6.99930612		25.3418438	0.23419521
r	5.00054199		2.80954434	11.99127993
n_x		0.1947970	0.88546719	−0.74569520
n_y		0.5449293	−0.4647002	−0.66073840
n_z		−0.81554037	−0.0012322	−0.08581051
d		31.748989	8,215.588	

The compared programs provided identical results

of the circle. The position and radius are obtained with the functional model of the sphere.

The cylinder has only one form parameter, too, the radius r. An implicit model of a cylinder with infinite length is given by

$$\|(P_i - P_0) \times n_0\| = r \quad (8)$$

A point $P_0 = [x_0\ y_0\ z_0]^T$ and the normalized direction vector $n_0 = [n_x\ n_y\ n_z]^T$ describe the cylinder axis' position and orientation. The radius is the distance of this axis to the cylinder's surface.

The results of the form fitting with the different implementations are identical. Table 2 shows the number of identical decimal places of the estimated values. Only the approximate GHM of the authors' implementation revealed significant differences. This proves that none of the tested programs estimates the form parameters with the approximate model. All the points representing the forms were introduced as uncorrelated with the same weights.

However, the above mentioned software packages differ in terms of available quality information on the estimates. The geometry fit report of SA presents the estimated parameters of the form (e.g. center point and radius of a sphere). Furthermore the report includes a list of the deviations of each point to the estimated form and a graphical presentation of the point distribution. Apart from that, no other parameters (i.e. standard deviations etc.) are available to assess the estimated form parameters in terms of quality or accuracy.

A simple stochastic model for some forms is implemented in LSGE. The points representing circles, spheres and cylinders can be weighted individually. A weighting of single coordinate values or of points representing lines or planes is not possible. Furthermore the user can retrieve a three by three covariance matrix for the center point. The variance of the radius of circles, spheres and cylinders is also available. For the normal vector of the circle and the direction vector of the cylinder, respectively, another three by three covariance matrix is available. LSGE calculates the deviations of all points to the estimated form as well.

In the Form Fitting Toolbox of JAG3D the coordinates of the points can be weighted separately by introducing a fully populated covariance matrix. The information on the accuracy of the estimated form parameters is available through a fully populated covariance matrix as well. The size of this matrix corresponds to the number of estimated parameters. For example, center point, radius, normal vector and distance to the point of origin of a circle are characterized by an eight by eight covariance matrix. Besides, the following information is presented for each point: standard deviation, redundancy number, estimation of gross error and whether or not the point is an outlier. This is inferred from two statistic tests with user defined levels of significance.

Conclusion

Based on the studies from Schwieger et al. (2010) and Lösler and Bähr (2010) this paper focuses on SpatiaAnalyzer as a tool for net adjustment and form fitting. In contrast to Lösler and Bähr (2010) a special network of small size was chosen, to minimize the influence of earth's curvature. The group of JAG3D, Gama and Netz3D provided identical results in the estimated coordinates as well as in the standard deviations. The differences of up to 0.4 mm of SA to the solution of the above mentioned group of programs are probably due to the different mathematical model of SA using concatenated similarity transformations. The differences of the standard deviations in SA to the above mentioned group could not be explained. By introducing the same gross error free data to all the tested programs, the authors ensured no observations being excluded. Therefore the different standard deviations cannot be explained, e.g., by reduced redundancy. Whether the differences in coordinates are significant with respect to the standard deviations has to be verified in further tests. However, a Monte-Carlo simulation in SA of the uncertainties of the adjusted network points provided values similar to the standard deviations of Gama, JAG3D and Netz3D. This leads to the cautiously optimistic assumption, that the differences in coordinates are not significant.

In terms of form fitting all the programs included in the present comparison provided identical results. It could be verified that all the programs obtain their least squares solution in a rigorous GHM. However, the programs differ in the available quality information on the estimated form parameters. JAG3D offers the widest range of information.

Acknowledgments The authors thank COS Geoinformatik GbR for providing the measurement data of the network. Furthermore they thank Alexander Hiller of VMT GmbH as well as Scott Sandwith and Joseph Calkins of New River Kinematics for the open discussion and help.

References

Calkins JM (2002) Quantifying coordinate uncertainty fields in coupled spatial measurement systems. PhD Thesis, The Virginia Polytechnic Institute and State University

Čepek A (2011) GNU Gama 1.11, Adjustment of geodetic networks. www.gnu.org

Drixler E (1994) Analyse der Form und Lage von Objekten im Raum. DGK series C, vol 409. C. H. Beck, Munich

Eschelbach C, Haas R (2003) The IVS-reference point at Onsala—high end solution for a real 3D-determination. In: Schwegmann W, Thorandt V (eds) Proceedings of the 16th working meeting on European VLBI for geodesy and astrometry, Leipzig, 9–10 May 2003. Federal agency for cartography and geodesy, Leipzig/Frankfurt am Main, pp 109–118

Grontmij Nederland bv (2011) move3 USER MANUAL, Version 4.0 (move3.com)

Illner I (1983) Freie Netze und S-Transformation. AVN 90(5):157–170

Jäger R, Müller T, Saler H, Schwäble R (2005) Klassische und robuste Ausgleichungsverfahren. Wichmann, Berlin

Lenzmann L, Lenzmann E (2004) Strenge Auswertung des nichtlinearen Gauß-Helmert-Modells. AVN, 2/2004. Wichmann, Berlin, pp 68–72

Lösler M, Bähr H (2010) Vergleich der Ergebnisse verschiedener Netzausgleichungsprogramme. In: Zippelt K (ed) Vernetzt und Ausgeglichen. Kit Scientific Publishing, Karlsruhe

Neitzel F (2010) Generalization of total least-squares on example of unweighted and weighted 2D similarity transformation. J Geod 84(12):751–762

Schwieger V, Foppe K, Neuner H (2010) Qualitative Aspekte zu Softwarepaketen der Ausgleichungsrechnung. In: Qualitätsmanagement geodätischer Mess- und Auswerteverfahren. Contributions to the 93rd DVW-seminar. Series of the DVW, vol 61

Späth H (2000a) Ein Verfahren zur Bestimmung des Least-Squares-Zylinders. AVN, 2/2000. Wichmann, Berlin, pp 65–67

Späth H (2000b) Ausgleich mit einem Kreis im Raum. AVN, 11–12. Wichmann, Berlin, pp 398–399

Witte B, Schmidt H (2000) Vermessungskunde und Grundlagen der Statistik für das Bauwesen. Wittwer, Stuttgart

Part III

Recursive State-Space Filtering

State-Space Filtering with Respect to Data Imprecision and Fuzziness

I. Neumann and H. Kutterer

Abstract

State-space filtering is an important task in geodetic science and in practical applications. The main goal is an optimal combination of prior knowledge about a (non-linear) system and additional information based on observations of the system state. The widely used approach in geodesy is the extended Kalman filter (KF), which minimizes the quadratic error (variance) between the prior knowledge and the observations. The quality of a predicted or filtered system state is only determinable in a reliable way if all significant components of the uncertainty budget are considered and propagated appropriately. But in the nowadays applications, many measurement configurations cannot be optimized to reveal or even eliminate non-stochastic error components.

Therefore, new methods and algorithms are shown to handle these non-stochastic error components (imprecision and fuzziness) in state-space filtering. The combined modeling of random variability and imprecision/fuzziness leads to fuzzy-random variables. In this approach, the random components are modeled in a stochastic framework and imprecision and fuzziness are treated with intervals and fuzzy membership functions. One example in KF is presented which focuses on the determination of a kinematic deformation process in structural monitoring. The results are compared to the pure stochastic case. As the influence of imprecision in comparison to random uncertainty can either be significant or less important during the monitoring process it has to be considered in modeling and analysis.

Keywords

Fuzziness • Fuzzy random variables • Imprecision • Monitoring • State-space filtering • Uncertainty

1 Introduction

In the nowadays applications, many measurement configurations cannot be optimized in order to reveal or even eliminate non-stochastic error components. This has mainly logistic and/or economic reasons. Typical examples are the monitoring of slide slopes or the measurements of mobile mapping systems. In both cases, polar measurements, e.g., with laserscanners or total stations are carried out which are not elementary controllable. Another example is the estimation of a trajectory with IMUs (especially without GPS) which usually have a systematic but unknown drift in the measurements. In order to be able to estimate uncertainty measures and to realize a reliable analysis over time, state space filtering of the data is carried out.

The main goal of filtering is an optimal combination of prior knowledge about the (non-linear) system and additional

I. Neumann (✉)
Geodetic Institute, Leibniz Universität Hannover, Nienburger Straße 1, 30167 Hannover, Germany
e-mail: neumann@gih.uni-hannover.de

H. Kutterer
Federal Agency for Cartography and Geodesy, Richard-Strauss-Allee 11, 60598 Frankfurt am Main, Germany

H. Kutterer et al. (eds.), *The 1st International Workshop on the Quality of Geodetic Observation and Monitoring Systems (QuGOMS'11)*, International Association of Geodesy Symposia 140,
DOI 10.1007/978-3-319-10828-5_13,

information based on observations of the system state. The quality of a predicted or filtered system state is only determinable in a reliable way if all significant components of the uncertainty budget are considered and propagated appropriately. Conventionally, all uncertainty components are modeled in a stochastic framework using random variables and stochastic processes. This implicitly assumes that non-stochastic measurement errors such as, e.g., systematic errors, are reducible or even eliminable due to changing measurement setups and/or calibration processes. But often this is not realistic and the reported uncertainty measures are too optimistic since the respective imprecision is neglected. Additionally, discretization errors and insufficient knowledge about the system (fuzziness) play an essential rule in nowadays applications. Especially in the identification of kinematic systems, it is not possible to change measurement setups in order to reveal or even eliminate non-stochastic error components. For this reason, more general methods and algorithms are required to consider and propagate imprecision and fuzziness within the uncertainty budget of state-space filters.

2 Modeling of Imprecision and Fuzziness

2.1 Sources of Uncertainty

In the classical geodetic data analysis, the uncertainty is described by stochastics (random variability). Random variability describes random deviations due to the laws of probability and can be seen in repeated data samples. It corresponds to the complexity of reality were the future behavior is uncertain at the moment (Bandemer 2006).

In contrast to that, imprecision is a non-stochastic part of the uncertainty that can be assigned to the observations, fuzziness is the non-stochastic uncertainty mainly assigned to the models (of Geodesy) and the observed object (see Fig. 1). The relevance of imprecision and fuzziness in the applications is meaningful due to many reasons which are described in the following.

For understanding and modeling of imprecision it is important that the original measurement results are typically preprocessed before they are introduced in the measurement equation of a state-space filter. These preprocessing steps comprise several factors **s** influencing the observations (see also Fig. 2):

- Physical parameters for the reduction and correction steps from the original to the reduced observations.
- *Sensor parameters* (e.g., remaining error sources that cannot be modelled).

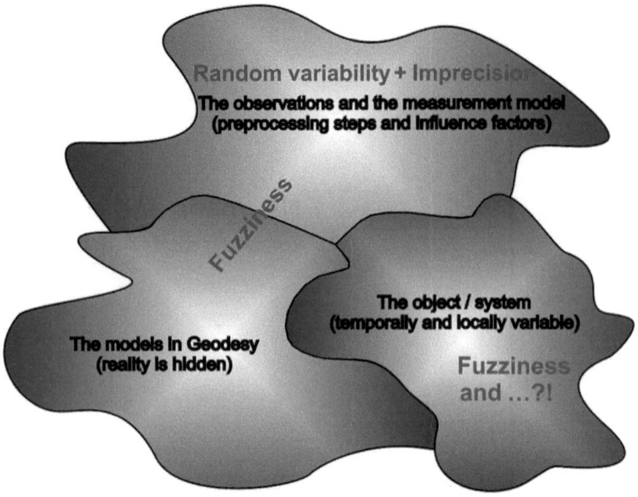

Fig. 1 Sources of uncertainty within the modeling process

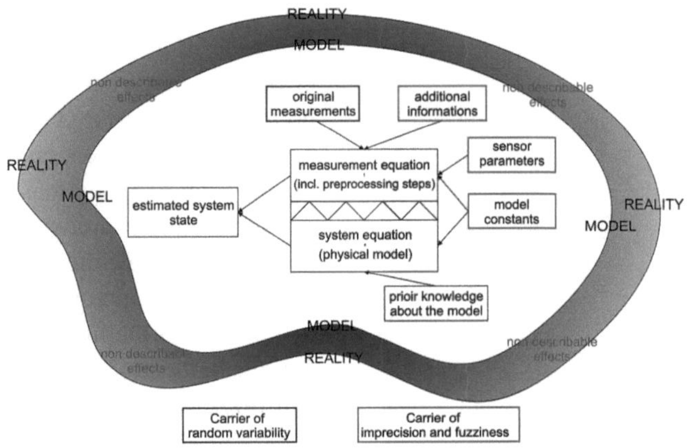

Fig. 2 Interaction between the system and the measurement equation and their influence factors

- *Additional information* (e.g., temperature and pressure measurements for the reduction steps of a distance measurement).

Most of these influence factors are uncertain realisations of random variables; their imprecision is meaningful by many reasons:

- The number of additional information (measurements) may be too small to estimate reliable distributions.
- Displayed measurement results are affected by rounding errors.
- Other non-stochastic errors of the reduced observations occur due to neglected correction and reduction steps.

Figure 2 shows the interaction between the system and measurement equation and their influence factors. While correction and reduction steps are systematic, the imprecision of the influence parameters is directly transferred to the reduced observations, which are now carrier of random variability and imprecision.

The fuzziness in the model is, e.g., due to insufficient knowledge about a priori introduced *model constants*. This concerns especially *model constants* that are only partially representative for the given situation (e.g., the model constants for the refraction index for distance measurements in the formula of Barrel and Sears). Additionally, the exact knowledge about reality is not existent which leads to effects that cannot be modeled (imperfect models). A strict separation between the measurement model and the models in Geodesy is not possible. Therefore the fuzzy component of the models in Geodesy does also partially affect the measurement model (see Fig. 1).

The fuzziness of the object can be explained by the unknown behavior of the object due to external influence factors. A typical example is the unknown temperature of the object during the measurements. This leads to an expansion (e.g. of the steel or concrete) during the measurement process and also in between different measurement epochs. A detailed discussion of the here presented extended uncertainty budget can be found in Kutterer (2002) and Neumann (2009).

2.2 Modeling of Uncertainty

Due to the above mentioned reasons, the quantification of the uncertainty budget of empirical measurements is often too optimistic. For this reason, one has to find (exact) enclosures for the non stochastic part of the influence factors. This step is based on expert knowledge and on error models concerning the deterministic behavior of these parameters (Neumann 2009). The here described procedure leads to so called fuzzy-random variables (FRV) which are a combination of probability and fuzzy-theory, see, e.g., Kwakernaak (1978).

Imprecision and fuzziness can be modeled with the aid of fuzzy theory (Zadeh 1965). Here, LR-fuzzy intervals according to Dubois and Prade (1980) are used.

An LR-fuzzy interval is a special case of a one-dimensional *fuzzy set* $\tilde{A}$ which is described by a *membership function* $m_{\tilde{A}}(x)$:

$$\tilde{A} := \left\{\left(x, m_{\tilde{A}}(x)\right) | x \in \Re\right\} \text{ with } m_{\tilde{A}} : \Re \to [0,1]. \quad (1)$$

The *core* of a fuzzy set is the classical set of elements of $\tilde{A}$ with membership degree equal to 1.

An *LR-fuzzy interval* is defined as a fuzzy set over $\Re$ with a non-empty core. Its membership function is constructed by monotonously decreasing *reference functions* L and R (see Fig. 3). The α-cut of a fuzzy-interval $\tilde{A}$ is defined by:

$$\tilde{A}_\alpha := \left\{x \in X \left| m_{\tilde{A}}(x) \geq \alpha\right.\right\} \quad (2)$$

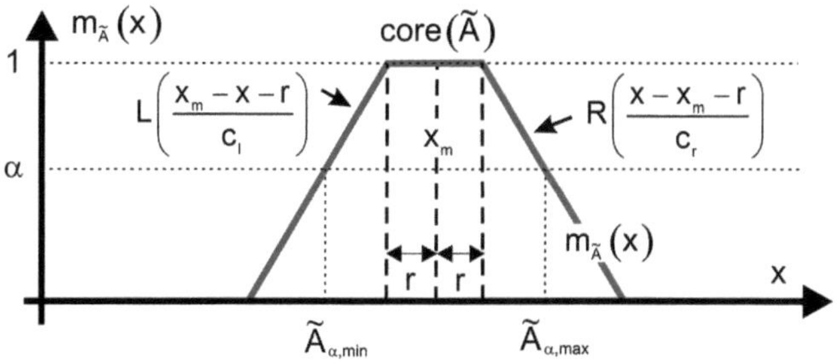

Fig. 3 LR-fuzzy interval with linear reference functions

with $\alpha \in (0, 1]$. Each α-cut represents in case of monotonously decreasing reference functions a classical interval. The lower bound $\tilde{A}_{\alpha,min}$ and upper bound $\tilde{A}_{\alpha,max}$ of an α-cut are obtained as:

$$\tilde{A}_{\alpha,min} = \min\left(\tilde{A}_\alpha\right) \text{ and } \tilde{A}_{\alpha,max} = \max\left(\tilde{A}_\alpha\right). \quad (3)$$

LR-fuzzy intervals can be represented by $\tilde{X} = (x_m, r, c_l, c_r)_{LR}$. The *midpoint* is denoted by x_m, and the *radius* of the interval representing the core is r. Together with the deterministic *spreads* c_l and c_r it serves as a measure of uncertainty. Strategies to construct fuzzy numbers or fuzzy intervals based on expert knowledge are given in Nguyen and Kreinovich (1996) and Neumann (2009). In this paper only symmetric reference functions are used. The main benefit of this is that the *midpoint* equals the result of the pure stochastic case within linear filtering processes.

Random variability is then introduced through the midpoint of an LR-fuzzy interval which is modeled as a random variable and hence treated by methods of stochastics. The combination of random variability and imprecision/fuzziness in uncertainty modeling leads to the theory of fuzzy-random variables (FRV, see Fig. 4). FRVs serve as basic quantities; they are an extension of the classical probability theory. For this reason, all statistical methods have to be extended to imprecise data and all statistical quantities are imprecise by definition.

Here, this yields an *LR-fuzzy-random interval* $\underline{\tilde{X}} = (\underline{X}_m, X_r, c_l, c_r)_{LR}$ with a stochastic midpoint $\underline{X}_m$; the underline indicates a random variable. Actually, $\underline{\tilde{X}}$ is a special case of a fuzzy-random variable, see also Möller and Beer (2004). In contrast to the general case only the expectation value is considered as superposed by imprecision/fuzziness but not the variance. Without non-stochastic errors the pure stochastic case is obtained ($X_r = c_l = c_r = 0$), see Dubois and Prade (1980) for examples.

In case of normal distributed values for the random part, the standard deviation σ_x is the carrier of the stochastic uncertainty, and the *radius* X_r and *spreads* c_l and c_r are the carrier of imprecision. A geometric interpretation of a FRV

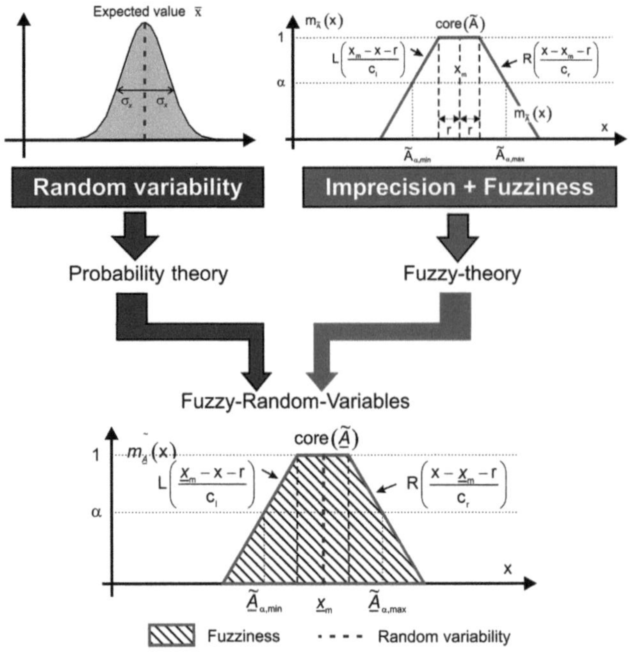

Fig. 4 Construction of a FRV

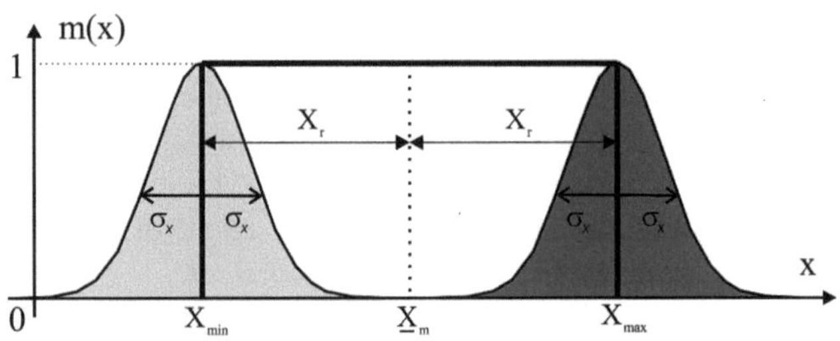

Fig. 5 FRV for $c_l = c_r = 0$ with the variation range of the random uncertainty component inside the core element

with $c_l = c_r = 0$ is given in Fig. 5. The lower and upper bound of the core define the variation range (in a deterministic manner) of the random midpoint.

3 Introduction to State-Space Filtering

3.1 System and Measurement Equation

In this section, the notation for the filtering process is introduced. Section 4 then extends the here given equations to the extended uncertainty budget. The knowledge about the system is given by the system equation with the state vector $\mathbf{x}_k \in \Re^{n_x}$, and the time index k.

$$\begin{aligned} &\mathbf{x}_{k+1} = \boldsymbol{f}(\mathbf{x}_k, \mathbf{u}_k, \mathbf{w}_k) \qquad \text{with} \\ &\boldsymbol{f} : \mathbf{x}_k \in \Re^{n_x}, \ \mathbf{u}_k \in \Re^{u}, \ \mathbf{w}_k \in \Re^{w} \rightarrow \mathbf{x}_{k+1} \in \Re^{n_x} \end{aligned} \tag{4}$$

where $\boldsymbol{f}$ is a known, in general nonlinear function, $\mathbf{u}_k$ is a vector of known (deterministic) input, and $\mathbf{w}_k$ is the process noise vector, which is caused by mismodeling effects and other disturbances in the motion model. The main goal of filtering is to estimate the optimal $\mathbf{x}_{k+1}$ from the observations $\mathbf{y}_{k+1} \in \Re^{n_y}$ and $\mathbf{x}_k \in \Re^{n_x}$, with the help of the measurement equation:

$$\begin{aligned} &\mathbf{y}_{k+1} = \boldsymbol{h}(\mathbf{x}_{k+1}, \mathbf{v}_{k+1}) \qquad \text{with} \\ &\boldsymbol{h} : \mathbf{x}_{k+1} \in \Re^{n_x}, \ v_{k+1} \in \Re^{v} \rightarrow \mathbf{y}_{k+1} \in \Re^{n_y} \end{aligned} \tag{5}$$

where $\boldsymbol{h}$ is a known, in general nonlinear function, and $\mathbf{v}_{k+1}$ is the measurement noise vector, which is assumed to be independent and uncorrelated with known PDF.

Different algorithms were development in order to find the optimal solution and to solve the computational problems. Some of the approaches approximate the nonlinear function (e.g. extended KF) and other approaches approximate the distribution of measurements and state (e.g. unscented KF and sequential Monte Carlo filter). The widely used approach in the applications is the extended KF. It is used in engineering geodesy especially in applications such as engineering navigation and deformation analysis. In order to extend the KF to imprecision and fuzziness, the basic algorithm is introduced in the following.

3.2 Extended Kalman Filter

Engineering navigation and deformation analysis requires a sequential estimation of the system state based on information coming from a (dynamic) model of the system and from external observations. The extended KF minimizes the quadratic error (variance) between the prior knowledge and the observations. The algorithm starts with the definition of the input data vector $\mathbf{L}$ and its associated variance–covariance matrix (VCM) $\boldsymbol{\Sigma}_{LL}$:

$$\mathbf{L} = \left[\mathbf{x}_k \ \mathbf{u}_k \ \mathbf{w}_k \ \mathbf{y}_{k+1}\right]^{\mathrm{T}}$$

$$\boldsymbol{\Sigma}_{LL} = \begin{bmatrix} \boldsymbol{\Sigma}_{xx,k} & \mathbf{0} & \mathbf{0} & \mathbf{0} \\ \mathbf{0} & \boldsymbol{\Sigma}_{uu,k} & \mathbf{0} & \mathbf{0} \\ \mathbf{0} & \mathbf{0} & \boldsymbol{\Sigma}_{ww,k} & \mathbf{0} \\ \mathbf{0} & \mathbf{0} & \mathbf{0} & \boldsymbol{\Sigma}_{yy,k+1} \end{bmatrix} \tag{6}$$

with the initial conditions of the state estimates $\mathbf{x}_k$, disturbances $\mathbf{w}_k$ for time t_k, as well as the vector of measurements $\mathbf{y}_{k+1}$ for time t_{k+1}. The vectors $\mathbf{x}_k$, $\mathbf{u}_k$ and $\mathbf{w}_k$ are assigned to the physical model, which represents the theoretical component of the filter (system equation) whereas the vector of measurements $\mathbf{y}_{k+1}$ represents the experimental component (measurement equation).

The evaluation continues with the extrapolation of the system equation (prediction) and the error propagation for time t_{k+1}. The predicted system state $\overline{\mathbf{x}}_{k+1}$, and its VCM $\boldsymbol{\Sigma}_{\overline{xx},k+1}$ are given by:

$$\begin{aligned}\overline{\mathbf{x}}_{k+1} &= \underbrace{[\mathbf{T}_{k+1,k}\ \mathbf{B}_{k+1,k}\ \mathbf{C}_{k+1,k}]}_{=:\,\mathbf{P}_{k+1,k}}\begin{bmatrix}\mathbf{x}_k\\ \mathbf{u}_k\\ \mathbf{w}_k\end{bmatrix} = \mathbf{P}_{k+1,k}\mathbf{x}_k^*\\ \boldsymbol{\Sigma}_{\overline{xx},k+1} &= \mathbf{T}_{k+1,k}\boldsymbol{\Sigma}_{xx,k}\mathbf{T}_{k+1,k}^T + \dots\\ &\dots \mathbf{B}_{k+1,k}\boldsymbol{\Sigma}_{uu,k}\mathbf{B}_{k+1,k}^T + \mathbf{C}_{k+1,k}\boldsymbol{\Sigma}_{ww,k}\mathbf{C}_{k+1,k}^T,\end{aligned} \tag{7}$$

with the transition matrix $\mathbf{T}_{k+1,k}$, the Jacobi matrix $\mathbf{B}_{k+1,k}$, and the coefficient matrix of disturbing variables $\mathbf{C}_{k+1,k}$. The matrices $\mathbf{T}_{k+1,k}$, $\mathbf{B}_{k+1,k}$, and $\mathbf{C}_{k+1,k}$ depend on the physical model to be observed and must be chosen accordingly.

The optimal estimation of the system state $\widehat{\mathbf{x}}_{k+1}$ at time t_{k+1} (filtering) in a Gauss–Markov-Model is now based on the predicted system state $\overline{\mathbf{x}}_{k+1}$ at time t_{k+1}, and on the (indirect) external measurements $\mathbf{y}_{k+1}$:

$$\begin{aligned}\widehat{\mathbf{x}}_{k+1} &= \overline{\mathbf{x}}_{k+1} + \mathbf{K}_{k+1}\mathbf{d}_{k+1},\\ \mathbf{K}_{k+1} &= \boldsymbol{\Sigma}_{\overline{xx},k+1}\mathbf{A}_{k+1}^T\left(\boldsymbol{\Sigma}_{yy,k+1} + \mathbf{A}_{k+1}\boldsymbol{\Sigma}_{\overline{xx},k+1}\mathbf{A}_{k+1}^T\right)^{-1}.\end{aligned} \tag{8}$$

With the Kalman gain matrix $\mathbf{K}_{k+1}$, and the vector of innovation $\mathbf{d}_{k+1} = \mathbf{y}_{k+1} - \boldsymbol{h}(\overline{\mathbf{x}}_{k+1})$ as well as the design matrix $\mathbf{A}_{k+1}$ which is the Jacobi matrix of the function $\boldsymbol{h}$ with respect to $\mathbf{x}$. This function models the relation between the predicted system state and the collected measurements at time t_{k+1}. In the linearized case considered here the vector of innovation $\mathbf{d}_{k+1}$ is given by $\mathbf{d}_{k+1} = \mathbf{y}_{k+1} - \mathbf{A}_{k+1}\overline{\mathbf{x}}_{k+1}$.

4 Imprecise Filter Extension

4.1 General Case

In this section the filtering process is extended to the above mentioned non-stochastic uncertainties. Therefore, the equations and algorithms from Sect. 3 have to be extended to the FRVs from Sect. 2. The system equation (2) with FRVs yields to:

$$\begin{aligned}&\underline{\tilde{\mathbf{x}}}_{k+1} = \tilde{\boldsymbol{f}}\left(\underline{\tilde{\mathbf{x}}}_k, \underline{\tilde{\mathbf{u}}}_k, \underline{\tilde{\mathbf{w}}}_k\right) \text{ with}\\ &\tilde{\boldsymbol{f}} : \Im\left(\Re^{n_x+u+w}\right) \to \Im\left(\Re^{n_x}\right),\end{aligned} \tag{9}$$

where $\Im(\Re)$ is the FRV-space over the real numbers. The extension of the measurement equation in the imprecise case is:

$$\begin{aligned}&\underline{\tilde{\mathbf{y}}}_{k+1} = \tilde{\boldsymbol{h}}\left(\underline{\tilde{\mathbf{x}}}_{k+1}, \underline{\tilde{\mathbf{v}}}_{k+1}\right) \quad \text{with}\\ &\tilde{\boldsymbol{h}} : \Im\left(\Re^{n_x+v}\right) \to \Im\left(\Re^{n_y}\right)\end{aligned} \tag{10}$$

This mapping is a fuzzy extension of the classical probability theory and the Eqs. (8) and (9) have to be solved with the extension principle according to Zadeh (1965). This leads, e.g. for Eq. (8) to:

$$\begin{aligned}\underline{\tilde{\mathbf{x}}}_{k+1} &:= m_{\underline{\tilde{\mathbf{x}}}_{k+1}}(\mathbf{x})\\ &= \sup_{\substack{(\mathrm{x}_{k,1},\dots,\mathrm{w}_j)\in\Re_1\times\dots\times\Re_{n_x+u+w}\\ \mathrm{x}_{k+1}=f(\mathrm{x}_k,\mathrm{u}_k,\mathrm{w}_k)}}\\ &\min\left(\mathrm{m}_{\underline{\tilde{\mathrm{x}}}_{k,1}}(\mathrm{x}_{k,1}), \dots, \mathrm{m}_{\underline{\tilde{\mathrm{w}}}_{k,w}}(\mathrm{w}_{k,w})\right)\end{aligned} \tag{11}$$

This equation is solvable by α-cut optimization (Möller and Beer 2004) which reduces Eq. (11) to its interval extension if the discussion is based on the α-cuts (see, e.g., Viertl 2011):

$$\begin{aligned}\underline{\tilde{\mathbf{x}}}_{m,k+1} &= \boldsymbol{f}\left(\underline{\tilde{\mathbf{x}}}_{m,k}, \underline{\tilde{\mathbf{u}}}_{m,k}, \underline{\tilde{\mathbf{w}}}_{m,k}\right),\\ \underline{\tilde{\mathbf{x}}}_{k+1_{\alpha,\min}} &= \min_{\mathbf{x}\in[\underline{\tilde{\mathbf{x}}}_\alpha],\ \mathbf{u}\in[\underline{\tilde{\mathbf{u}}}_\alpha],\ \mathbf{w}\in[\underline{\tilde{\mathbf{w}}}_\alpha]} \boldsymbol{f}\left(\underline{\tilde{\mathbf{x}}}_k, \underline{\tilde{\mathbf{u}}}_k, \underline{\tilde{\mathbf{w}}}_k\right),\\ \underline{\tilde{\mathbf{x}}}_{k+1_{\alpha,\max}} &= \max_{\mathbf{x}\in[\underline{\tilde{\mathbf{x}}}_\alpha],\ \mathbf{u}\in[\underline{\tilde{\mathbf{u}}}_\alpha],\ \mathbf{w}\in[\underline{\tilde{\mathbf{w}}}_\alpha]} \boldsymbol{f}\left(\underline{\tilde{\mathbf{x}}}_k, \underline{\tilde{\mathbf{u}}}_k, \underline{\tilde{\mathbf{w}}}_k\right),\end{aligned} \tag{12}$$

and

$$\begin{aligned}&\mathrm{m}_{\underline{\tilde{\mathbf{x}}}_{k+1}}(\mathrm{x}) = \sup_{\alpha\in(0,1]} \alpha \cdot \mathrm{i}_{\underline{\tilde{\mathbf{x}}}_{k+1_\alpha}}(\mathrm{x})\\ &\text{with } \mathrm{i}_{\underline{\tilde{\mathbf{x}}}_{k+1_\alpha}} = \left[\underline{\tilde{\mathbf{x}}}_{k+1_{\alpha,\min}}, \underline{\tilde{\mathbf{x}}}_{k+1_{\alpha,\max}}\right]\end{aligned} \tag{13}$$

The multidimensional mapping of FRVs leads to wrapping effects and therefore for an overestimation of the true membership function of the resulting FRVs. A detailed discussion of this topic can be found in Schön and Kutterer (2005) as well as in Kutterer and Neumann (2009). For this reason, the general solution of the system and measurement equation with imprecision and fuzziness is at the moment not available. In the following, the solution for the KF is presented.

4.2 KF with Imprecise and Fuzzy Data

The imprecise/fuzzy extension of the extended KF can be reduced to its interval extension if the discussion is based on the α-cuts, see Eqs. (12) and (13). The interval extension of the system state $\widehat{\mathbf{x}}_{k+1}$ at time t_{k+1} is obtained by the evaluation of Eq. (8) using the fundamental arithmetic rules of interval mathematics. The formulation of imprecision/fuzziness is directly referred to the (basic) influence factors $\mathbf{s}$ in order to reduce the overestimation effect which is caused by the sub-distributivity property of the intervals (intervals are a special case of fuzzy sets). This is possible if the filter algorithm is formulated in a non-recursive way. Note that interval extension is only required for the propagation of imprecision/fuzziness to the estimated parameters since in the present approach the precise midpoint of the estimated system state equals the result of the pure stochastic case.

In order to avoid this overestimation the recursion can be resolved as introduced by Neumann and Kutterer (2007). For this purpose, Eq. (14) is reformulated in full equivalence to Eq. (8) as a linear mapping of independent quantities which are not updated within the recursion, i.e. the initial state vector and all system state measurements up to time t_{k+1}:

$$\widehat{\mathbf{x}}_{k+1} = \left[\mathbf{K}_{k+1}\ (\mathbf{E} - \mathbf{K}_{k+1}\mathbf{A}_{k+1})\,\mathbf{P}_{k+1,k}\mathbf{H}_k\right] \begin{bmatrix} \mathbf{y}_{k+1} \\ \vdots \\ \mathbf{y}_1 \\ \widehat{\mathbf{x}}_0 \end{bmatrix}, \tag{14}$$

with $\mathbf{H}_k = \left[\mathbf{K}_k\ (\mathbf{E} - \mathbf{K}_k\mathbf{A}_k)\,\mathbf{P}_{k,k-1}\mathbf{H}_{k-1}\right]$, and $\mathbf{H}_0 = \mathbf{E}$ the identity matrix. Here, for the respective midpoints the interval-mathematical evaluation yields:

$$\widehat{\mathbf{x}}_{m,k+1} = \left[\mathbf{K}_{k+1}\ (\mathbf{E} - \mathbf{K}_{k+1}\mathbf{A}_{k+1})\,\mathbf{P}_{k+1,k}\mathbf{H}_k\right] \begin{bmatrix} \mathbf{y}_{m,k+1} \\ \vdots \\ \mathbf{y}_{m,1} \\ \widehat{\mathbf{x}}_{m,0} \end{bmatrix}, \tag{15}$$

which equals the classical formulation if the actual measurement values are taken as interval midpoints. For a specific description of imprecision the interval radii of the influence parameters $[\mathbf{s}] = [\mathbf{s}_m - \mathbf{s}_r, \mathbf{s}_m + \mathbf{s}_r]$ are introduced. This yields:

$$\widehat{\mathbf{x}}_{r,k+1} = \left|\left[\mathbf{K}_{k+1}\ (\mathbf{E} - \mathbf{K}_{k+1}\mathbf{A}_{k+1})\,\mathbf{P}_{k+1,k}\mathbf{H}_k\right]\mathbf{MF}\right|\mathbf{s}_r \tag{16}$$

without specifying the composition of $\mathbf{s}$ in detail. With the matrix of partial derivatives $\mathbf{F} = \partial\mathbf{y}/\partial\mathbf{s}$ that describes the functional relationship between the observations and the influence factors. The matrix $\mathbf{M}$ models the influence of the measurement method on the propagation process.

The interval vector of the filtered system state estimate $\widehat{\mathbf{x}}_{k+1}$ is then composed as:

$$\left[\widehat{\mathbf{x}}_{k+1}\right] = \left[\widehat{\mathbf{x}}_{m,k+1} - \widehat{\mathbf{x}}_{r,k+1}, \widehat{\mathbf{x}}_{m,k+1} + \widehat{\mathbf{x}}_{r,k+1}\right]. \tag{17}$$

In contrast to Eq. (12), this representation of the parameter vector is exact component by component. Hence, it is the tightest interval inclusion of the range of values in the given parameterization, although it still overestimates the correct range of values which is a convex polyhedron (a.k.a. zonotope); see, e.g., Schön and Kutterer (2005) for a detailed discussion.

In Sect. 5 some exemplarily results of the presented algorithm of the extended KF are depicted.

5 Example for Imprecise Kalman Filtering

This numerical example demonstrates an application of the extended KF in a geodetic deformation analysis for the monitoring of a lock (see Fig. 6). Due to changing water levels (from 42 to 65 m) inside the lock, different deformations of the lock chamber occur. We focus on the periodical expansion of the lock chamber during the time when ships are passing through the lock.

Therefore, the 3d-coordinates of different points at the right position of the lock chamber were observed with a total station (Leica TPS 1101), measuring horizontal directions (a), zenith angles (b), and distances (c). The main influence factors for the imprecision of the measurements are given in Table 1. Additionally, the imprecision of the measurement period (time step) must be considered. The measurements of the temperature and the pressure are only partially representative for the given situation, because they are measured at the position of the instrument, only. On this account, their imprecision is stronger than their uncertainty determined in terms of standard deviations. Table 2 shows the theoretical standard deviations (from the manufacturer) and interval radii of the measurements (for the α-cut equal to zero). Due to the definition of the geodetic reference system, the expansion of the lock chamber is visible in the y-coordinate of the point positions. On that account, the numerical examples were mainly reduced to the y-coordinates of the points. The time series consists of about 300 epochs observed in periods of 75 s. There are four nearly identical expansions of the lock chamber during the 300 epochs. For this reason only the first 120 epochs with two expansions are shown. The goal is to compute the displacements, velocities and accelerations of the y-coordinates. The uncertainties of the y-coordinate are mainly influenced by the direction measurements and the uncertainties of the x-coordinate are mainly influenced by the distance measurements (see Fig. 6).

Figure 7 shows the estimated system states (y-coordinate) and Fig. 8 the stochastic uncertainty (standard deviations) of the system state. The characteristic of the implemented KF in this example is as follows: if the innovation is determined as significant, the standard deviations of the measurements are increased. This leads to maximum standard deviations of the system state in case of strong innovations.

Due to the lack of space, the determination of the significance of the innovation with statistical tests in case of imprecise data cannot be shown; the reader is referred to Neumann (2009) for the general procedure.

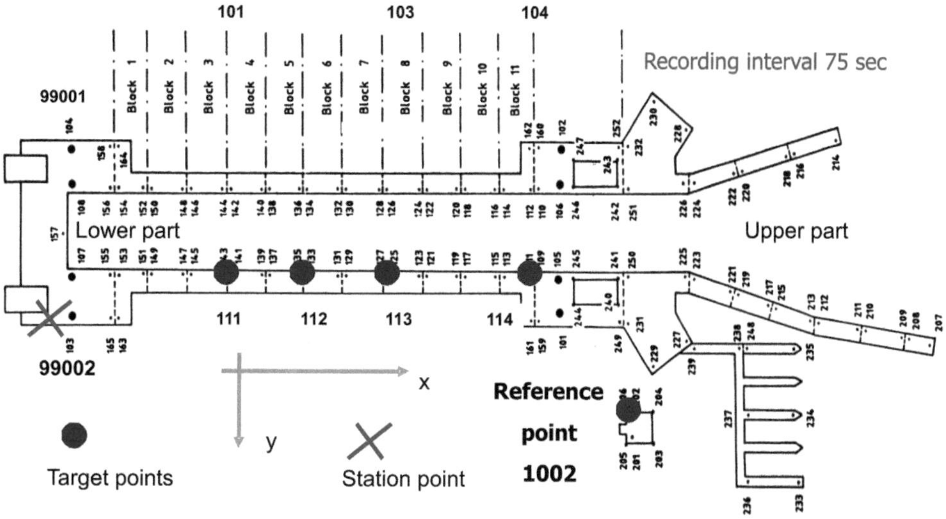

Fig. 6 The lock Uelzen and the measurement configuration

Table 1 An assortment of the most important influence factors

Influence factor	Interval radii (imprecision and fuzziness)	Affected measurements
Temperature	1.0 °C	(c)
Pressure	1.0 hPa	(c)
Visual axis error	0.1 mgon	(a)
Collimation error	0.1 mgon	(a)
Vertical axis error	0.2 mgon	(a) and (b)
Time step	3 s	(a), (b) and (c)

Table 2 Standard deviations and interval radii of the observations

Measurement	Standard deviations	Interval radii
(a) Horizontal direction	0.7 mgon	0.5 mgon
(b) Zenith angle	0.7 mgon	0.5 mgon
(c) Distance	2.0 mm	1.5 mm

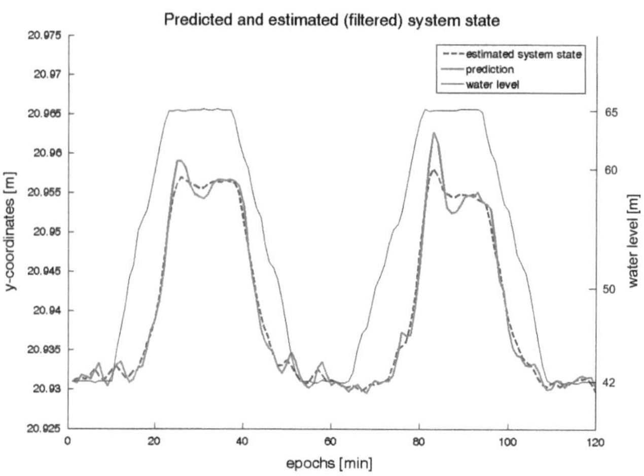

Fig. 7 Y-coordinates of the filtered 3d-point positions and the corresponding water levels

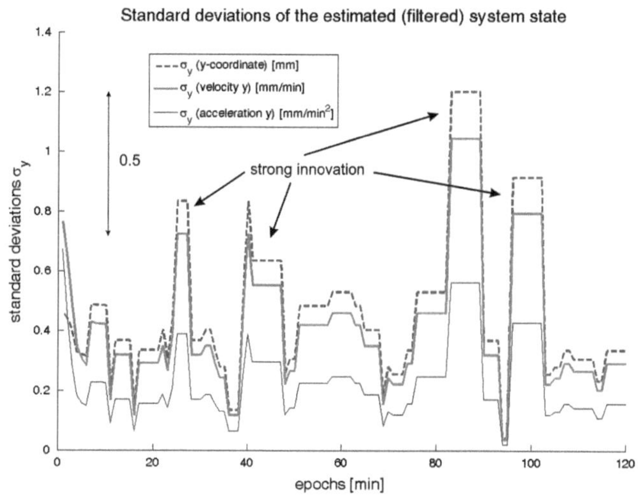

Fig. 8 Standard deviations of the estimated state vector

The fuzziness of the model is mainly due to possible time delays of the measurements (time step). The importance of the fuzzy component is strongly regulated by the characteristic of the deformations. These effects become more and more important for higher velocities and accelerations (see Fig. 9). This is due to the proceeding expansion of the lock during a time delay, which is stronger for high velocities and accelerations.

Whereas the chosen measurement configuration leads to a stronger imprecision/fuzziness of the point positions, the first (velocity) and second (acceleration) derivatives of the process (in the y-coordinates) are well determined (see Fig. 9). This is in full accordance with the theoretical expectations because the imprecision/fuzziness of the inner geometry (derivatives) of processes is significantly reduced by observing only process differences.

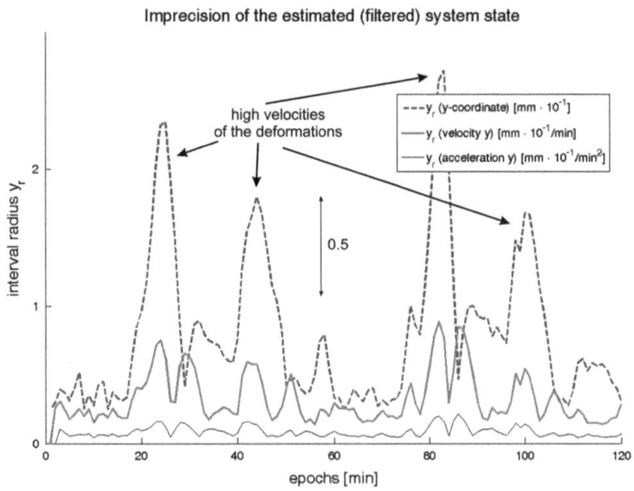

Fig. 9 Imprecision of the estimated state (y-coordinate)

Conclusions

The presented paper shows some ideas for the extension of state-space filtering to imprecise data and fuzziness. The consideration of imprecision and fuzziness leads to an extended uncertainty budget which can be explained by the nowadays applications, where measurement configurations cannot be optimized in order to reveal or even eliminate non-stochastic error components. Especially in the identification of kinematic systems, it is not possible to change measurement setups in order to reveal or even eliminate non-stochastic error components.

Whereas random variability can be modeled in terms of variance–covariance matrices, imprecision and fuzziness are modeled by fuzzy theory. If imprecision and fuzziness are absent, the presented approach leads to the same results than in the stochastic case.

Both, the evaluation of statistical tests in case of imprecision and fuzziness, e.g., for the determination of the significance of the innovation, and the application of the presented approach to adaptive KF techniques are possible. Further work has to deal with the reduction of the computational complexity. Therefore the recursive propagation of imprecision and fuzziness is required. This is already possible within recursive estimation (Kutterer and Neumann 2011) and should be extended to the filtering process. Additionally, the treatment of non-linear functions within the propagation and estimation process is required in order to handle all of the practical applications.

References

Bandemer H (2006) Mathematics of uncertainty – ideas, methods, application problems. Springer, Berlin

Dubois DJ, Prade HM (1980) Fuzzy sets and systems: theory and applications. Academic, New York

Kutterer H (2002) Zum Umgang mit Ungewissheit in der Geodäsie – Bausteine für eine neue Fehlertheorie. DGK, Reihe C, Nr. 553, München

Kutterer H, Neumann I (2009) Fuzzy extensions in state-space filtering. In: Proceedings of ICOSSAR 2009. Taylor and Francis Group, London, pp 1268–1275, ISBN 978-0-415-47557-0

Kutterer H, Neumann I (2011) Recursive least-squares estimation in case of interval observation data. Int J Reliab Saf 5(3/4):229–249

Kwakernaak H (1978) Fuzzy random variables: definitions and theorems. Inf Sci 15:1–29

Möller B, Beer M (2004) Fuzzy randomness – uncertainty in civil engineering and computational mechanics. Springer, Berlin

Neumann I (2009) On the modeling of an extended uncertainty budget in parameter estimation and hypothesis testing (in German). PhD thesis, German Geodetic Commission, No. C-364

Neumann I, Kutterer H (2007) A Kalman filter extension for the analysis of imprecise time series. In: Proceedings of the 15th European Signal Processing Conference (EUSIPCO), Poznan, 2007 (CD-ROM)

Nguyen HT, Kreinovich V (1996) Nested intervals and sets: concepts, relations to fuzzy sets, and applications. In: Kearfott RB, Kreinovich V (eds) Applications of interval computations. Kluwer Publishers, Dordrecht, pp 245–290

Schön S, Kutterer H (2005) Using zonotopes for overestimation-free interval least-squares – some geodetic applications. Reliab Comput 11(2):137–155

Viertl R (2011) Statistical methods for fuzzy data. Wiley, London

Zadeh LA (1965) Fuzzy sets. Inf Control 8:338–353

Unscented Kalman Filter Algorithm with Colored Noise and Its Application in Spacecraft Attitude Estimation

Lifen Sui, Zhongkai Mou, Yu Gan, and Xianyuan Huang

Abstract

The accuracy and reliability of the estimation and prediction of satellite attitude are affected by not only the random noise and systematic errors, but also the colored noise related to time. Any theory or technique based on the hypothesis of Gaussian white noise ignoring the colored noise cannot guarantee the actual reliability of the parameter estimates. On the basis of Unscented Kalman Filter (UKF), the paper regards colored noise as pseudo white noise and considers colored noise as ARMA model, calculates its variance by polynomial-quotient which expresses colored noise model as form of progression. The random model can be corrected with this method. Then the new UKF is formulated by time series analysis theory. In order to verify the validity and rationality of this method, a simulated experiment is showed which validates that the method can restrain effectively the influence of colored noise for satellite attitude estimation.

Keywords

ARMA model • Attitude estimation • Colored noise • Quaternion • Unscented Kalman Filter

1 Introduction

The state-space model of satellite attitude determination is seriously nonlinear model in which the Extended Kalman Filter algorithm is commonly used in all mature data processing methods (Lefferts 1982). However, EKF transforms the nonlinear problem into a linear problem through series expansion which introduces the model error because of ignoring the high order terms. While the initial is not accurate, the filter is divergent easily. So UKF is used in satellite attitude determination (Vandyke 2004; Crassidis 2003). Compared with EKF, UKF can achieve more than second-order accuracy. However, they are built on the basis of Gaussian white noise. In the practical problems of attitude determination, the measurement error and kinetic model error usually do not belong to Gaussian white noise, but to colored noise which characteristics are unusual or time-space related. The presence of colored noise seriously influences the accuracy and reliability of UKF filter. A few filter methods of controlling colored noise have been proposed to handle colored observation noise. The classical approach is to use the group difference of the adjacent observations to transform the observation equations which can transfer the observed colored noise into white noise. Then the EKF is used to solve it (Yang 2006; Xiong 2007).

Unlike the traditional approach, colored noise may be modeled by ARMA or AR model, and be treated as a virtual white noise, which is expanded into series expression by the polynomial long division and get its variance. Then the filter is built by modern time series analysis method (Huang 2008). The simulation results show that the filtering method can effectively control the impact of the colored noise on the results of UKF filtering and improve filtering accuracy and reliability to some extent.

L. Sui (✉) • Z. Mou • Y. Gan • X. Huang
Zhengzhou Institute of Surveying and Mapping, 66th Longhai Middle Road, Zhengzhou 450052, China
e-mail: suilifen@163.com

H. Kutterer et al. (eds.), *The 1st International Workshop on the Quality of Geodetic Observation and Monitoring Systems (QuGOMS'11)*, International Association of Geodesy Symposia 140,
DOI 10.1007/978-3-319-10828-5_14,

2 Mathematical Model of Satellite Attitude

2.1 Quaternion in Satellite Attitude

Since 1980s, the quaternion becomes the most widely used attitude parameter because of its simple kinematical equation and conversion with the attitude angles. Quaternion is a four-dimensional vector, defined as

$$q = \begin{bmatrix} q_{13} \\ q_4 \end{bmatrix} = \begin{bmatrix} n \sin(\theta/2) \\ \cos(\theta/2) \end{bmatrix} \quad (1)$$

Where $q_{13} = \begin{bmatrix} q_1 \; q_2 \; q_3 \end{bmatrix}^T$ is the vector part of quaternion, n, θ is the rotation axis and rotation angle respectively. It can be found from the above expression that its four elements meet the normalized constraints.

And the quaternion attitude matrix is:

$$A(q) = \left(q_4^2 - |q_{13}|^2\right) I_3 - 2q_4 [q_{13}\times] + 2q_{13}q_{13}{}^T \quad (2)$$

The kinematical equation of satellite based on the quaternion (Crassidis 2007) is:

$$\dot{q} = \frac{1}{2}\Omega(\omega)\, q = \frac{1}{2}\Xi(q)\omega \quad (3)$$

Where $\omega = \begin{bmatrix} \omega_1 \; \omega_2 \; \omega_3 \end{bmatrix}^T$ is the satellite attitude angular velocity.

$$\Omega(\omega) = \begin{bmatrix} -[\omega\times] & \omega \\ -\omega^T & 0 \end{bmatrix}$$
$$\Xi(q) = \begin{bmatrix} q_4 I_{3\times3} + [q_{13}\times] \\ -q_{13}^T \end{bmatrix}$$
$$[\omega\times] = \begin{bmatrix} 0 & -\omega_3 & \omega_2 \\ \omega_3 & 0 & -\omega_1 \\ -\omega_2 & \omega_1 & 0 \end{bmatrix} \quad (4)$$

Assuming a sampling interval $\begin{bmatrix} t_k \; t_k + T \end{bmatrix}$, Eq. (3) closed-form solution can be obtained with fixed ω direction (Zhang 2004):

$$q_{k+1} = \left(\cos\frac{|\Theta_k|}{2} I_{4\times4} - \frac{\sin\frac{\Theta_k}{2}}{|\Theta_k|} \begin{bmatrix} 0_{3\times3} & \Theta_k \\ \Theta_k^T & 0 \end{bmatrix}\right) q_k$$

Where, $\Theta_k = \int_{t_k}^{t_k+T} \omega(\tau)d\tau$, $|\Theta| \le \pi$. When ω remains unchanged, $\Theta_k = \omega_k T$.

It is worth noting that quaternion is redundant for representing the global attitude, because quaternion represents three-dimensional attitude through four parameters. So the four parameters of quaternion must satisfy the normalization constraint, while the commonly used numerical integration algorithm cannot guarantee its normalization constraint.

2.2 Nonlinear Model of the Satellite Attitude

Using quaternion as the attitude parameters, the nonlinear state equation of satellite attitude is:

$$\dot{x}(t) = \begin{bmatrix} f_1[q(t), \omega(t)] \\ f_2[\omega(t)] \end{bmatrix} + \begin{bmatrix} 0_{4\times3} \\ J^{-1} \end{bmatrix} \omega(t) \quad (5)$$

Where f_1 and f_2 are kinematical equations and dynamic equations respectively.

$$f_2[w(t)] = J^{-1}\left[T - \dot{h} - [\omega\times](J\omega + h)\right]$$

Among them, T is the total external torque (including the control torque, moment of atmospheric drag, solar pressure torque and the other external disturbance torque), h is the total angular momentum, J is the inertia matrix.

Assuming Δt is the sampling period, when the attitude angular velocity ω remains constant during $[t_k, t_k + T]$, the discrete attitude quaternion propagation equation is expressed as follows in accordance with Eq. (3) closed-form solution:

$$\begin{aligned} q_{k+1} &= \Omega(\omega_k)\, q_k \\ &= \begin{bmatrix} \cos(0.5\,||\,\omega_k\,||\,\Delta t)\, I_{3\times3} & \psi_k \\ -\psi_k^T & \cos(0.5\,||\,\omega_k\,||\,\Delta t) \end{bmatrix} q_k + w_k \end{aligned} \quad (6)$$

Where, $\psi_k = \sin(0.5||\omega_k||\Delta t)\omega_k/||\omega_k||$, w_k is the process noise introduced in the calculation of angular velocity through dynamic equation.

Attitude measurement sensors usually used star sensor because it is not only most accurate in all attitude determination sensors but also can provide the full range attitude information (Zhang 2004). For convenience of calculation, assuming that the star sensor coordinate system coincides with the star coordinate system, the measurement direction vectors of two star sensors are represented as unit reference vectors in the inertial reference coordinate system at the same time. So the measurement equation provided by star sensors is:

$$y_k = h(q_k) + v_k = \begin{bmatrix} A(q_k) r_1 \\ A(q_k) r_2 \end{bmatrix} + v_k \qquad (7)$$

Where, $A(q)$ is attitude matrix, v_k is Gaussian white noise.

3 UKF Algorithm

For the nonlinear model, the extended Kalman filtering (EKF) which ignores high-order terms based on Taylor series expansion can be a good approximation, but it applied only to weakly nonlinear model and may also cause filter divergence when the initial state estimation error is large. The Unscented Kalman Filter (UKF) is proposed based on the idea that it is easier to approximate Gaussian distribution than to approximate nonlinear functions (Pan 2005; Julier 1995; Fredrik 2005). It can achieve more than second order accuracy based on a set of deterministic sampled Sigma points to approximate the nonlinear distribution which can be propagated directly through the nonlinear model and get the mean and variance of the state vector. Compared to EKF algorithm, it can give more higher-order nonlinear system state estimation.

Suppose the state space model of nonlinear system:

$$\widehat{X}_{k+1} = f\left(\widehat{X}_k\right) + w_k \qquad (8a)$$

$$L_k = h\left(\widehat{X}_k\right) + v_k \qquad (8b)$$

Where $f(\bullet)$ and $h(\bullet)$ are nonlinear function, $\widehat{X}_k$ and L_k are the state estimation and measurement vector respectively in k moment. w_k is the process noise arising from the disturbance and model error. $\boldsymbol{v_k}$ is the measurement noise. $w(k)$ and $v(k)$ are zero mean and satisfy the following relations.

$$E\left[w(i)w^T(j)\right] = \delta_{ij} Q(i), \quad E\left[v(i)v^T(j)\right] = \delta_{ij} R(i),$$
$$E\left[v(i)w^T(j)\right] = 0$$

UKF algorithm is calculated as follows:

(1) Initialization: According to the state mean and covariance, the following equations are obtained

$$\widehat{X}(t_0) = E\left(\widehat{X}_0\right)$$

$$\Sigma_{X_0} = E\left\{\left(X(t_0) - \widehat{X}_0\right)\left(X(t_0) - \widehat{X}_0\right)^T\right\}$$

The Sigma points are sampled by the following equation:

$$\chi_{k-1} = \left[\widehat{X}_{k-1} \; \widehat{X}_{k-1} + \gamma\sqrt{\Sigma_{x_{k-1}}} \; \widehat{X}_{k-1} - \gamma\sqrt{\Sigma_{x_{k-1}}}\right] \qquad (9)$$

Where $\gamma = \sqrt{L + \lambda}$.

(2) State propagation: The Sigma points are spread through the state equation,

$$\chi_{k|k-1} = f(\chi_{k-1}) \qquad (10)$$

$$\overline{X}_k = \sum_{i=0}^{2L} W_i^m \chi_{i,k|k-1},$$

$$\overline{\Sigma}_{X_k} = \sum_{i=0}^{2L} W_i^c \left(\chi_{i,k|k-1} - \overline{X}_k\right)\left(\chi_{i,k|k-1} - \overline{X}_k\right)^T \qquad (11)$$

(3) Measurement update:

$$\Upsilon_{k|k-1} = h\left(\chi_{k|k-1}\right), \quad \overline{L}_k = \sum_{i=0}^{2L} W_i^m \Upsilon_{i,k|k-1} \qquad (12)$$

$$\Sigma_{L_k L_k} = \sum_{i=0}^{2L} W_i^c \left(\Upsilon_{i,k|k-1} - \overline{L}_k\right)\left(\Upsilon_{i,k|k-1} - \overline{L}_k\right)^T + R_k \qquad (13)$$

$$\Sigma_{X_k L_k} = \sum_{i=0}^{2L} W_i^c \left(\chi_{i,k|k-1} - \overline{X}_k\right)\left(\Upsilon_{i,k|k-1} - \overline{L}_k\right)^T \qquad (14)$$

$$K_{x_k} = \Sigma_{X_k L_k} \Sigma_{L_k L_k}^{-1} \qquad (15)$$

$$\widehat{X}_k = \overline{X}_k + K\left(L_k - \overline{L}_k\right)$$

$$\Sigma_{X_k} = \overline{\Sigma}_{X_k} - K \Sigma_{L_k L_k} K^T \qquad (16)$$

From the above formula, it can be seen that UKF considers observation noise as Gaussian white noise with variance R. But in the practical problem, the noise from the different time is likely correlated each other, that is the colored noise. When the observation noise is colored noise, it is necessary to consider the colored noise as virtual white noise and obtain the prior distribution information. Then UKF is applied to solve it.

In addition, although the attitude kinematical equation based on quaternion is simple and easy to convert with the attitude angles, the quaternion need to satisfy normalized constraints. In the prediction process, UKF cannot guarantee quaternion normalized constraints in the form of weighting. So this paper uses error quaternion δq_k for UKF filter in which the satellite attitude is obtained by the expression (Xiong 2007) $\widehat{q}_{k+1} = \left[\delta \widehat{q}_{k+1}{}^T \; \sqrt{1 - \delta \widehat{q}_{k+1}{}^T \delta \widehat{q}_{k+1}}\right]^T \otimes \widehat{q}_k$, which cannot only guarantee the quaternion normalization constraint but also simplify the quaternion propagation in the calculation process.

4 Series Representation of Colored Noise and Its Variance Calculation

It is necessary for any filter to know the noise prior distribution. Usually, measurement noise and process noise are seen as Gaussian white noise which priori statistical properties are known in order to simplify the calculation and not to affect the estimation accuracy. However this approach ignores the correlation of noise between adjacent times which would greatly affect the estimation accuracy. In order to take into account the relevance of the adjacent time, the colored noise model is seen as ARMA or AR model which coefficients will be expanded into a series form by polynomial long division. And then its variance is calculated for approximately virtual white noise.

Assuming that colored measurement noise v_k and white noise ξ_k with known statistical properties satisfy the following relationship:

$$C\left(q^{-1}\right) v_k = D\left(q^{-1}\right) \xi_k \tag{17}$$

The following equation is obtained by deforming (17):

$$v_k = D\left(q^{-1}\right) / C\left(q^{-1}\right) \xi_k \tag{18}$$

Where $D(q^{-1})$ and $C(q^{-1})$ are the polynomial of unit delay operators q^{-1}

$$C\left(q^{-1}\right) = 1 + C_1 q^{-1} + \cdots + C_{n_c} q^{-n_c}$$

$$D\left(q^{-1}\right) = 1 + D_1 q^{-1} + \cdots + D_{n_d} q^{-n_d}$$

Using polynomial long division to expand the coefficients $D(q^{-1})/C(q^{-1})$ of Eq. (18) into series form:

$$D\left(q^{-1}\right) / C\left(q^{-1}\right) = 1 + f_1 q^{-1} + f_2 q^{-2} + \cdots + f_j q^{-j} + \cdots$$

Assuming that the expression of $C(q^{-1})$, $D(q^{-1})$ is known and the order n_f is determined. Based on $\xi_k q^{-1} = \xi_{k-1}, \cdots, \xi_k q^{-n_f} = \xi_{k-n_f}$, the Eq. (18) can be expressed as follows:

$$v_k = \xi_k + f_1 \xi_{k-1} + \cdots + f_{n_f} \xi_{k-n_f}$$

The coefficients $f_1, \cdots, f_{n_f}$ can be obtained based on the last equation. According to knowledge of mathematical statistics, $\xi_k, \xi_{k-1}, \ldots, \xi_{k-n_f}$ are independent and identically distributed under $n_f << t$. Therefore, the colored noise variance can be obtained by variance propagation law.

$$\Sigma_v = \left(1 + f_1^2 + \cdots + f_{n_f}^2\right) \Sigma_\xi$$

Therefore, for a single epoch, the colored state noise v_k can be seen as the virtual white noise with zero mean and Σ_v variance.

5 Test Computation and Analysis

The Eqs. (5) and (7) are state equation and observation equation of satellite attitude. In order to analyze the impact of color observation noise on the attitude estimation, we suppose observation noise as colored noise which belongs to AR model. Its expression is:

$$v_k = 0.8 v_{k-1} + \xi_k$$

The simulation parameters of satellite attitude estimation are as follows:Two reference vectors: $r_1 = \left[1\,0\,0\right]^T$, $r_2 = \left[0\,1\,0\right]^T$

Initialization of quaternion:

$$q(0) = \widehat{q}(0) = \left[1/\sqrt{3}\sin(5°)\ 1/\sqrt{3}\sin(5°)\ 1/\sqrt{3}\sin(5°)\ \cos(5°)\right]^T$$

Initialization of angle velocity:

$$\omega(0) = \widehat{w}(0) = \left[0\,0\,0\right]^T (°/s)$$

Initial error covariance:

$$P_0 = diag\left(10^{-6} \cdot I_{3\times3}, 10^{-12} \cdot I_{3\times3}\right)$$

The inertia matrix is selected according to (Pisiaki 1990):

$$J = \begin{bmatrix} 198918 & -210 & -1821 \\ -210 & 265365 & 55 \\ -1821 & 55 & 67406 \end{bmatrix} \circ$$

To analyze the impact of colored measurement noise on satellite attitude estimation and to verify the effectiveness of treatment method of colored noise, the paper uses the following three programs to estimate attitude and compares the results of each program with the true value.

Scenario 1: observation noise is white noise with mean 0 and variance R, UKF is used;

Scenario 2: observation noise is colored noise, but the paper still consider the observation noise as white noise with mean 0 and variance R, UKF is used;

Scenario 3: observation noise is colored noise, the paper considers it as virtual white noise based on AR model which coefficients are known and obtain its variance using polynomial long division. UKF is used.

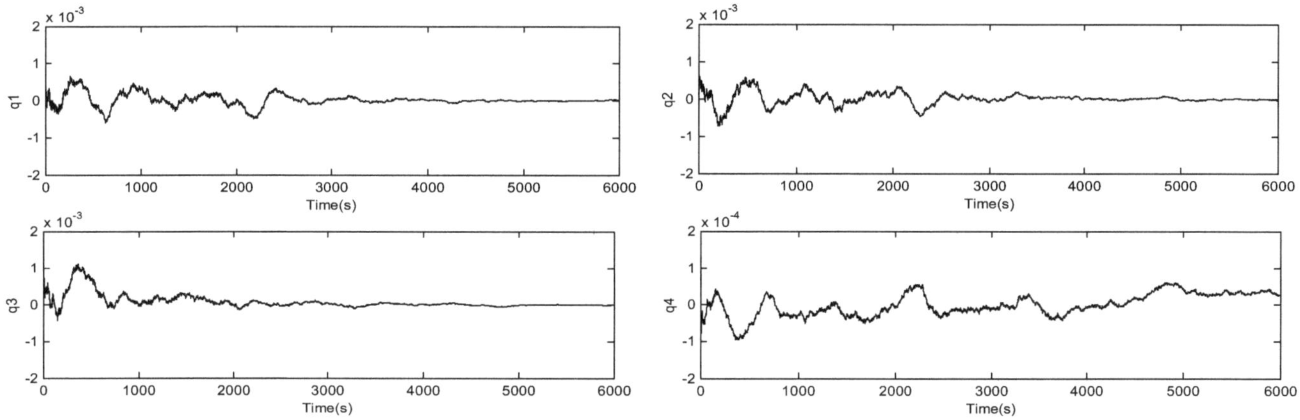

Fig. 1 Scenario 1

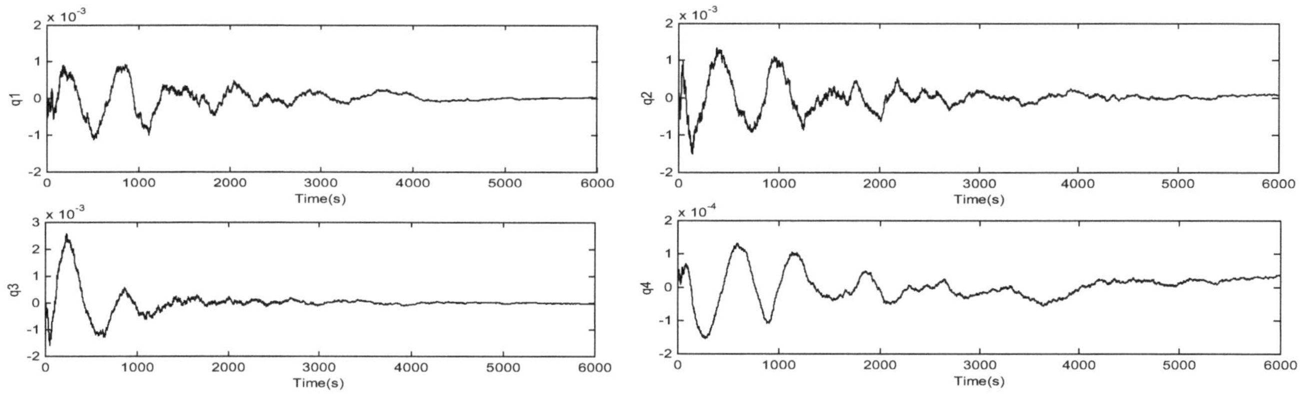

Fig. 2 Scenario 2

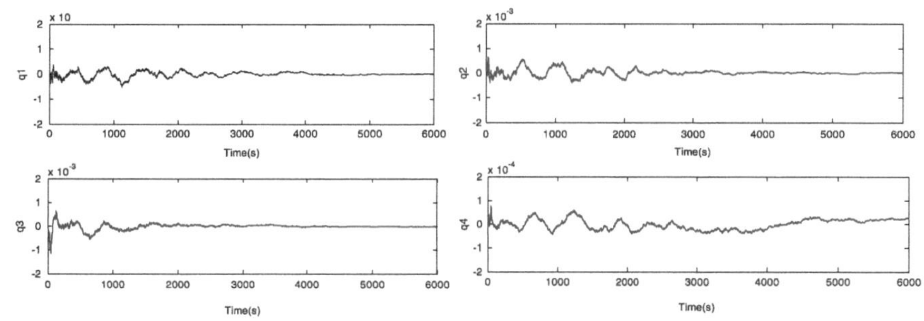

Fig. 3 Scenario 3

The results are shown from Figs. 1 to 3. Figure 1 is UKF error figure when the observation noise is white noise. Figure 2 is error figure when the observation noise is colored noise. Figure 3 is UKF error figure when the colored noise is treated.

The following conclusions can be gotten from the above results: when the observation noise contains colored noise, the convergence becomes slow, and the estimation error increases significantly under colored noise due to the presence of colored noise which makes the correlation between states more complex. However, the filtering process has ignored the correlations and affected the filtering accuracy. It can be seen from Fig. 3 that the use of polynomial long division can be good for the suppression of the influence of colored noise.

Concluding Remarks

The colored noise greatly influenced the accuracy of the nonlinear Filtering. However UKF algorithm cannot effectively suppress the impact of colored observation noise on the filter estimation accuracy. Under the conditions of the new technology, the colored noise can be seen as a virtual white noise and its prior information can be obtained by the series expansion method. Next, UKF is used to determinate satellite attitude with colored noise, which can reduce the influence of colored noise to attitude estimation to some extent. It should also be noted that the colored noise is considered as the AR model with known model coefficients. So the further research is that how to establish different models based on the different characteristics of colored noise.

Acknowledgment This paper has been supported by the National Natural Science Foundation of China (Grant Nos. 41 274016, 40974010, 40971306).

References

Crassidis JL, Markley FL (2003) Unscented filtering for spacecraft attitude estimation. J Guid Control Dynam 26(4):536–542

Crassidis JL, Markley FL, Cheng Y (2007) Survey of nonlinear attitude estimation methods. J Guid Control Dynam 30(1):12–28

Fredrik Orderud (2005) Comparison of Kalman filter estimation approaches for state space models with nonlinear measurements. In: Proceedings of Scandinavian conference on simulation and modeling (SIMS)

Julier SJ, Uhlmann JK, Durrant Whyten HF (1995) A new approach for filtering nonlinear system. In: Proceedings of the American control conference. Washington, pp 1628–1632

Lefferts EJ, Markley FL, Shuster MD (1982) Kalman filtering for spacecraft attitude estimation. J Guid Control Dynam 5(5):417–429

Pan Q, Yang F, Ye L et al (2005) Survey of a kind of nonlinear filters—UKF (in Chinese). Control Decis 20(5):481–489

Pisiaki ML, Martel F, Pal PK (1990) Three-axis attitude determination via Kalman filtering of magnetometer data. J Guid 13(3):506–514

Vandyke MC, Schwartz JL, Hall CD (2004) Unscented Kalman filtering for spacecraft attitude state and parameter estimation [DB/OL]. In: Proceedings of the 14th AAS/AIAA space flight mechanics meeting. Maui, Hawaii, pp 217–228

Wei X, Li-kui C, You H, Jingwei Z (2007) Unscented Kalman filter with colored noise (in Chinese). J Electron Inf Technol 29(3):598–600

Xianyuan H, Lifen S, Pengpai F (2008) A new approach for colored measurement noises by correcting random model (in Chinese). Geomat Informat Sci Wuhan Univ 33(6):644–647

Yang Y (2006) Adaptive navigation and kinematic positioning (in Chinese). Surveying Mapping Press, Beijing, pp 146–160

Zhang H (2004) Nonlinear filtering and its applications in satellite attitude determination (in Chinese). Harbin Industry University

Principles and Comparisons of Various Adaptively Robust Filters with Applications in Geodetic Positioning

Yuanxi Yang, Tianhe Xu, and Junyi Xu

Abstract

The quality of kinematic positioning and navigation depends on the quality of the kinematic model describing the vehicle movements and the reliability of the measurements. A series of adaptive Kalman filters have been studied in recent years. The main principles of four kinds of adaptive filters are summarized, i.e. fading Kalman filter, adaptive Sage windowing filter, robust filter and adaptively robust filter. Some of the developed equivalent weight functions and the adaptive factors including the fading factors are also introduced. Some applications are mentioned.

Keywords

Adaptive filter • Functional model error • Kalman filter • Navigation • Robust estimation

1 Introduction

The adaptive Kalman filtering can be classified into two families, functional model adaptation and stochastic model adaptation. For functional model adaptation, the multiple model based adaptive estimation (MMAE) studied in some references (Mohamed and Schwarz 1999; Moore and Wang 2001) has been analyzed. It has been pointed out by Yang (1999) that the functional model adaptation cannot actually adapt the complicated kinematic model errors.

For stochastic model adaptation, an adaptive Sage windowing filter makes the covariance matrices of the observation vectors and the model predicted state errors adapted to the observation information (Mohamed and Schwarz 1999), this kind of adaptive filter uses the residuals from the previous m epochs to compute the covariance matrices of the observation vectors and the state errors to replace those at present epoch. It is hard to precisely describe the online physical reality or measurement reality by the windowing residual information (Yang et al. 2001a, b).

Other kinds of stochastic model adaptation methods called the robust M–LS, LS–M and M–M filters (Yang 1991) were developed for controlling the outlier influences on the kinematic state estimates (Yang 1997). Furthermore, Koch and Yang (1998) derived the robust Kalman filter for the rank deficient observational model. These robust Kalman filters, which adapt the weights or variance-covariance elements of measurements to their actual errors, belong to the adaptive filter.

A new adaptively robust filter was set up in 2001 for controlling the influences of both the stochastically kinematic model errors and the measurement outliers on the kinematic state estimates (Yang et al. 2001a). After that several adaptive factors have been developed (Yang et al. 2001a, b, 2004). Three kinds of learning statistics for judging the kinematic model errors have been set up, (Xu and Yang 2000; Yang and Gao 2004; Yang et al. 2001a, b) An optimal adaptive Kalman filtering was also given (Yang and Gao 2006).

Y. Yang (✉)
China National Administration of GNSS and Applications, Beijing 100088, China

Xian Research Institute of Surveying and Mapping, Xian 710054, China
e-mail: yuanxi_yang@163.com

T. Xu
Xian Research Institute of Surveying and Mapping, Xian 710054, China

J. Xu
Technic Division of Surveying and Mapping, Xian 710054, China

H. Kutterer et al. (eds.), *The 1st International Workshop on the Quality of Geodetic Observation and Monitoring Systems (QuGOMS'11)*, International Association of Geodesy Symposia 140,
DOI 10.1007/978-3-319-10828-5_15,

An adaptive filter with multiple adaptive factors has been studied (Ou et al. 2004; Ren et al. 2005), it was valid when the state parameters belong to different groups with different characteristics. Furthermore, the adaptive Kalman filters with classified adaptive factors and multiple adaptive factors have been studied (Cui and Yang 2006; Yang and Cui 2008).

In order to illustrate the progress of the adaptive filtering we start from the principle of standard Kalman filter.

Let the linear dynamic system be given by

$$\mathbf{X}_k = \mathbf{\Phi}_{k,k-1}\mathbf{X}_{k-1} + \mathbf{W}_k \tag{1}$$

where $\mathbf{X}_k$ denotes $u \times 1$ state vector at epoch t_k, $\mathbf{\Phi}_{k,k-1}$ the $u \times u$ transition matrix, and $\mathbf{W}_k$ the state noise vector.

At epoch t_k, the observation model reads

$$\mathbf{L}_k = \mathbf{A}_k\mathbf{X}_k + \mathbf{\Delta}_k \tag{2}$$

where $\mathbf{L}_k$ represents $n_k \times 1$ observation vector, $\mathbf{A}_k$ the $n_k \times u$ design matrix, and $\mathbf{\Delta}_k$ the observational noise vector. Let the covariance matrices of $\mathbf{W}_k$ and $\mathbf{\Delta}_k$ be taken as $\mathbf{\Sigma}_{\mathbf{W}_k}$ and $\mathbf{\Sigma}_k$ respectively, and $\mathbf{W}_k$, $\mathbf{W}_j$, $\mathbf{\Delta}_k$ and $\mathbf{\Delta}_j$ be mutually uncorrelated. The predicted state vector is denoted as

$$\overline{\mathbf{X}}_k = \mathbf{\Phi}_{k,k-1}\widehat{\mathbf{X}}_{k-1} \tag{3}$$

$$\mathbf{\Sigma}_{\overline{X}_k} = \mathbf{\Phi}_{k,k-1}\mathbf{\Sigma}_{\widehat{X}_{k-1}}\mathbf{\Phi}^T_{k,k-1} + \mathbf{\Sigma}_{W_k} \tag{4}$$

The error equation is

$$\mathbf{V}_k = \mathbf{A}_k\widehat{\mathbf{X}}_k - \mathbf{L}_k \tag{5}$$

$$\mathbf{V}_{\overline{X}_k} = \widehat{\mathbf{X}}_k - \overline{\mathbf{X}}_k \tag{6}$$

where $\mathbf{V}_\mathbf{k}$ and $\mathbf{V}_{\overline{X}_\mathbf{k}}$ are the residual vectors of $\mathbf{L}_\mathbf{k}$ and $\overline{\mathbf{X}}_\mathbf{k}$ respectively, $\widehat{\mathbf{X}}_\mathbf{k}$ is the estimated state vector. It is well known that the standard Kalman filter is based on the following loss function

$$\mathbf{\Omega}\ (\mathbf{k}) = \mathbf{V}_k^T\mathbf{\Sigma}_k^{-1}\mathbf{V}_k + \mathbf{V}^T_{\overline{X}_k}\mathbf{\Sigma}^{-1}_{\overline{X}_k}\mathbf{V}_{\overline{X}_k} = \mathbf{min} \tag{7}$$

By derivation we can easily get the estimator of the standard Kalman filter.

2 Principle of Adaptive Filtering

In order to analyze the characteristics of various adaptive filters we need not only to have their estimators but also to know their principles and their corresponding loss functions.

2.1 Principle of Fading Kalman Filter

Fading Kalman filter aims at forgetting the previous state information, and weakening its contribution in the present estimates of the state vector. The loss function is like

$$\mathbf{\Omega}\ (\mathbf{k}) = \mathbf{V}_k^T\mathbf{\Sigma}_k^{-1}\mathbf{V}_k + \frac{\mathbf{1}}{\boldsymbol{\lambda}_\mathbf{k}}\mathbf{V}^T_{\widehat{X}_{k-1}}\mathbf{\Sigma}^{-1}_{\widehat{X}_{k-1}}\mathbf{V}_{\widehat{X}_{k-1}} + \widehat{\mathbf{w}}_k^T\mathbf{\Sigma}^{-1}_{w_k}\widehat{\mathbf{w}}_k = \mathbf{min} \tag{8}$$

where $\boldsymbol{\lambda}_k$ ($\boldsymbol{\lambda}_k \geq \mathbf{1}$) is a fading factor (Xia et al. 1990), $\mathbf{V}_{\widehat{X}_{k-1}}$ and $\widehat{\mathbf{w}}_k$ are the residual vectors of $\widehat{\mathbf{X}}_{k-1}$ and the model error vector respectively.

2.2 Principle of Sage Windowing Filter

Sage windowing filter are divided into two types, i.e. IAE (Innovation-based adaptive estimation), and RAE (Residual-based adaptive estimation). These two adaptively windowing estimations can be seen in many literatures (Mohamed and Schwarz 1999; Yang et al. 2001a; Yang and Xu 2003). The loss function of the Sage filter can be deduced as

$$\mathbf{\Omega}\ (\mathbf{k}) = \mathbf{V}_k^T\widehat{\mathbf{\Sigma}}_k^{-1}\mathbf{V}_k + \mathbf{V}^T_{\widehat{X}_{k-1}}\mathbf{\Sigma}^{-1}_{\widehat{X}_{k-1}}\mathbf{V}_{\widehat{X}_{k-1}} + \widehat{\mathbf{w}}_k^T\widehat{\mathbf{\Sigma}}^{-1}_{w_k}\widehat{\mathbf{w}}_k = \mathbf{min} \tag{9}$$

where $\widehat{\mathbf{\Sigma}}_k$ and $\widehat{\mathbf{\Sigma}}_{W_k}$ are evaluated by windowing method.

2.3 Principle of Robust Filter

In order to control the influence of measurement outlier on the kinematic state estimates, a robust filter principle is set up (Yang 1991, 1997). A loss function of M–LS filter is like

$$\mathbf{\Omega}\ (\mathbf{k}) = \sum_{i=1}^{n_k}\mathbf{P}_{k_i}\rho\left(\mathbf{V}_{k_i}\right) + \mathbf{V}^T_{\overline{X}_k}\mathbf{\Sigma}^{-1}_{\overline{X}_k}\mathbf{V}_{\overline{X}_k} = \mathbf{min} \tag{10}$$

where ρ is a continuous and convex function (Yang 1991). This kind of robust filter uses robust maximum (M) likelihood principle on measurement vector and uses least squares (LS) principle on the predicted state vector. It is why the method is named M–LS. There are two other kinds of robust filters which are called LS–M filter and M–M filter respectively (Yang 1991, 1997).

2.4 Adaptively Robust Filter

In order to robustly adjust the contribution of the measurements and the kinematic model information, a new adaptively robust filter motivated by robust estimation principle is developed (Yang et al. 2001a, b). The loss function is as follows

$$\boldsymbol{\Omega}(\mathbf{k}) = \sum_{i=1}^{n_k} \mathbf{P}_{k_i}\rho\left(\mathbf{V}_{k_i}\right) + \alpha_k \mathbf{V}_{\overline{X}_k}^T \boldsymbol{\Sigma}_{\overline{X}_k}^{-1} \mathbf{V}_{\overline{X}_k} = \mathbf{min} \tag{11}$$

where $\boldsymbol{\alpha_k}$ $(\mathbf{0} \leq \boldsymbol{\alpha_k} \leq \mathbf{1})$ is an adaptive factor which balances the contributions of the measurements and the predicted state vector at $\boldsymbol{t_k}$. The first part of (11) is robust loss function and the second part is an adaptive loss function.

3 Design of the Adaptive Factors

The key problem of the adaptive filters is to design the adaptive factors. A reasonable adaptive factor should reflect the measurement errors and/or the kinematic model errors.

3.1 Fading Factors

There are many fading factors in applications. Following are the two factors (Xia et al. 1990)

$$\lambda_k = \max\left\{1, \frac{1}{n} tr\left(\mathbf{N}_k \mathbf{M}_k^{-1}\right)\right\} \tag{12}$$

where $\boldsymbol{tr}[\cdot]$ denotes the trace of a matrix, $\boldsymbol{M_k}$ and $\boldsymbol{N_k}$ are expressed as

$$\mathbf{M}_k = \mathbf{A}_k \boldsymbol{\Phi}_{k,k-1} \boldsymbol{\Sigma}_{\widehat{X}_{k-1}} \boldsymbol{\Phi}_{k,k-1}^T \mathbf{A}_k^T \tag{13}$$

$$\mathbf{N}_k = \boldsymbol{\Sigma}_{\overline{V}_k} - \mathbf{A}_k \boldsymbol{\Sigma}_{W_k} \mathbf{A}_k^T - \boldsymbol{\Sigma}_k \tag{14}$$

where $\boldsymbol{\Sigma}_{\overline{V}_k}$ is the covariance matrix of the predicted residual vector $\overline{\mathbf{V}}_k$, with

$$\overline{\mathbf{V}}_k = \mathbf{A}_k \overline{\mathbf{X}}_k - \mathbf{L}_k \tag{15}$$

3.2 Robust Weight Element

In robust filter, the robust equivalent weight elements may determined by following weight function (Yang 1999; Yang et al. 1999)

$$\overline{p}_i = \begin{cases} p_i & |V_i/\sigma_{v_i}| \leq c_0 \\ p_i \frac{c_0 \sigma_{v_i}}{|V_i|}\left(\frac{c_1 - |V_i/\sigma_{v_i}|}{c_1 - c_0}\right)^2 & c_0 < |V_i/\sigma_{v_i}| \leq c_1 \\ 0 & |V_i/\sigma_{v_i}| > c_1 \end{cases} \tag{16}$$

where $\boldsymbol{c_0}$ and $\boldsymbol{c_1}$ are two constants, which are usually chosen as 1.0–1.5 and 3.0–4.5 respectively (Yang 1999).

3.3 Adaptive Factors

We have set up several adaptive factors for the adaptive filtering. A three segment function is like (Yang et al. 2001a, b)

$$\alpha_k = \begin{cases} 1 & \left|\Delta\tilde{\mathbf{X}}_\mathbf{k}\right| \leq c_0 \\ \frac{c_0}{\left|\Delta\tilde{\mathbf{X}}_\mathbf{k}\right|}\left(\frac{c_1 - \left|\Delta\tilde{X}_k\right|}{c_1 - c_0}\right)^2 & c_0 < \left|\Delta\tilde{\mathbf{X}}_\mathbf{k}\right| \leq c_1 \\ 0 & \left|\Delta\tilde{\mathbf{X}}_\mathbf{k}\right| > c_1 \end{cases} \tag{17}$$

where

$$\Delta\tilde{\mathbf{X}}_k = \left\|\tilde{\mathbf{X}}_k - \overline{\mathbf{X}}_k\right\| / \sqrt{tr\left(\boldsymbol{\Sigma}_{\overline{X}_k}\right)} \tag{18}$$

and $\tilde{\mathbf{X}}_k$ is the robust estimated state vector only by measurements at epoch $\boldsymbol{t_k}$, $\boldsymbol{c_0}$ and $\boldsymbol{c_1}$ are two constants, which are usually chosen as 1.0–1.5 and 3.0–4.5 respectively.

Three learning statistics and three kinds of adaptive factors have been constructed (Yang et al. 2001a, b; Yang and Gao 2004).

4 Computations and Comparisons

A set of kinematic GPS observation data using two Trimble receivers is used in the computation. A rover receiver is mounted in an aircraft, and a reference receiver is fixed at a site about 1 km from the initial aircraft location. The aircraft takes off for a flight time of about 90 min. The measurements used are C/A code and L1 and L2 carrier phases.

In order to analyze the influences of the vehicle disturbances on various Kalman filtering results, the highly precise results from double-differenced carrier phase measurements are used as reference values to compare with the results from the code measurements. The constant velocity model is employed in all the filters. The initial variances for position, velocity and C/A code measurements are selected as 0.2 m^2, 0.001 m^2 s^{-2} and 1 m^2, respectively. The spectral density for velocities is chosen to be 0.01 m^2 s^{-3}. The dynamic model covariance matrix is the same as that in Yang et al. (2001a, b).

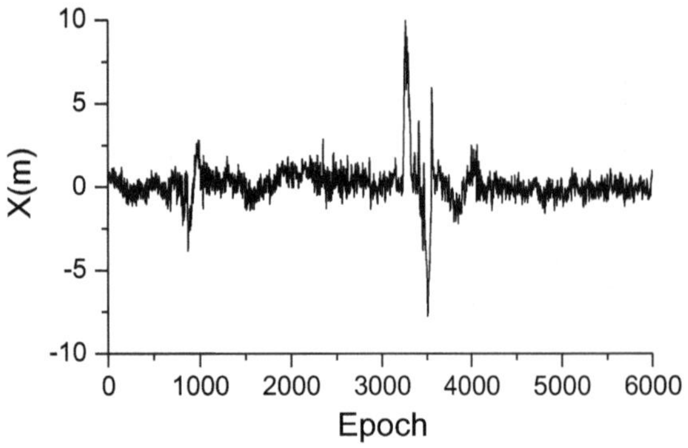

Fig. 1 Standard Kalman filtering

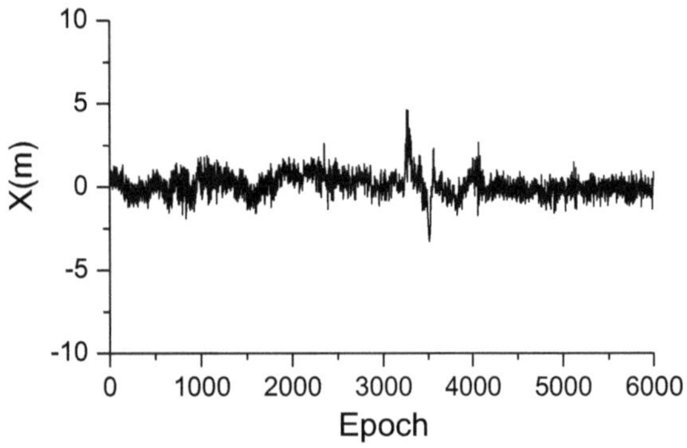

Fig. 2 Robust LS–M Kalman filtering

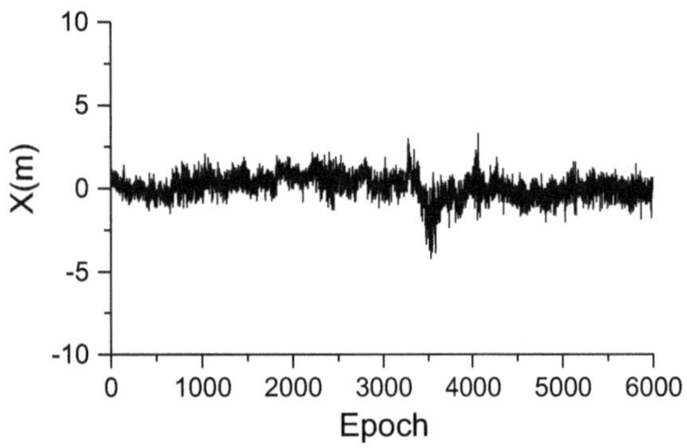

Fig. 3 Adaptive Sage filtering

The following five schemes are adopted:
(1) Scheme 1: standard Kalman filtering;
(2) Scheme 2: robust LS–M Kalman filtering;
(3) Scheme 3: adaptive Sage filtering;
(4) Scheme 4: fading filtering;
(5) Scheme 5: adaptively robust filtering.

The position differences for the *X* component between the results from the five computation schemes and the reference values are shown in Figs. 1, 2, 3, 4 and 5. The RMSs for X, Y and Z components are given in Table 1.

From the computed results and comparisons, the following facts can be stated.

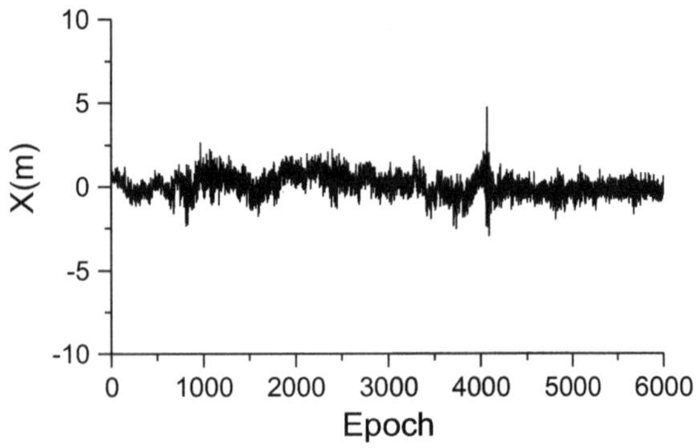

Fig. 4 Fading filtering

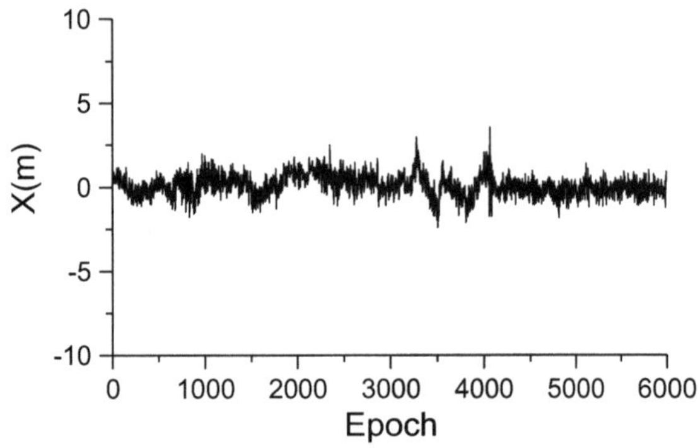

Fig. 5 Adaptively robust filtering

Table 1 The RMSs for X, Y and Z components

Scheme	X/m	Y/m	Z/m
1	1.246	1.481	1.562
2	0.664	0.440	1.015
3	0.721	0.522	1.053
4	0.640	0.646	0.895
5	0.611	0.431	0.870

(1) From Fig. 1, we can see that the two disturbances in the flight have serious effects on the standard Kalman filtering.
(2) Figure 2 shows that the robust LS–M Kalman filter is robust in controlling the influences of kinematic disturbing, whose results are remarkably superior to the standard Kalman filtering.
(3) Sage adaptive filter is better than the standard Kalman filter, to some extent, in controlling the influences of the dynamic model errors (see Fig. 3). However, Sage adaptive filter, requires the historical information and the width of moving windows is uncertain.
(4) Fading filter can control the influence of dynamic disturbances to some extent, and its results are equivalent to those of robust LS–M filter or Sage adaptive filter (see Fig. 5). One disadvantage of the fading filter is that the formula of fading factor is too complex.

(5) Figure 5 and Table 1 illustrate that the adaptively robust filtering can realize a real adaptation. Its results are obviously superior to other filterings, especially in the notable sudden changes of the flight. It can not only flexibly control the influences of the kinematic model disturbances but also resist the influences of the measurement outliers.

Concluding Remarks

The main principle of four kinds of adaptive filters as well as their advantage and disadvantages are summarized elaborately. The adaptive factors play significant roles in adaptive Kalman filter, thus the design of adaptive factor was also discussed. From the application of those adaptive filters, it is shown that the adaptively robust filter cannot only resist the influences of the measurement outliers, but also effectively control the influences of the state disturbances. So we recommend the adaptively robust filter be used in application.

Acknowledgments The project is sponsored by Natural Science Foundations of China (Grant Nos. 41374019 and 41020144004 as well as National 863 project No. 2013AA122501).

References

Cui X, Yang Y (2006) Adaptively robust filtering with classified adaptive factors. Proc Natl Acad Sci U S A 16(8):846–851

Koch KR, Yang Y (1998) Robust Kalman filter for rank deficient observation model. J Geophys Res 72(8):436–441

Mohamed AH, Schwarz KP (1999) Adaptive Kalman filtering for INS/GPS. J Geophys Res 73(4):193–203

Moore M, Wang J (2001) Adaptive dynamic modeling for kinematic positioning. IAG Assembly, Budapest

Ou J, Chai Y, Yuan Y (2004) Adaptive filter for kinematic positioning by selection of the parameter weights. In: Progress in geodesy and geodynamics, Hubei Science & Technology Press, pp 816–823

Ren C, Ou J, Yuan Y (2005) Application of adaptive filtering by selecting the parameter weight factor in precise kinematic GPS positioning. Proc Natl Acad Sci U S A 15(1):41–46

Xia Q, Sun Y, Zhou C (1990) An optimal adaptive algorithm for fading Kalman filter and its application. Acta Automatic Sinica 16(3):210–216 (in Chinese)

Xu T, Yang Y (2000) Modified Sage adaptive filtering. Sci Surv Mapping 25:22–24

Yang Y (1991) Robust Bayesian estimation. Bull Geodesique 65:145–150

Yang Y (1997) Robust Kalman filter for dynamical system. J Zhengzhou Inst Surv Mapping 14(2):79–84 (In Chinese)

Yang Y (1999) The basis in theory and comparisons of various robust filter models. In: Collection papers of geodesy, Surveying and Mapping Press, Beijing, pp 51–57 (in Chinese)

Yang Y, Cui X (2008) Adaptively robust filter with multi adaptive factors. Surv Rev 40(309):260–270

Yang Y, Gao W (2004) Integrated navigation by using variance component estimates of multi-sensor measurements and adaptive weights of kinematic model information. In: Selected Papers for English Edition, Acta Geodaetica et Cartographica Sinica, pp 8–13

Yang Y, Gao W (2006) An optimal adaptive Kalman filter. J Geophys Res 80:177–183

Yang Y, Xu T (2003) An adaptive Kalman filter based on Sage windowing weights and variance components. J Navigation 56(2):231–240

Yang Y, Cheng MK, Shum CK et al (1999) Robust estimation of systematic errors of satellite laser range. J Geodesy 73:345–349

Yang Y, He H, Xu G (2001a) Adaptively robust filtering for kinematic geodetic positioning. J Geophys Res 75(2):109–116

Yang Y, Xu T, He H (2001b). On adaptively kinematic filtering. In: Selected Papers for English of Acta Geodetica et Cartographica Sinica, pp 25–32

Yang Y, Cui X, Gao W (2004) Adaptive integrated navigation for multi-sensor adjustment outputs. J Navigation 57(2):287–285

Alternative Nonlinear Filtering Techniques in Geodesy for Dual State and Adaptive Parameter Estimation

H. Alkhatib

Abstract

In many fields of geodesy applications, state and parameter estimation are of major importance within modeling of on-line processes. The fundamental block of such processes is a filter for recursive estimation. Kalman Filter is the well known filter, a simple and efficient algorithm, as an optimal recursive Bayesian estimator for a somewhat restricted class of linear Gaussian problems. However, in the case that state and/or measurement functions are highly non-linear and the density function of process and/or measurement noise are non-Gaussian, classical filters do not yield satisfying estimates. So it is necessary to adopt alternative filtering techniques in order to provide almost optimal results. A number of such filtering techniques will be reviewed in this contribution, but the main focus lays on the sequential Monte Carlo (SMC) estimation. The SMC filter (well known as particle filter) allows to reach this goal numerically, and works properly with nonlinear, non-Gaussian state estimation. The main idea behind the SMC filter is to approximate the posterior PDF by a set of random particles, which can be generated from a known PDF. These particles are propagated through the nonlinear dynamic model. They are then weighted according to the likelihood of the observations. By means of the particles the true mean and the covariance of the state vector are estimated. However, the computational cost of particle filters has often been considered as their main disadvantage. This occur due to the large, sufficient number of particles to be drawn. Therefore a more efficient approach will be presented, which is based on the combination of SMC filter and the Kalman Filter. The efficiency of the developed filters will be demonstrated through application to a method for direct georeferencing tasks for a multi-sensor system (MSS).

Keywords

Nonlinear filtering in state space • Kalman filter • Bayesian filter • Sequential Monte Carlo filter • Multi-sensor system

1 Introduction

The Kalman filtering technique is used in geodesy especially in applications such as engineering navigation and deformation analysis. Both kinds of application require a sequential estimation of the system state based on information coming from a (dynamic) model of the system and from external observations. Over the years refined models were introduced for the system description (i.e., the trajectory of a vehicle), and a great variety of sensors was used to observe the system state. This yielded often nonlinearities in the equations of the Kalman filter. In the field of engineering navigation for example, Aussems (1999) describes the vehicle's trajectory by a refined model of consecutive arcs. In this model the

H. Alkhatib (✉)
Geodetic Institute Leibniz Universität Hannover, Nienburger Str. 1, 30167 Hannover, Germany
e-mail: alkhatib@gih.uni-hannover.de

H. Kutterer et al. (eds.), *The 1st International Workshop on the Quality of Geodetic Observation and Monitoring Systems (QuGOMS'11)*, International Association of Geodesy Symposia 140,
DOI 10.1007/978-3-319-10828-5_16, © Springer International Publishing Switzerland 2015

vehicle coordinates are nonlinearly related to other state parameters like the angular velocity in the horizontal plane, the pitch angle or the tangential component of the linear velocity. Another approach that may lead to nonlinear state equations is the adaptive enhancement of the Kalman filter. For the estimation of the thermal diffusivity of an aluminum column Eichhorn (2008) introduces this physical parameter into the state vector. Thus, the initially linear state equations become nonlinear at the order of reciprocal exponentials. A majority of the performed observations of the system are nonlinearly related to its state parameters; refer to Sternberg (2000) for a disquisition on observation equations in engineering navigation. Non-linearities can occur directly due to the nature of the relation between measurements like distances or angles and point coordinates or due to the transformation of the observations in a different coordinate system as is often the case in navigation applications. Hence over the years a multitude of approximate nonlinear filters has been proposed, see e.g. Bar-Shalom et al. (2001), and Simon (2006). These methods can be loosely grouped into the following two main categories:

- *Gaussian approximate methods:* A well known analytical approximation to handle a nonlinear system is to linearize the measurement and the system equations using Taylor series expansions; see Gelb (1974). However, as pointed out in Doucet et al. (2001) this type of nonlinear filter which includes the first-order and the higher-order extended Kalman filter (EKF), is tended to diverge if the system equations are highly nonlinear. Another method to overcome the linearization process is the Unscented Kalman filter (UKF). The UKF has been introduced in Julier and Uhlmann (1997) and approximates the first and the second moment of the posterior PDF rather than to approximate nonlinear functions.
- *Sequential Monte-Carlo methods:* The SMC filter (also known as particle filter) is a suboptimal filter for implementing the recursive Bayesian filter by Monte Carlo (MC) techniques, see e.g. Doucet et al. (2001) and Ristic et al. (2004). The main idea behind the SMC filter is to approximate the posterior PDF by a set of random samples, which can be generated from a known PDF. These generated particles are propagated through the dynamic model. They are then weighted according to the likelihood of the observations. In a resampling step, the new weighted particles are drawn with a probability proportional to the likelihood of the observation. By means of the drawn particle the true mean and the covariance of the state vector are estimated. If we assume that the number of drawn particles is very large, the MC approximation becomes an equivalent representation to the functional description of the posterior PDF.

In this paper nonlinear filtering approaches are considered. We focus only on the second category, namely the SMC filter. The problem of high computational cost due to the large required number of particles has been solved in Alkhatib et al. (2012). The approach that has been proposed in Alkhatib et al. (2012) to overcome the computational effort and improve the performance of the filtering process was to combine the SMC filter with one of the filter from the Gaussian approximate methods such as EKF. In this paper, an extension of the aforementioned algorithm [proposed in Alkhatib et al. (2012)] to the estimation of state and static or minimal time-varying unknown model parameters (adaptive parameters) is introduced. The estimation of both the dynamic state and the static parameters is commonly known as the dual estimation. Numerous papers have been dealt with the developing of estimation algorithms based on SMC methods, refer, e. g., to Storvic (2002), which considered models with sufficient statistics for the parameters and applied particle filters to an augmented vector of states and sufficient statistics. The used strategy here is based on Storvic (2002) by adding random walk to the parameters, and then expand the state space with the extended parameters for the dual estimation.

The paper is organized as follows. Section 2.1 describes the mathematical model of the system which includes both the system dynamics and the measurement model. Section 2.2 briefly introduces the EKF and the UKF. The SMC filter is presented in Sect. 2.3. Two numerical simulations are demonstrated in Sect. 3. Finally, section "Conclusion" summarizes the results and gives an outlook for future work.

2 Nonlinear State Estimation

2.1 The Probabilistic Inference

Probabilistic inference is the problem of estimating states or parameters of a system in an optimal and consistent approach (using probability theory) given noisy observations. This general framework is shown in Fig. 1. In particular, we will be addressing the sequential (recursive) probabilistic inference problem within discrete-time nonlinear dynamic systems that can be described by a dynamic state-space model (DSSM). The hidden system state x_k, with initial probability density $p(x_0)$, evolves over time (k is the discrete time index) as an indirect or partially observed first order Markov process according to the conditional probability density $p(x_k|x_{k-1})$. The observations y_k are conditionally independent given the state and are generated according to the conditional probability density $p(y_k|x_k)$.

The DSSM can also be written as a set of nonlinear system equations

$$\mathbf{x}_k = f(\mathbf{x}_{k-1}, \mathbf{u}_{k-1}, \mathbf{w}_{k-1}) \quad (1)$$

$$\mathbf{y}_k = h(\mathbf{x}_k, \mathbf{v}_k) \quad (2)$$

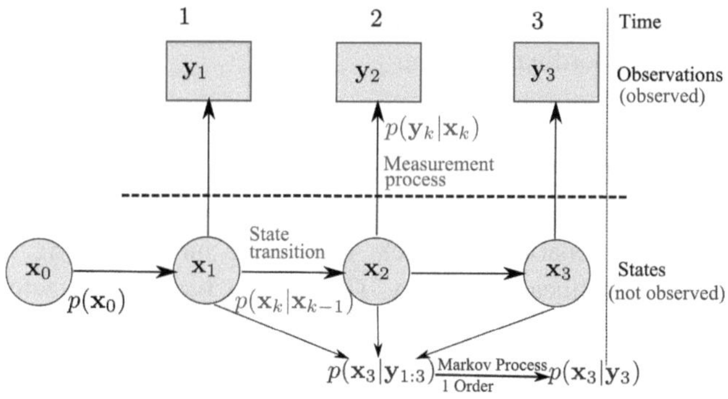

Fig. 1 Graphical model of a probabilistic dynamic state-space model

where $\mathbf{v_k}$ is the process noise that drives the dynamic system through the nonlinear state transition function f, and $\mathbf{n_k}$ is the observation or measurement noise corrupting the observation of the state through the nonlinear observation function h. The state transition density $p(\mathbf{x_k}|\mathbf{x_{k-1}})$ is fully specified by f and the process noise distribution $p(\mathbf{v_k})$, whereas h and the observation noise distribution $p(\mathbf{v_k})$ fully specify the observation likelihood $p(\mathbf{y_k}|\mathbf{x_k})$. The exogenous input to the system, $\mathbf{u_k}$, is assumed known.

2.2 The Bayes Filter

From a Bayesian perspective, the filtering problem is to estimate the state $\mathbf{x}_{k+1}$ recursively given the data $\mathbf{y}_{1:k+1}$ up to time $k+1$. Thus, it is required to evaluate the joint posterior PDF given the hole data. That is:

$$p(\mathbf{x}_{k+1}|\,\mathbf{y}_{1:k+1}) = \frac{p(\mathbf{y}_{k+1}|\,\mathbf{x}_{k+1}) \cdot p(\mathbf{x}_{k+1}|\,\mathbf{y}_{1:k})}{p(\mathbf{y}_{k+1}|\,\mathbf{y}_{1:k})} \quad (3)$$

where the posterior PDF at time k, $p(\mathbf{x}_k|\,\mathbf{y}_{1:k})$, is first projected forward in time in order to calculate the prior PDF at time $k+1$. This is done by using the probabilistic process model (cf. Simon 2006, pp. 464):

$$p(\mathbf{x}_{k+1}|\,\mathbf{y}_{1:k}) = \int p(\mathbf{x}_{k+1}|\,\mathbf{x}_k) \cdot p(\mathbf{x}_k|\,\mathbf{y}_{1:k}) d\mathbf{x}_k. \quad (4)$$

The probabilistic model of the state evolution $p(\mathbf{x}_{k+1}|\,\mathbf{x}_k)$ is defined by the system described in Eq. (1) and the known PDF of the noise vector $\mathbf{w}_k$. The term $p(\mathbf{y}_{k+1}|\,\mathbf{y}_{1:k})$ in Eq. (3) is a normalizing factor.

2.3 The Extended Kalman Particle Filter

On approach that has been proposed for improving particle filtering is described in Alkhatib et al. (2012) which combine the generic particle filter with the extended Kalman filter. In this approach each particle is updated ate the measurement time using the EKF, and the resampling is performed using the measurement equation. The main goal in this approach was the improving the numerical efficiency of the SMC filter.

The main drawback of the SMC filter is its computational cost which leads to increasing the sufficient number of particles to approximate the statistical moments of the state parameter and the adaptive parameter as well. In Ristic et al. (2004) and Simon (2006), several implementation issues are considered for improving the PF algorithm, including degeneracy, the selection of the importance density, and particle filters with an improved sample diversity. Due to the lack of space we discuss here only the developed approach for enhancement of convergence based on combination with the well known Kalman filter such as the extended Kalman filter (EKF). The novelty of the proposed EKPF algorithm was the update of each particle at every time step k using the EKF, when a new measurement $\mathbf{y}_k$ arrives. In other words, we are running an extra EKF step for every particle i:

$$\begin{aligned}
\mathbf{P}^-_{k+1,i} &= \mathbf{F}_{k,i}\mathbf{P}^+_{k,i}\mathbf{F}^T_{k,i} + \mathbf{Q}_k \\
\mathbf{K}_{k+1,i} &= \mathbf{P}^-_{k+1,i}\mathbf{H}^T_{k+1,i}\left(\mathbf{H}_{k+1,i}\mathbf{P}^-_{k+1,i}\mathbf{H}^T_{k+1,i} + \mathbf{R}_{k+1}\right)^{-1} \\
\mathbf{x}^+_{k+1,i} &= \mathbf{x}^-_{0,i} + \mathbf{K}_{k+1,i}\left[\mathbf{y}_{k+1} - \mathbf{h}(\mathbf{x}^-_{k+1,i})\right] \\
\mathbf{P}^+_{k+1,i} &= (\mathbf{I} - \mathbf{K}_{k+1,i}\mathbf{H}_{k+1,i})\,\mathbf{P}^-_{k+1,i}.
\end{aligned} \quad (5)$$

$\mathbf{K}_{k+1}$ represents the Kalman gain of the i-th particle, and $\mathbf{P}_{k+1,i}$ is the appropriate estimation of the state covariance matrix. We distinguish in Eq. (5) between the a priori $\mathbf{P}^-_{k+1,i}$ and the a posteriori $\mathbf{P}^+_{k+1,i}$. The transition and design matrices $\mathbf{A}_k$ and $\mathbf{H}_k$ in Eq. (5) are defined as:

$$\mathbf{A}_k = \frac{\partial f}{\partial \mathbf{x}}\,|_{\mathbf{x}=\mathbf{x}^+_k} \quad \text{and} \quad \mathbf{H}_k = \frac{\partial h}{\partial \mathbf{x}}\,|_{\mathbf{x}=\mathbf{x}^-_k}, \quad (6)$$

respectively. The key idea behind this approach is the substitution of the possibly non-linear model given by Eq. (1) with a linearized model to reduce the variance of the drawn particles in order to get short computing times without increasing the number of samples.

The generated prior particles $\mathbf{x}^-_{k,i}$ would be transformed to a new set of particles $\mathbf{x}^+_{k,i}$ using the EKF step given by Eq. (5). Based on the transformed particles $\mathbf{x}^+_{k,i}$ and their covariance matrix $\mathbf{P}^+_{k+1,i}$ we generate and propagate a new set of particles using the Gaussian importance PDF:

$$\mathbf{x}_{k+1,i} \sim p(\mathbf{x}^i_{k+1}|\mathbf{x}^i_k, \mathbf{y}_{k+1}) \sim \mathcal{N}(\mathbf{x}^+_{k+1,i}, \mathbf{P}^+_{k+1,i}). \quad (7)$$

where the symbol $\sim$ in Eq. (7) means that the particles are generated from a specific PDF (in this case the Gaussian PDF). The remaining computational steps of the EKPF are similar to the generic PF, see Alkhatib et al. (2012).

2.4 Combined Parameter and State Estimation in EKPF Algorithm

The algorithm in Sect. 3.2 has to be developed in order to estimate the static unknown parameters. The maximum likelihood parameter estimation should be then performed based on PF and an effective stochastic approximation gradient algorithm is used to optimize cost function. The estimation of static parameters and dynamic state variables is performed simultaneously. The algorithm shown in this paper is adapted from Yang et al. (2008).

The state-space model should be extended by the static parameters $\boldsymbol{\vartheta}$:

$$\mathbf{x}_k \sim p\left(\mathbf{x}_k | \mathbf{x}_{k-1}, \boldsymbol{\vartheta}\right) \tag{8}$$

$$\mathbf{y}_k \sim p\left(\mathbf{y}_k | \mathbf{x}_k, \boldsymbol{\vartheta}\right) \tag{9}$$

The static parameter $\boldsymbol{\vartheta} \in \mathbb{R}^m$ with m dimensional unknown static parameters vector. The method introduced here focuses rather on the estimation of $\boldsymbol{\vartheta}$ directly by the maximum likelihood method. In other words, the dynamic state parameters are be estimated by the SMC filter and static parameters are estimated by recursive ML method online.

The cost function to estimate the likelihood is given by:

$$\begin{aligned} f(\vartheta) = & p\left(\mathbf{y}_k | \mathbf{y}_{0:k-1}, \boldsymbol{\vartheta}\right) \\ & \int p\left(\mathbf{y}_k | \mathbf{x}_k, \boldsymbol{\vartheta}\right) p\left(\mathbf{x}_k | \mathbf{y}_{0:k-1}, \boldsymbol{\vartheta}\right) d\mathbf{x}_k. \end{aligned} \tag{10}$$

A closed-form of the integral given in (10) expect for only fey special cases is impossible. Numerical methods will be approximate the optimal solution. In every filter step, the cost function will be maximized. The problem here is to find the zeros of the gradient $\nabla f(\boldsymbol{\vartheta})$:

$$\vartheta_k = \boldsymbol{\vartheta}_{k-1} + \gamma_k \hat{\nabla} f(\boldsymbol{\vartheta}_{k-1}). \tag{11}$$

In (11) $\hat{\nabla} f(\boldsymbol{\vartheta}_{k-1})$ denotes the estimated value of the gradient in the point $\boldsymbol{\vartheta}_{k-1}$ and γ_k denotes a sequence of decreasing step size. After a sufficient number of iterations, the true value of ϑ_k will be estimated. The gradient estimate in (11) is obtained by numerical approximation, for more details see Yang et al. (2008).

3 Numerical Applications

We compare in this section the performance of the different nonlinear filter techniques presented in Sect. 2 by two numerical experiments. The first one is a simple target tracking problem, which is typically arises in navigation, and the second is to derive position and orientation parameters for the transformation of a local sensor-defined coordinate system to a global coordinate system.

3.1 Tracking of a Nonlinear Trajectory

In this scenario we track a vehicle moving along a highly nonlinear trajectory. Consider a base state vector $\mathbf{x} = [x_k \; y_k \; \dot{x}_k \; \dot{y}_k]$, where x_k and y_k specify the position of the vehicle; and $\dot{x}_k$ and $\dot{y}_k$ are the velocities in the Cartesian plane. The system is given by a linear kinematic model and a nonlinear measurement model.

Dynamic system: The dynamic system (cf. Eq. 1) of the vehicle can be modeled with a discretized Wiener velocity model, cf. Bar-Shalom et al. (2001) and Särkkä (2006):

$$\mathbf{x}_{k+1} = \begin{bmatrix} x_{k+1} \\ y_{k+1} \\ \dot{x}_{k+1} \\ \dot{y}_{k+1} \end{bmatrix} = \begin{bmatrix} 1 & 0 & \Delta t & 0 \\ 0 & 1 & 0 & \Delta t \\ 0 & 0 & 1 & 0 \\ 0 & 0 & 0 & 1 \end{bmatrix} \tag{12}$$

$$\boldsymbol{\Sigma}_{\mathbf{w}_k\mathbf{w}_k} = \begin{bmatrix} \frac{\Delta t^3}{3} & 0 & \frac{\Delta t^2}{2} & 0 \\ 0 & \frac{\Delta t^3}{3} & 0 & \frac{\Delta t^2}{2} \\ \frac{\Delta t^2}{2} & 0 & \Delta t & 0 \\ 0 & \frac{\Delta t^2}{2} & 0 & \Delta t \end{bmatrix} \cdot q, \tag{13}$$

where $\mathbf{w}_k$ is Gaussian with $E(\mathbf{w}_k) = \mathbf{0}$ and $q = 0.1[m^2/sec^3]$ fixes the spectral density of the noise.

Measurement model: We assume that two sensors provide measurements of distance s_i and horizontal angle ϑ_i to the moving vehicle with a time discretization step of $\Delta t = 0.01\ sec$:

$$s_{i,k+1} = \sqrt{(x_{k+1} - X_0^i)^2 + (y_{k+1} - Y_0^i)^2} \tag{14}$$

$$\vartheta_{i,k+1} = \arctan\left(\frac{y_{k+1} - Y_0^i}{x_{k+1} - X_0^i}\right) \tag{15}$$

where X_0^i, Y_0^i are the locations of the sensors with $i \in \{1, 2\}$. The noise vector $\mathbf{v}_k \sim \mathcal{N}(\mathbf{0}, \boldsymbol{\Sigma}_{\mathbf{vv}})$ is characterized by $\boldsymbol{\Sigma}_{\mathbf{vv}} = diag[\sigma_{s_1}^2 \, \sigma_{\vartheta_1}^2 \, \sigma_{s_2}^2 \, \sigma_{\vartheta_2}^2]$.

Figure 2 shows the final tracking results for three different filtering techniques compared to the real trajectory. The prior standard deviation of the measurements given by $\sigma_{s_i} = 0.05[m]$ for distances and $\sigma_{\vartheta_i} = 0.01[rad]$ for angles. As we can see in Fig. 2 all filtering methods for such optimal measurement accuracies give almost identical estimates of the system states. However, Low Cost sensors have become prevalent in geodetic applications in the last years. Therefore high variances of the measurements have to be taken into account in the analysis process. If a more rigorous analysis of

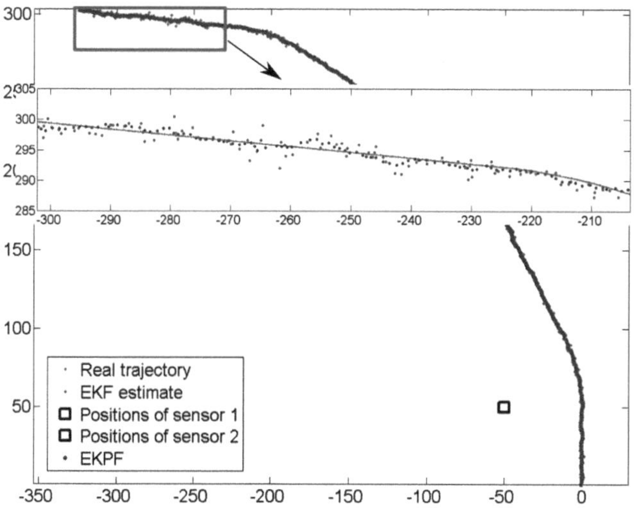

Fig. 2 Nonlinear filtering results of nonlinear trajectory

the measurements of low-cost sensors can be carried out, the acquisition costs of a sensor system can be decreased. From a theoretical point of view, the higher the variances the more important is a very good approximation of the nonlinear function (EKF) or of the PDF of the system (PF, EKPF). One would expect that in cases of very high variances the EKPF produces significantly more accurate results of the estimated system states. For this reason, the variances in this simulation study are increased by a factor of four. The RMSE values for all 500 runs are shown in Fig. 3. It is clearly noticeable that nearly all runs of the PF have a smaller RMSE in comparison to the EKF and UKF.

Naturally, a more realistic dynamic model may reduce the mismodelling effects due to approximation errors of the function or the PDF, but this is not in the scope of the study.

3.2 Tracking of a Nonlinear Trajectory

In this section an application of the algorithm presented in Sect. 2 is shown and the results are discussed. The main goal of the numerical investigation is to derive position and orientation parameters for the transformation of a local sensor-defined coordinate system to a global coordinate system. This is a typical task within the direct georeferencing procedure of static terrestrial laser scans. For this purpose, an adapted sensor-driven method based on a multi-sensor system (MSS) has been developed at the Geodetic Institute of the Leibniz Universität Hannover (GIH). The MSS is established by a sensor fusion of a phase-based terrestrial laser scanner (TLS) and additional navigation sensors to observe the parameters.

The above mentioned transformation parameters include the position of the MSS, which is equal to the translation vector and a rotation matrix, which contains the orientation of the three axes of the MSS—roll, pitch and yaw angle known from aeronautics. The MSS as well as the mathematical modeling in form of a Kalman filtering approach are presented in details in Paffenholz et al. (2010).

This approach uses the constant rotation of the TLS about its vertical axis with combination of kinematic GNSS measurements to estimate 4 of the 6° of freedom of the transformation—the position vector as well as the orientation in the horizontal plane. Therefore one GNSS antenna is mounted eccentrically on the TLS. In order to optimize the direct georeferencing strategy the MSS is enhanced with additional navigation sensors—inclinometers—to estimate the residual spatial rotation angles about the x- and y-axis of the TLS.

In this MSS application the trajectory can be described by a circle in 3D space. This parameterization is due to the orbital motion of the antenna reference point (ARP) caused by the constant rotation of the TLS about its vertical axis, as already mentioned. The orientation change of the ARP within two time steps is given by the circular arc segment s divided by the radius r_k.

The state vector is expressed by the following components:

$$\mathbf{x}_k^G = \left[\mathbf{X}_k^G \; \alpha_{S,k}^G \; \beta_{S,k}^L \; \gamma_{S,k}^L\right]^T \tag{16}$$

where $\mathbf{X}_k^G$ is the global position of the ARP at the epoch k, $\alpha_{S,k}^G$ describes the azimuth orientation of the MSS, $\beta_{S,k}^L$ the inclination in scan direction and $\gamma_{S,k}^L$ is perpendicular to the scan direction. The indices G and L in Eq. (16) are refereed to the global or to the local coordinate system, respectively. The space state model leads to:

$$\mathbf{x}_{k+1} = \begin{bmatrix} \mathbf{X}_k^G + \mathbf{R}_L^G(\lambda,\varphi)\cdot\mathbf{R}_{SN}^{aP}\left(\alpha_{S,\,k}^G\right)\cdot\boldsymbol{\Delta X}_k^L \\ \alpha_{S,k}^G + \frac{s_k}{r_k} \\ \beta_{S,k}^L \\ \gamma_{S,k}^L \end{bmatrix} \mathbf{w}_k. \tag{17}$$

The term $\Delta\mathbf{X}_k^L$ in Eq. (17) is given by:

$$\Delta\mathbf{X}_k^L = \begin{bmatrix} r_k\cdot\cos\left(\gamma_{S,k}^L\right)\cdot\sin\left(\frac{s_k}{r_k}\right) \\ r_k\cdot\cos\left(\gamma_{S,k}^L\right)\cdot\cos\left(\frac{s_k}{r_k}\right) \\ s_k\cdot\sin\left(\beta_{S,k}^L\right) \end{bmatrix} - \boldsymbol{\Delta X}_{SN,k}^{GNNS} \tag{18}$$

where $\boldsymbol{\Delta X}_{SN,k}^{GNNS}$ represents the eccentric position of the GNSS antenna. $\Delta\mathbf{X}_k^L$ in Eq. (17) is responsible for the high-nonlinearity in the space state model. It should be pointed out, that in Paffenholz et al. (2010) additional adaptive

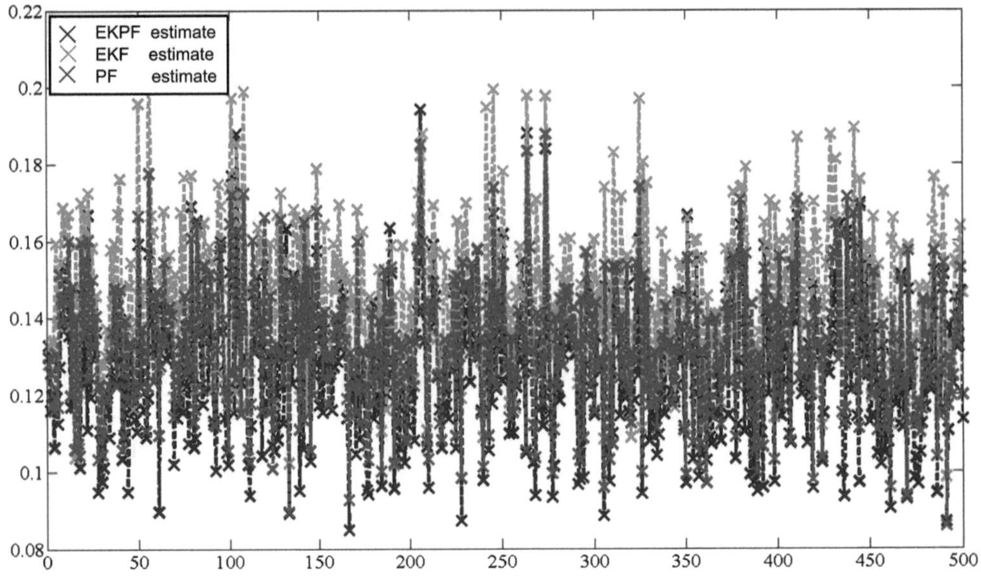

Fig. 3 RMSE values for the nonlinear filtering results of the 500 runs

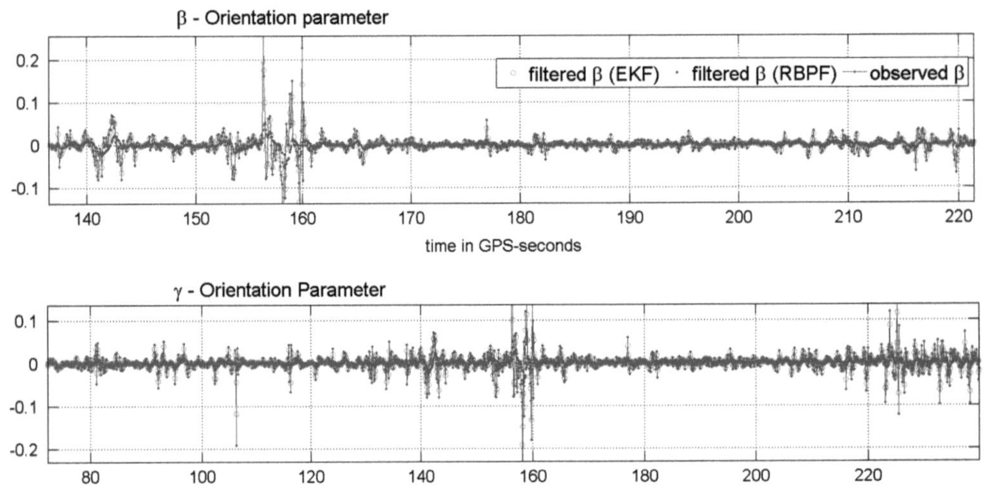

Fig. 4 Filtering results of EKF and the EKPF; *top* the residuals obtained within a linear regression of the orientation α_S^G; *middle and bottom* the filtered inclinations β_S^L and γ_S^L

parameters are considered in the space state model. However, the consideration of such adaptive parameters in the EKPF algorithm (refer to Sect. 3.2) needs a significant modification, which will be shown in future works.

The measurement model is characterized by the position of the GNSS antenna $\mathbf{X}_k^G$, and the measurements of the inclination sensor $\beta_{S,k}^L$ and $\gamma_{S,k}^L$. This yields to:

$$\mathbf{y}_{k+1} = \begin{bmatrix} \mathbf{X}_{k+1}^G \\ \beta_{S,k+1}^L \\ \gamma_{S,k+1}^L \end{bmatrix} = \mathbf{A}_{k+1} \cdot \mathbf{x}_{k+1}^G + \mathbf{v}_{k+1}. \quad (19)$$

The state vector [refer to Eq. (17)] has been initialized on $\mathbf{x}_0 = \mathbf{0}$ with the initial covariance matrix which has been chosen equally to the noise covariance matrix in Alkhatib et al. (2012):

$$\mathbf{P}_0 = \text{diag}\,[100\,\text{mm}^2 \quad 100\,\text{mm}^2 \quad 900\,\text{mm}^2 \ldots$$
$$0.01\,\text{mg rad}^2 \quad 1\,\text{mg rad}^2 \quad 1\,\text{mg rad}^2].$$

As start value for the EKPF, we randomly draw 500 particles from $\mathcal{N}(\mathbf{x}_0, \mathbf{P}_0)$.

Figure 4 presents a subsample of the estimated state parameters by classical EKF algorithm and EKPF. The upper part of this figure shows the residuals obtained within a linear regression of the orientation α_S^G. Due to the constant rotation of the TLS about its vertical axis, we expect a linear relationship between α_S^G and time. Therefore, the residuals are quality indicators. The residuals are in both algorithms comparable, and lead to a metric uncertainty of about 1.5 cm for the azimuth calculation at a distance up to 35 m. The middle and lower part of Fig. 4 show a comparison between the filtered inclinations β_S^L and γ_S^L, respectively. Here again, the EKPF effect is noticeable for the filtered inclinations, mainly in case of higher noise level.

Conclusion

In this paper, the newly developed EKPF algorithm for non-linear dynamic systems with adaptive static parameter. The algorithm is based on the combination of the newly developed EKPF algorithm and the gradient techniques. The algorithm has been applied to derive transformation parameters for the direct georeferencing of terrestrial laser scans. The EKPF algorithm is based on a combination of the SMC techniques and an EKF step, which guarantees a faster convergence. The results of the developed filter show an improvement of the filter effect. The EKPF with adaptive parameter shows a better performance in case of high-nonlinear space state equations.

References

Alkhatib H, Paffenholz J-A, Kutterer H (2012) Sequential Monte Carlo Filtering for nonlinear GNSS trajectories. In: Nico S (ed) VII Hotine-Marussi Symposium on mathematical geodesy. Proceedings of the Symposium in Rome, 6–10 June, 2009, pp.81–86. Springer (International Association of Geodesy Symposia, 137), Berlin/New York

Aussems T (1999) Positionsschätzung von Landfahrzeugen mittels KALMAN-Filterung aus Satelliten- und Koppelnavigationsbeobachtungen. Veröffentlichungen des Geodätischen Instituts derRheinisch-Westfälischen Technischen Hochschule Aachen, Nr. 55, Aachen

Bar-Shalom Y, Li XR, Kirubarajan T, Li X-R (2001) Estimation with applications to tracking and navigation. Theory algorthims and software. Wiley, New York

Doucet A, Freitas N, Gordon N (2001) Sequential Monte Carlo methods in practice. Springer, New York

Eichhorn A (2008) Analysis of dynamic deformation processes with adaptive Kalman-filtering. J Appl Geodesy 1(1):9–15

Gelb A (1974) Applied optimal estimation. MIT, Cambridge

Julier SJ, Uhlmann JK (1997) A new extension of the Kalman filter to nonlinear systems. In: SPIE Proceedings of AeroSense. The 11th International Symposium on Aerospace/Defense Sensing, Simulation and Controls. SPIE, Orlando, FL, USA

Paffenholz J-A, Alkhatib H, Kutterer H (2010) Direct georeferencing of a static terrestrial laser scanner. J Appl Geodesy 4(3):115–126

Ristic B, Arulampalam S, Gordon N (2004) Beyond the Kalman filter. Particle filters for tracking applications. Artech House, Boston

Särkkä S (2006) Recursive Bayesian inference on stochastic differential equations. Ph.D. thesis, Helsinki University of Technology

Simon D (2006) Optimal state estimation. Kalman, H infinity, and nonlinear approaches // Kalman, H [infinity] and nonlinear approaches. Wiley, Hoboken

Sternberg H (2000) Zur Bestimmung der Trajektorie von Landfahrzeugen mit einem hybriden Messsystem. Schriftenreihe des Studienganges Geodäsie und Geoinformation, Universität der Bundeswehr Mänchen, No. 67, Neubiberg

Storvic G (2002) Particle filters in state space models with the presence of unknown static parameters. IEEE Trans Signal Process 90(2):281–289

Yang X, Xing K, Shi K, Pan Q (2008) Joint parameter and state estimation in particle filtering and stochastic optimization. J Control Theory Appl 6(2):215–220

Part IV

Sensor Networks and Multi Sensor Systems in Engineering Geodesy

Parametric Modeling of Static and Dynamic Processes in Engineering Geodesy

A. Eichhorn

Abstract

In this paper, the main focus is set on the utilization of parametric methods for the quantification of causative relationships in static and dynamic deformation processes. Parametric methods are still 'exotic' in engineering geodesy but state of the art e.g. in civil and mechanical engineering. Within this context, an essential part is the physical (parametric) modeling of the functional relationships based on partial or ordinary differential equations using the corresponding numerical solutions represented by finite element (FE) or finite difference (FD) models.

The identification of a physical model is realised by combination with monitoring data. One important part of the identification includes establishing the deterministic model structure and estimating a priori unknown model parameters and initial respectively boundary conditions by filtering (e.g. adaptive Kalman-filtering). Major challenges are establishing the parametric model structure, quantifying disturbances and the identifiability of the model parameters which are possibly non-stationary. These challenges are discussed with the help of a practical example from engineering geology.

Keywords

Descriptive and causative view • GB-SAR • Mass movement • Parametric modeling • Static and dynamic deformation processes

1 Introduction

Modern engineering geodesy often deals with the observation, analysis and interpretation of time-dependent processes, e.g. in vehicle positioning and navigation, acquisition of geodata and monitoring of deformations. This may be performed from a 'descriptive point of view' describing only the temporal behaviour of the process (e.g. concerning trend, signal, noise, discussion of stationarity and error distribution etc.) or from a more 'causative point of view', taking also into account the causes (triggers) for the temporal behaviour of the process (see Fig. 1). Especially the second case aims to explaining and understanding the underlying mechanisms, which are responsible for the structure of the observed process and enables (within boundaries) the prediction of the future behaviour also with significantly changing triggers.

According to Heunecke (1995) and Welsch et al. (2000), causative relationships can be mathematically quantified either by

- Static models, where the time is not explicitly modeled (the process is modeled as a series of succeeding balanced states without considering the transition)
- or by dynamic models, where the time is explicitly modeled and also the transition between balanced states is described.

Dynamic models represent the most realistic and universal quantification of time-dependent processes. Nevertheless, they are often difficult to prepare. Especially in deformation

A. Eichhorn (✉)
TU Darmstadt Institute of Geodesy, Franziska-Braun Str. 7, 64287 Darmstadt, Germany
e-mail: eichhorn@geod.tu-darmstadt.de

H. Kutterer et al. (eds.), *The 1st International Workshop on the Quality of Geodetic Observation and Monitoring Systems (QuGOMS'11)*, International Association of Geodesy Symposia 140,
DOI 10.1007/978-3-319-10828-5_17, © Springer International Publishing Switzerland 2015

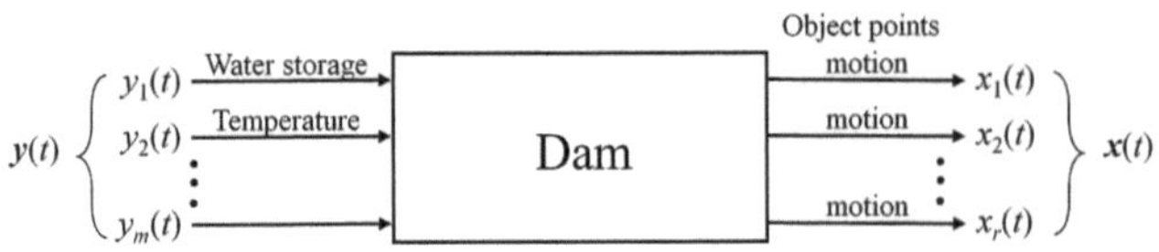

Fig. 1 Example: causative view of dam deformations

analysis, in many cases also the static representation of a process is sufficient, dependent on the goals of process analysis and interpretation. In the following, the focus is set on the quantification of parametric static and dynamic deformation processes. With the help of a practical example from engineering geology (quantification of a mass movement), the main differences between static and dynamic modeling and identification are presented.

2 Parametric Modeling of Deformation Processes

The quantification of causative deformation models can be performed either non-parametric or parametric (e.g. Isermann 1988; Heunecke 1995; Welsch et al. 2000). In general, non-parametric modeling is performed in a mathematically abstract way using regression or correlation relationships between the influences (triggers, e.g. temperature and mechanical loads) and the deformation quantities (e.g. movements of selected object points, see Fig. 1). The model parameters enable no or only a very restricted interpretation of the physical structure of the deformation process. Consequently, non-parametric models are called 'black box'- or 'grey box'-models (see Welsch et al. 2000). Typical examples are convolution integrals, ARMA, Artificial Neural Networks and Fuzzy approaches. Non-parametric models are directly derived from the monitoring data (pure experimental system analysis) and miss a theoretical system analysis based on physical justified laws (see Isermann 1988). Consequently,

- Their validity is restricted to a small range around the monitored temporal progress of influence and deformation quantities (in some cases, the processes are even presumed to be stationary, which means that no significant change in their stochastical properties is admissible),
- they show a very limited capability for prediction respectively simulation (theoretical experiments with significantly changing triggers)
- and they do not explicitly explain the physical mechanisms which are responsible for the deformations.

In contrast to non-parametric models, parametric deformation models base on physical laws derived in the framework of a theoretical system analysis. The common representation is performed by finite element (FE, e.g. Gallagher 1976) or finite difference (FD, e.g. Smith 1985) models. They are generated as numerical solutions from continuum mechanics and represent the physical structure of the deformation process ('white box'-models, see Welsch et al. 2000). Consequently, numerical models like FE and FD

- Can be also valid in ranges which are significantly different from the monitored deformation process,
- are capable to perform predictions also with significantly changing trigger data and to calculate simulations, this means to perform computational experiments, which could not be observed at the real object
- and are helpful to understand the physical mechanisms which are underlying the deformation process.

Quantifying a parametric model, major challenges are

- Designing the model structure itself (especially the load functions which connect triggers like air temperature, wind, rainfall with the physical structure of the object),
- to establish the initial and boundary conditions of the model
- and to determine the numerical values of a priori insufficiently known model parameters (e.g. material properties like rigidity or damping parameters).

In general, the quantification of model structure and initial respectively boundary conditions is part of the theoretical system analysis and is performed by an interdisciplinary cooperation, e.g. with experts from civil-engineering, geotechnics or engineering geology. In a second step, the monitoring data obtained at the real object is used to improve the theoretical model. One goal is to minimize the residuals between calculated and measured deformations adapting the model further to reality (including the numerical values of model parameters). This step is called 'parametric identification' and is one important subject of geodetic research in deformation analysis (e.g. Heunecke 1995; Welsch et al. 2000; Eichhorn 2005; Lienhart 2007).

One suitable tool for parametric identification is given by the application of adaptive Kalman-filtering techniques (e.g. Gelb et al. 1974; Heunecke 1995), which (principally) enable the optimal estimation of a priori insufficiently known model parameters. Until now, adaptive Kalman-filtering was successfully used in civil engineering in combination with static FE-models from artificial structures like dams (Gülal 1997), pylons (Heunecke 1996) and shell structures (Hesse 2002). In Lienhart (2007), the static FE-model of a monolithic bridge is embedded in the environmental soil structure.

Nevertheless until now, real dynamic applications for parametric identification are only performed for small and well described structures under lab conditions like thermal deformations of machine elements in mechanical engineering (Eichhorn 2005). The transfer to complex buildings and natural objects is still missing. Especially the quantification and identification of natural deformation processes like mass movements is still a big challenge.

3 Case Study: The Mass Movement 'Steinlehnen'

The mass movement 'Steinlehnen' (see Fig. 2) is located near Gries im Sellrain (Northern Tyrol, Austria), approximately 20 km southwest of Innsbruck and is part of the austroalpine polymetamorphic Ötztal-Stubai crystalline complex. The unstable mass consists of paragneisses, amphibolites and granodiorite-gneisses. The slope strikes to east with a dip angle between 30° in the lower and 40° in the upper part (Zangerl et al. 2007; Mair am Tinkhof et al. 2010). In its high active zone, the slope currently performs movements up to 2–3 dm per year (Schmalz et al. 2010).

The mass movement is investigated in the project 'KASIP' (knowledge-based alarm system with identified deformation predictor), which is funded by the FWF (Austrian Science Fund). The project contains an interdisciplinary cooperation between TU Vienna (Engineering Geodesy and Engineering Geology), TU Darmstadt (Engineering Geodesy and Physical Geodesy) and alpS—Center for Natural Hazard and Risk Management (Innsbruck). One important goal is the creation of a realistic numerical slope model including its parametric identification using collected monitoring data (see Kasip 2011).

Fig. 2 Mass movement 'Steinlehnen'

Fig. 3 GB-SAR system IBIS-L facing towards the mass movement 'Steinlehnen' (S. Rödelsperger)

3.1 Monitoring System

The monitoring data from 'Steinlehnen' consists of

- Airborne laser scanning data from 2003 (performed by alpS, when the slope movement started with rockfalls, now only used for DTM creation),
- tacheometer measurements to discrete slope points (prisms, mean accuracy $\sigma_p = 5$–8 mm) which are periodically performed (since 2003, ca. all 6 months) by alpS and TU Darmstadt from a fixed pillar which is mounted on the opposite slope,
- continuous radar measurements (all 7 min, accuracy of the phase measurements strongly depends on the atmospheric conditions, especially humidity, and may rise up to $\sigma_\phi = \pi/2 = 4$ mm with a wavelength $\lambda = 17.4$ mm) performed by the GB-SAR (ground based synthetic aperture radar) system IBIS-L (e.g. Rödelsperger et al. 2010) within two 5 and 4 weeks measuring campaigns in 2010 and 2011 (see Fig. 3)
- and meteorological trigger data (e.g. daily temperature and rainfall) from a weather station in Gries.

In contrast to the tacheometer measurements which are restricted to selected (and accessible) slope points, the radar measurements enable a contact-free laminar acquisition of deformations with a high temporal and spatial resolution.

In Fig. 4 some results from the radar measurements are shown. The figure shows the integrated displacements (in the direction of the line of sight to the radar system) of the slope surface after 10.5 days in June 2010. The displacements are overlaid to a DTM (digital terrain model) of the slope derived from airborne laser scanning in 2003. The blue zone indicates local movements up to 15 mm (see also Rödelsperger 2011).

3.2 Parametric Slope Model

Taking a closer look to the monitoring data collected by IBIS-L, it is possible to associate a certain dependency between meteorological conditions (triggers) and the slope deformations. In Fig. 5 this dependency is exemplarily presented for the temporal progress of five radar dissolution cells extracted from a high active area (blue zone in Fig. 4). During a phase of significant change of the meteorological conditions from heavy rainfall over slight snowfall to sunshine, the displacements are gradually integrated (integration steps $\Delta t = 7$ min) over a period of 25 days. Starting from balanced

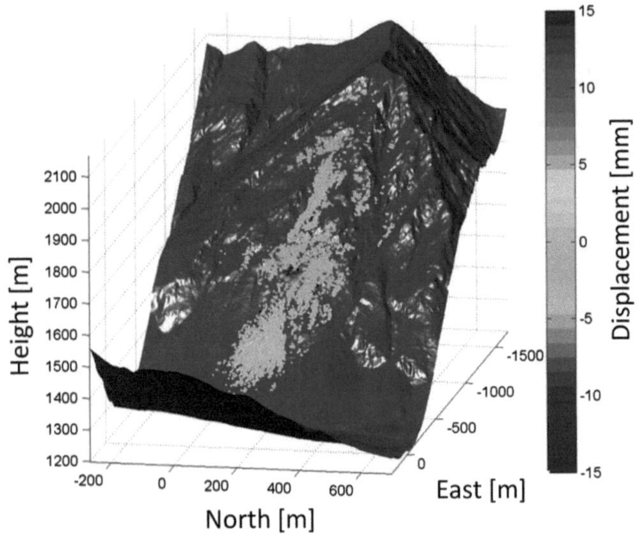

Fig. 4 DTM of 'Steinlehnen' with integrated displacements (KASIP measuring campaign 2010, S. Rödelsperger)

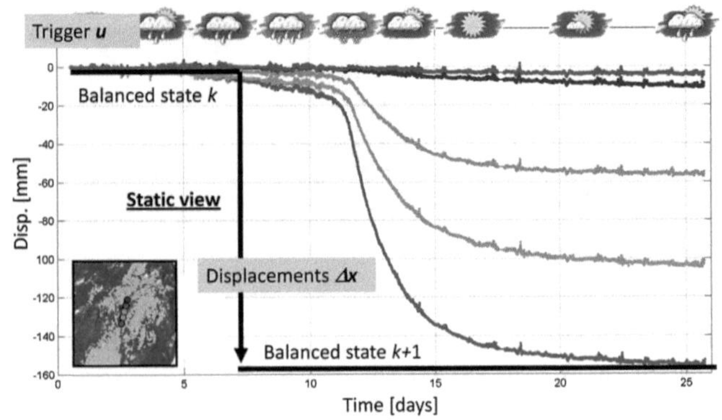

Fig. 5 Time series of five resolutions cells (KASIP measuring campaign 2010, S. Rödelsperger): integrated displacements over max. 25 days. Static view

states k, the displacements perform first accelerated and then decelerated motions to new balanced states $k + 1$.

The time series give a good impression concerning the typical behaviour of slope points as reaction to changing weather conditions. Learning from the observations, it is possible to establish two different views (see also Sect. 1). The 'static view' is linked with a resulting static deformation model, where only successive balanced states ($\ldots, k, k + 1, \ldots$) are considered, without modeling the time-dependent transition between the states (see Fig. 5). The static deformation model contains the quantification of the displacements $\boldsymbol{\Delta x}$ as reaction to the triggers $\boldsymbol{u}$.

The 'dynamic view' is linked with a resulting dynamic deformation model, where not only successive balanced states are modeled but also the time-dependent transition ($t_1 \ldots t_n$) between the states (see Fig. 6). The dynamic deformation model contains the quantification of the displacements

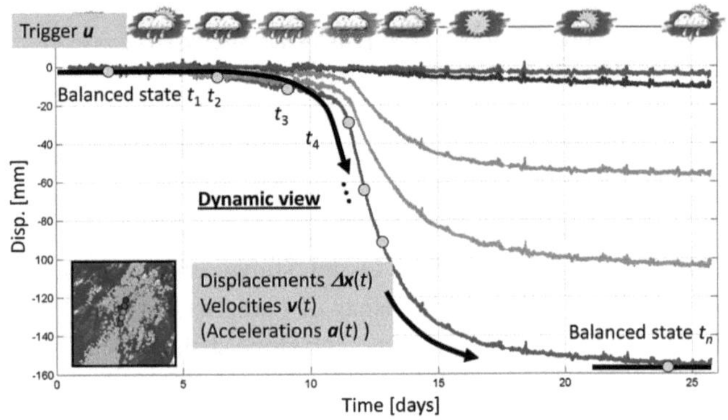

Fig. 6 The same time series but a dynamic view

$\boldsymbol{\Delta x}(t_k)$, their velocities $\boldsymbol{v}(t_k)$ and (if necessary) accelerations $\boldsymbol{a}(t_k)$ as reaction to the triggers $\boldsymbol{u}(t_k)$.

Considering the monitored motion of the slope surface, the possibilities for parametric static and dynamic modeling and identification are investigated in KASIP. The creation of a realistic parametric model of the mass movement is here realized on the base of finite differences (FD, see Smith 1985). The numerical modeling is performed by Engineering Geology (TU Vienna) with the FD-software 'FLAC3D' (Fast Lagrangian Analysis of Continua in three Dimensions) from Itasca Consulting Group Inc. (Itasca 2006). The software enables the calculation of 3D-continuum-models, using different elastic and plastic material models. The integration of dynamic extensions like material creeping makes FLAC3D suitable for static and for dynamic modeling.

In detail, the numerical model of the mass-movement is created in a theoretical system analysis based on the following pre-information.

- The geometry of the surface of the slope is derived from ALS in 2003.
- Movements at the surface of the slope are monitored (tacheometer measurements and visual inspection) by alpS since 2003.
- The geological structure of the slope (e.g. rock layers and zones of assumed homogenous material properties) are derived from geological mapping (Zangerl et al. 2007; Mair am Tinkhof et al. 2010).
- In the case of 'Steinlehnen' there are no material samples and lab experiments available (especially from the deeper layers), so the (initial) theoretical numerical values for the material parameters are taken from literature.

The dynamic numerical model of 'Steinlehnen' is just finished (see Fig. 7). Its geometrical dimensions are $1.4 \times 2.2 \times 1.2$ km (width $\times$ depth $\times$ height). It consists of seven homogenous zones with amphibolites, ortho- and paragneisses respectively rock mixtures. The discretization of the model is realized by finite meshes with a mean

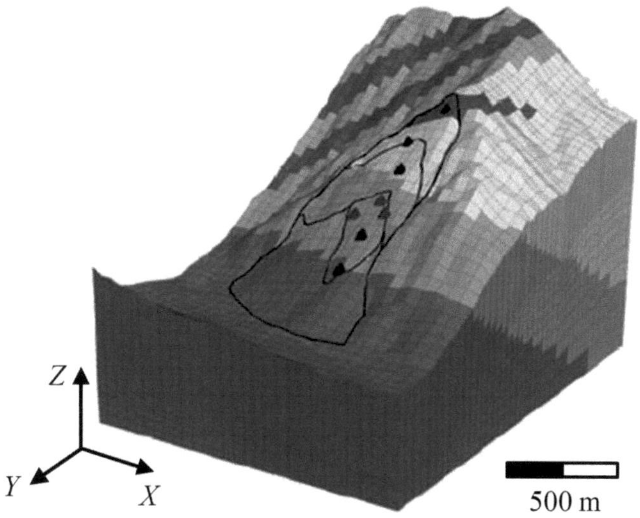

Fig. 7 FD-model of the mass movement 'Steinlehnen'

size of 25 m in the high active regions of the slope and up to 100 m in the boundary areas. The 3D-positions of their nodes represent the displacements. Each mesh is characterized by five mechanical parameters: density ρ, Young's modulus E, Poisson's ratio ν, angle of inner friction φ and cohesion c. These parameters are combined in a Mohr–Coulomb material model (e.g. Schmidt 2006) which quantifies amongst others the relationship between normal and critical shear stresses that may cause local or even global failure.

The failure mechanism is additionally 'supported' by a slide face in an assumed depth of approximately 100 m (Mair am Tinkhof et al. 2010). The integration of this mechanism is motivated by monitored shear strain rates at the surface. Its depth and extension is only hypothetical and may cause a significant model uncertainty. The model uncertainty respectively model quality represents one of the main restrictions for parametric identification. In the case of adaptive Kalman-filtering, it directly influences the parameter estimation. In the worst case, Kalman-filtering only performs a 'local' adaptation to the monitored deformation process and the identified model remains valid only for the monitoring data comparable with a non-parametric model. In the case of 'Steinlehnen', one method to proof the validity of the model could be to change the depth of the failure mechanism and to check the model sensitivity concerning this variation. Until now, this is not realized due to the extreme time-consuming modeling process.

One still unsolved challenge is the quantification of a direct functional relationship between external triggers (e.g. temperature and rainfall) and the mechanical properties of the model. Possibilities are given on the base of the integration of a groundwater table or pore-water pressure. Nevertheless, there's currently no monitoring data available.

In the existing model, the strength parameters (inner friction φ and cohesion c) of each mesh are assumed as indirect triggers. As a consequence of changing meteorological conditions, they change their numerical values and significantly influence the stability of the slope. To simulate a slope failure, the strength parameters are marginally reduced under certain boundary values (where the slope is just in a balanced state). This is very common in soil mechanics and called the 'strength reduction technique' (Mair am Tinkhof et al. 2010).

The determination of numerical values for the strength parameters φ and c is usually done by 'trial and error methods' (Mair am Tinkhof et al. 2010), visually matching the effects of the step by step performed strength reduction with available monitoring data. In the following, it is investigated how a parametric identification of the numerical model in terms of an optimal estimation of the strength parameters may support this process. As the FD-model of the full slope is just finished, no identification results with real monitoring data are currently available. So, the presented example deals with a simulation.

4 Parametric Identification of a Simplified Scarp

The following example for parametric identification deals with a simplified cut-out from the full slope model. The geometrical dimensions of the scarp are $30 \times 70 \times 40$ m (width $\times$ depth $\times$ height). To reduce the calculation time, the model is very rough, this means the displacements are calculated in only 700 nodes (in comparison to approximately 100,000 nodes in the full model). The scarp is assumed to be homogenous, this means there's only one set of material parameters needed. Like the full model, the motion of the scarp is 'triggered' by the strength parameters φ and c. The simplified dynamic FD-model is shown in Fig. 8.

Using reference values for the mechanical parameters, the displacements in the surface nodes of the front wall (red points in Fig. 8) are calculated and used as simulated 'monitoring' data. As shown in Sects. 2 and 3.2, the parametric identification can be performed under a static and a dynamic view.

4.1 Static Example

In the static example, the dynamic FD-model is used in a static way, this means only the total displacements between successive balanced states are calculated (see Fig. 9, green arrows). This procedure requires the variation of at least one of the two strength parameters as a sequence of static triggers u_k. In this example, the stepwise reduction of the

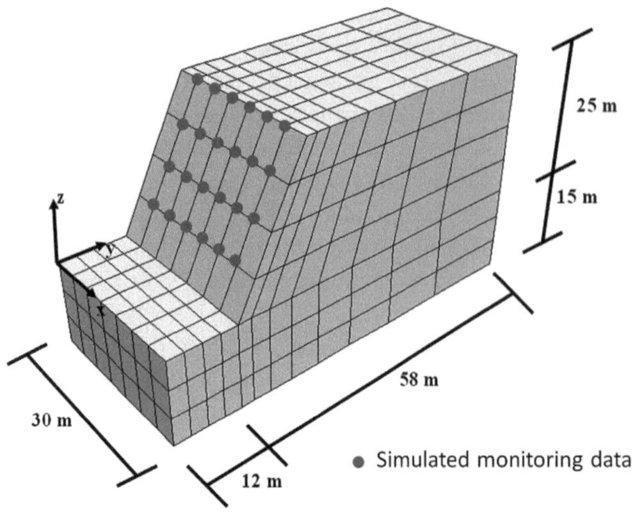

Fig. 8 Example for parametric identification: simplified scarp

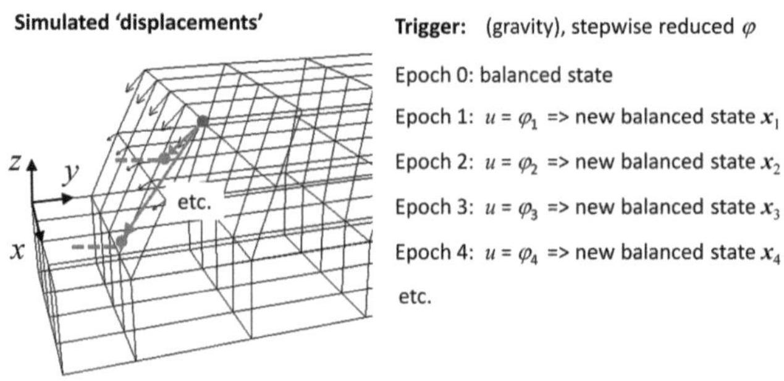

Fig. 9 Static view: simulation of displacements

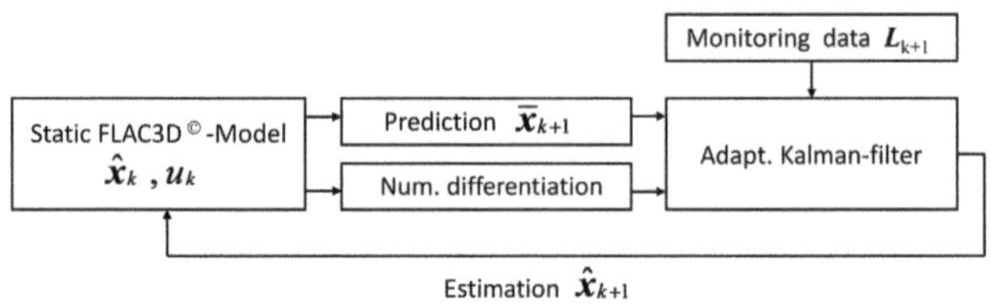

Fig. 10 Adaptive Kalman-filter for the static case

inner friction φ is chosen. Consequently, the balanced states $\boldsymbol{x}_k$ of the nodes are calculated by the sequence $u_k = \varphi_1, \varphi_2, \ldots, \varphi_n$, where k indicates the epoch. The state vector $\boldsymbol{x}_k$ contains the 3D-coordinates (x, y, z) of all nodes at epoch k. The observation (monitoring) vector $\boldsymbol{L}_k$ is a subset of $\boldsymbol{x}_k$, including only the surface nodes (see also (1)).

The second strength parameter c is assumed to be a priori unknown and is the target for the identification task. Its estimation is realized with an adaptive Kalman-filter, which contains the 'static' FD-model in the system equations and the simulated surface displacements $\boldsymbol{L}_k$ in the measuring equations (see also Fig. 10).

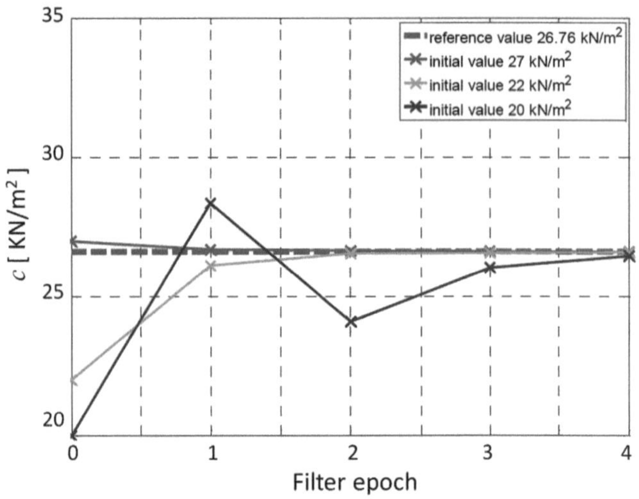

Fig. 11 Estimation progress of the cohesion c (according to Schmalz et al. 2010)

The state vector $\boldsymbol{x}_k$ is extended with an adaptive part, which contains the strength parameter c.

$$\begin{aligned} \boldsymbol{x}_k &= (x_1, \ldots, x_{700},\ y_1 \ldots y_{700}, z_1 \ldots z_{700} \mid c)_k^{\mathrm{T}} \\ \boldsymbol{L}_k &= (x_1, \ldots, x_{24},\ y_1 \ldots y_{24}, z_1 \ldots z_{24})_k^{\mathrm{T}} \end{aligned} \tag{1}$$

In each filter epoch, the prediction $\overline{\boldsymbol{x}}_{k+1}$ is calculated with the non-linear FD-model. The required matrices for the error propagation are obtained by numerical differentiation of the full FD-model.

Some results from the parametric identification of the static model are shown in Fig. 11.

Starting from different randomly initialized values c_0, the estimations $\widehat{c}$ of the cohesion converge against the reference value $c_{\mathrm{ref}} = 26.76$ kN/m^2. In the present case, the standard deviations of the different c_0 are derived from the (in the simulation) known deviation to the reference value and have the same order. In the case-study 'Steinlehnen', the relative error of material parameters derived from geological literature is generally assumed to be approximately 30 % (Zangerl et al. 2007). For each c_0, the estimation progress shows a stable behavior. After four filter epochs, the relative errors are <5 ‰ for all initializations. The precision of the results is sufficient for further stability investigations, e.g. calculating of realistic factors of safety (e.g. Roth 1999) for the scarp.

4.2 Dynamic Example

In the dynamic example, the time-dependent transition between two successive balanced states of the scarp is

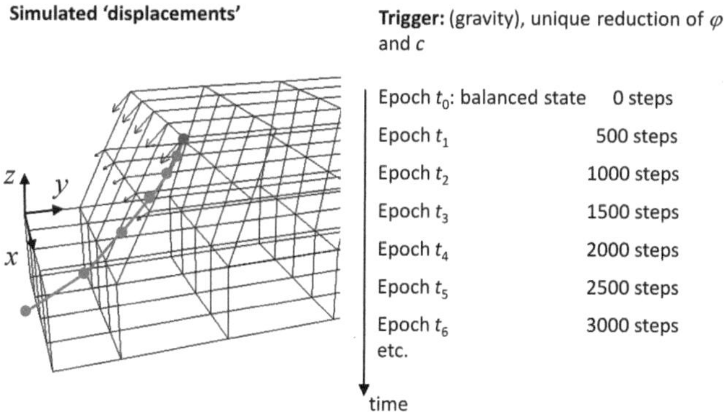

Fig. 12 Dynamic view: simulation of displacements

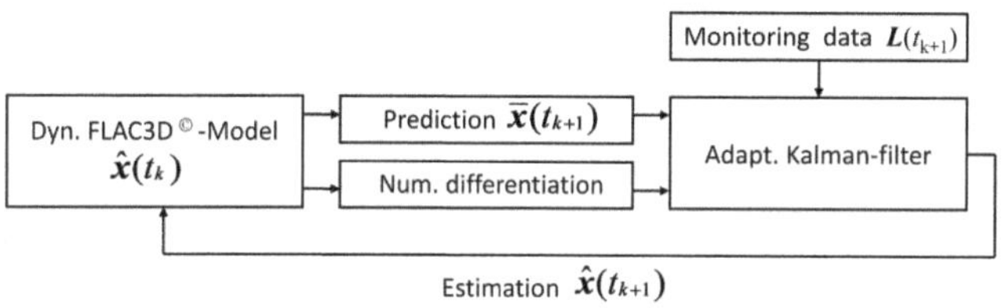

Fig. 13 Adaptive Kalman-filter for the dynamic case

calculated (see Fig. 12, green line). The motion is triggered by a unique reduction of φ and c. It has to be mentioned that FLAC3D calculates with an abstract time-unit expressed in 'steps'.

In the dynamic case, the state vector $\boldsymbol{x}(t_k)$ contains the 3D-coordinates (x, y, z) and the velocities (v_x, v_y, v_z) of all nodes at epoch t_k. In the present example, the accelerations of the nodes are small and can be neglected. The observation (monitoring) vector $\boldsymbol{L}(t_k)$ is a subset of $\boldsymbol{x}(t_k)$, including only the 3D-coordinates of the surface nodes [see also (2)].

$$\begin{aligned}\boldsymbol{x}\,(t_k) = &\ \Big(x_1,\ldots,x_{700},\ y_1\ldots y_{700}, z_1\ldots z_{700},\\ &\quad v_{x1},\ldots v_{x700}, v_{y1},\ldots v_{y700}, v_{z1},\ldots v_{z700}\ \big|\ \varphi,\ c\Big)^{\mathrm{T}}_{tk}\\ \boldsymbol{L}\,(t_k) = &\ (x_1,\ldots,x_{24},\ y_1\ldots y_{24}, z_1\ldots z_{24})^{\mathrm{T}}_{tk}\end{aligned} \tag{2}$$

As there is no definition of a varying trigger sequence like in the static example, both strength parameters can be assumed to be a priori unknown and are the targets for the identification task. Consequently, the adaptive part in (2) may contain both strength parameters φ and c. The filter strategy is shown in Fig. 13 (see also Sect. 4.1).

Some results from the parametric identification of the dynamic model are shown in Fig. 14. The estimated parameters show in each case a stable behaviour.

Like in the static example, the initialization c_0 and φ_0 of the strength parameters is performed randomly and their standard deviations are derived from the (in the simulation) known deviations to the reference values. After three filter epochs ($t_3 = 1{,}500$ time steps), the relative errors of the estimation results of the cohesion c are again within the range of some per mill. In contrast to c, the inner friction φ shows a more 'inertial' behaviour. The convergence process happens more slowly and the relative errors after four epochs ($t_4 = 2{,}000$ time steps) are within a range of 0.4–3 %.

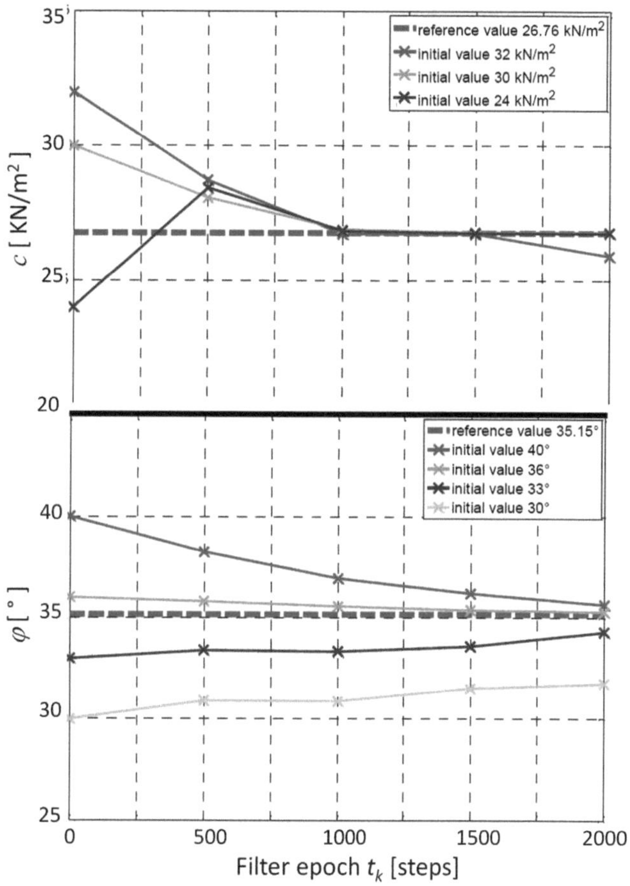

Fig. 14 Estimation progress of inner friction φ and cohesion c (according to Schmalz et al. 2010)

These results show clearly a worse identifiability of the friction in comparison to the cohesion. One reason for this is given by the monitoring data, which is (in this example) only available at the surface of the scarp and the slow progress of the deformations. This statement is supported comparing the monitored and predicted displacements of surface nodes in the progress of the filter. In Fig. 15, this is exemplarily shown for one surface node in z. Already after three filter epochs ($t_3 = 1{,}500$ time steps), the predicted and monitored z-coordinates of the node are fitting together within a range of some sub-mm, even if the estimation of φ does not 'hit' its reference value. This shows the low sensitivity of the monitoring data to changes in φ.

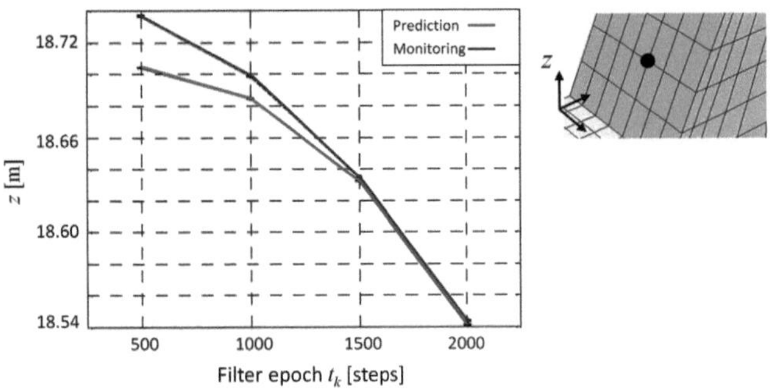

Fig. 15 Predicted versus monitored displacement (in z) in the progress of the filter (according to Schmalz et al. 2010)

Conclusions

The simulations performed in Sect. 4 show that a parametric identification of a numerical model is principally possible as well under static as under dynamic conditions. The next steps will be to leave the simulations and to integrate the real monitoring data (e.g. IBIS-L) in the numerical model. For this case, a further extension of the model structure with a combination of creeping and failure will be necessary which reflects the temporal behavior of the monitoring data shown in Fig. 6. This is currently performed by integrating a visco-elastic–plastic material model.

The decision for a static or dynamic Kalman-filter depends on the temporal density and progress of the available monitoring data. It is obvious, that the static case is more constraining because it requires a sequence of successive balanced states. Especially in the field of natural deformation processes, this may require a long-term monitoring and in the case of failure scenarios, there possibly will not exist such a behaviour. So the dynamic case is more realistic and flexible and in principle enables a real-time identification of the deformation process, because there is no requirement for the object to be in a balanced state.

Nevertheless, parametric modeling and identification of dynamic deformation processes is more complex and time consuming than in the static case. Using a numerical model with a high number of nodes (e.g. >10,000), the computation time for the Kalman-filter may rise from some hours for the static case to some days for the dynamic case. This is mainly effected by the numerical differentiation (Figs. 10 and 13) for the error propagation of the prediction, which is not only performed for coordinates but also for velocities and if necessary accelerations. This relativizes the term 'real-time' from above, especially in the context of alarm systems.

The creation of the functional relationships between triggers and the physical structure of the object is another big challenge in parametric modeling. In the case-study, this is done by a 'trick', using model parameters as indirect triggers. This approach cannot be assumed to be universal. The optimal case would be the direct use of monitored trigger data (e.g. temperature values and rainfall) as input for the Kalman-filter and should be one goal of the modeling process.

Acknowledgment The author thanks the FWF (Austrian Science Fund) for the financial support of the project 'KASIP', project number P20137.

References

Eichhorn A (2005) Ein Beitrag zur Identifikation von dynamischen Strukturmodellen mit Methoden der adaptiven Kalman-Filterung. PhD thesis, IAGB, Uni Stuttgart

Gallagher RH (1976) Finite-element-analysis. Springer, Berlin

Gelb A, Kasper JF, Nash RA, Price CF, Sutherland AA (1974) Applied optimal estimation. The M.I.T Press, Cambridge

Gülal E (1997) Geodätische Überwachung einer Talsperre; eine Anwendung der KALMAN-Filtertechnik. In: Wiss. Arbeiten der Fachrichtung Verm.wesen der Uni Hannover, Nr. 224, Hannover

Hesse C (2002) Deformation analysis of a shell structure under varying loads with Kalman-filter techniques. In: Proceedings of 2nd symposium for geotechnology and structural engineering, Berlin

Heunecke O (1995) Zur Identifikation und Verifikation von Deformationsprozessen mittels adaptiver KALMAN-Filterung (Hannoversches Filter). In: Wiss. Arbeiten der Fachrichtung Verm.wesen der Uni Hannover, Nr. 208, Hannover

Heunecke O (1996) Einige Gedanken zur fachübergreifenden Untersuchung von Deformationsvorgängen, dargestellt am Beispiel der Filterung der Biegelinie eines Pylons. In: Wiss. Arbeiten der Fachrichtung Verm.wesen der Uni Hannover, Hannover, pp 75–92

Isermann R (1988) Identifikation dynamischer Systeme. Springer, Berlin

Itasca (2006) Fast Lagrangian analysis of continua in three dimensions. Version 3.1. manual

Kasip (2011) Project homepage. http://info.tuwien.ac.at/ingeo/research/kasip/index.html

Lienhart W (2007) Analysis of Inhomogenous Structural Monitoring Data. PhD thesis, Engineering Geodesy, TU Graz

Mair am Tinkhof K, Preh A, Tentschert E, Eichhorn A, Schmalz T, Zangerl C (2010) FLAC^{3D} and adaptive Kalman-filtering – a new way to install effective alarm systems for landslides? In: Eurock rock mechanics symposium 2010, Lausanne

Rödelsperger S (2011) Real-time processing of ground based synthetic aperture radar (GB-SAR) measurements. PhD thesis, TU Darmstadt, Schriftenreihe der Fachrichtung Geodäsie, Heft 33

Rödelsperger S, Läufer G, Gerstenecker C, Becker M (2010) Terrestrische Mikrowelleninterferometrie – Prinzip und Anwendungen. In: AVN, 10/2010, pp 324–333

Roth W (1999) Entwicklung von Sicherheitsfaktoren mittels des kontinuumsmechanischen FD-Codes FLAC. Master thesis, Engineering Geology, TU Vienna

Schmalz T, Buhl V, Eichhorn A (2010) An adaptive Kalman-filtering approach for the calibration of FD-models of mass movements. J Appl Geod 3:127–135
Schmidt H-H (2006) Grundlagen der Geotechnik. Teubner Verlag, Wiesbaden
Smith GD (1985) Numerical solution of partial differential equations – FD methods. Oxford University Press, New York
Welsch W, Heunecke O, Kuhlmann H (2000) Auswertung geodätischer Überwachungsmesungen. Handbuch der Ingenieurgeodäsie, Wichmann
Zangerl C, Eberhardt E, Schönlaub H, Anegg J (2007) Deformation behaviour of deep-seated rockslides in crystalline rock. In: Rock mechanics

Land Subsidence in Mahyar Plain, Central Iran, Investigated Using Envisat SAR Data

M. Davoodijam, M. Motagh, and M. Momeni

Abstract

In recent decades land subsidence and its associated fissures have been observed in many plain aquifers of Iran. Knowledge of the deformation field in groundwater basins is of basic interest for understanding the cause and mechanism of deformation phenomenon, and for mitigating hazard related to it. In this paper the result of Envisat InSAR time-series analysis for monitoring land subsidence in Mahyar Plain, Central Iran, is presented. Long-term extraction of groundwater, which started in 1970 with the development of agriculture in this area, has caused substantial subsidence and formation of many earth fissures in Mahyar. Our analysis indicates significant subsidence bowl south of Mahyar plain with an elliptical pattern directed northwest–southeast along the axis of the plain. The velocity map obtained by the time-series analysis of InSAR data shows a maximum subsidence velocity of ~9 cm/year in the line of sight from the ground to the satellite in the year 2003–2006.

Keywords

Ground water • Interferometry • Land subsidence • Mahyar plain

1 Introduction

Subsidence is defined as the downward motion of a surface relative to a datum such as sea level (Waltham 1989). Generally, compaction of sediments, extraction of ground water, oil and gas result in compression of the clay layers beneath the land surface and subsequently elevation of the land surface is lowered. In recent years, with the growing population of the world and lack of water caused by global warming there has been an increased request for groundwater to supply domestic, industrial and agricultural needs. Excess exploitation of this resource causes underground fluid pressure to decrease. Thus the supportive effective stress on the rock matrix increases and rock compaction or land surface subsidence occurs (Murk et al. 1995; Jachens and Holzer 1982).

Several problems are associated with land subsidence including damages to underground infrastructure and civil engineering structures such as buildings, roads, canals and bridges, and increasing inland flooding along streams and waterways due to changes in stream gradient. Decreasing water level also increases pumping cost, as deeper wells are required to be built to extract water. It has been addressed in the literature that land subsidence caused by over extraction of ground water is increasing around the world. It is also a widespread problem in Iran (Motagh et al. 2008). This paper presents the result of an InSAR time-series method, called Small BAseline Subset (SBAS) (Berardino et al. 2002) for analyzing temporal and spatial variability of land subsidence in Mahyar plain, south of Isfahan province in Central Iran. The analysis done in this paper is based on 15 C-band ASAR radar images acquired by the Envisat satellite during 2003–2006.

M. Davoodijam (✉) • M. Momeni
Department of Geomatics, University of Isfahan, Isfahan, Iran
e-mail: ma.davodi@gmail.com

M. Motagh
Department of Geodesy and Remote Sensing, GFZ German Research Center for Geosciences 14473 Potsdam, Germany

H. Kutterer et al. (eds.), *The 1st International Workshop on the Quality of Geodetic Observation and Monitoring Systems (QuGOMS'11)*, International Association of Geodesy Symposia 140,
DOI 10.1007/978-3-319-10828-5_18,

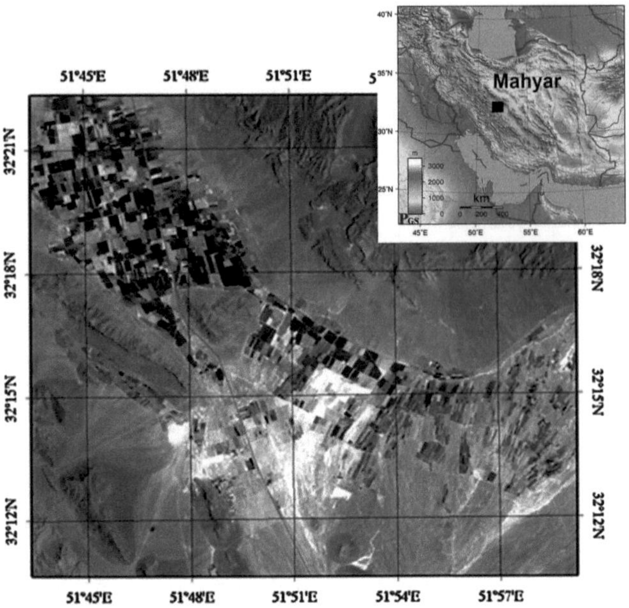

Fig. 1 Geographical location of Mahyar plain in Iran

2 Study Area

Located in an arid and semiarid region of Iran with an unconfined aquifer of an area of 158 km^2 (Ajalloein et al. 2006), Mahyar Plain is a NW–SE trending plain south of Isfahan province in central Iran (Fig. 1).

Long-term extraction of groundwater, which started about 30 years ago with the development of agriculture in this area, has caused subsidence and formation of many earth fissures in Mahyar, in turn leading to substantial damages to houses, farms, channels and roads in the region. Land subsidence monitoring is important in order to investigate the temporal and spatial extent of subsidence and for mitigating hazard related to it.

3 Methodology

To assess the temporal and spatial scale of land subsidence in Mahyar, we use the SBAS method as implemented in the STAMPS software (http://radar.tudelft.nl/~ahooper/stamps/index.html)

This method uses a large number of SAR acquisitions and implements a combination of a properly chosen set of multilooked DInSAR interferograms computed from these data. The unwrapped interferograms are input data for the least-squares method (Berardino et al. 2002), which is used to produce spatio-temporal evolution of ground surface deformation.

The following steps are done by STAMPS in order to generate time series analysis:

1. Generating subsets of small baseline interferograms
2. Selection of slowly decorrelating filtered phase (SDFP) pixels, defined as pixels whose interferometric phase after azimuth and range filtering shows slow decorrelation over short time intervals
3. 3-D phase unwrapping on SDFP pixels
4. Generating time series and mean velocity maps according to coherent pixels (Hooper 2008).

4 Experimental Results and Analysis

4.1 InSAR Result

In this study 15 radar images of the study area, acquired by the European Space Agency (ESA)'s Envisat satellite, are used. The dataset consists of ASAR images recorded in a descending trajectory spanning the year 2003–2006. Thirty-seven differential interferograms were generated and analyzed using the SBAS method. The topographic correction was done using the 90-m SRTM DEM. Figure 2 illustrates the final network of interferograms that we used for the time-series analysis.

Figure 3 presents an example of a geocoded differential interferogram in our study area. The interferogram covers the time interval between April 2003 and November 2004. One complete cycle of color from blue to yellow corresponds to half a wavelength of apparent range change (about 28 mm for the C-band radar onboard the Envisat satellite) between the spacecraft and the Earth's surface.

The obtained result indicates significant land subsidence occurring south of Mahyar. The subsidence area delineates an elliptical pattern directed northwest–southeast along the axis of the Mahyar plain.

Figure 4 illustrates the average subsidence rate in Mahyar obtained by the time-series analysis of Envisat SAR data. The negatives values in the map represent subsidence signals in the LOS direction (motion away from the satellite). We observe a maximum rate of LOS subsidence of about 9 cm/year in the southern part of Mahyar plain. Because of the steep incidence angle of the Envisat satellite ($\sim$23°) interferometric results are more sensitive to vertical than horizontal displacement (Hanssen 2001). This, coupled with field and hydrological evidence for aquifer compaction in the valley (next section), led us to interpret the LOS displacement as indicating mainly land subsidence in the valley.

Figure 5 shows displacement time series for four selected points, A, B, C, and D in Fig. 4. Point A lies in the area of maximum subsidence. The maximum amount of Line of Sight (LOS) displacement is about 23 cm during 2003–2006 (Fig. 5a). Point B lies on the main road that connects two metropolitan Isfahan and Shiraz. Here we obtain $\sim$9.3 cm displacement in the LOS direction. Point C and D are

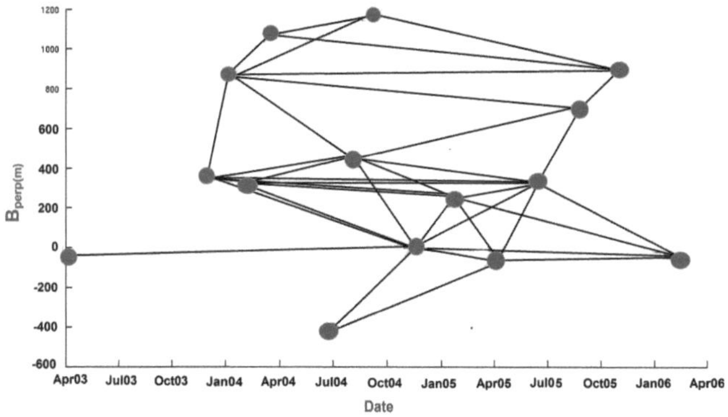

Fig. 2 The network of interferograms used in this study

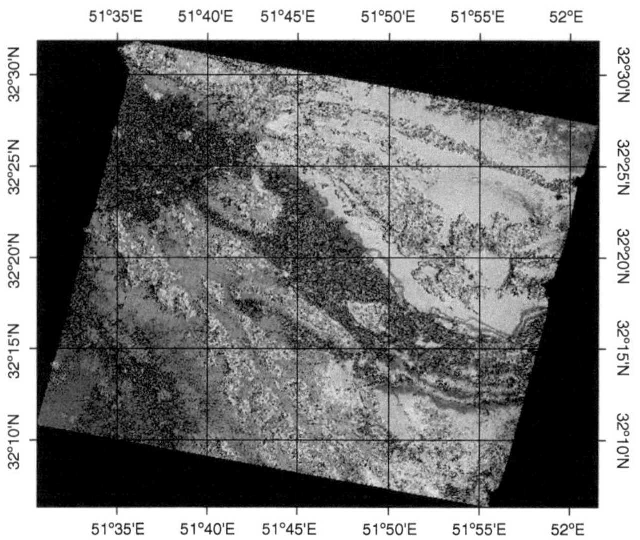

Fig. 3 An example of a geocoded differential interferogram in Mahyar between April 2003 and November 2004

two arbitrary points and their time-series results confirm that most parts of the Mahyar plain sustained significant subsidence in the year 2003–2006.

4.2 Earth Fissures

After InSAR observations a field survey of area for positioning of the fissures caused by land subsidence was done in the summer 2010. The observed fissures have been overlaid a descending interferogram in Fig. 6. As shown in Fig. 6 no fissure was observed to the south part of the southern Mahyar plain. In contrast, several parallel fissures were detected in other parts of region.

Many reasons contribute to the generation of fissures in areas affected by land subsidence. Factors such as decline in the water table, faults, aquifer heterogeneities, bedrock knob protruding into the aquifer system and presence of

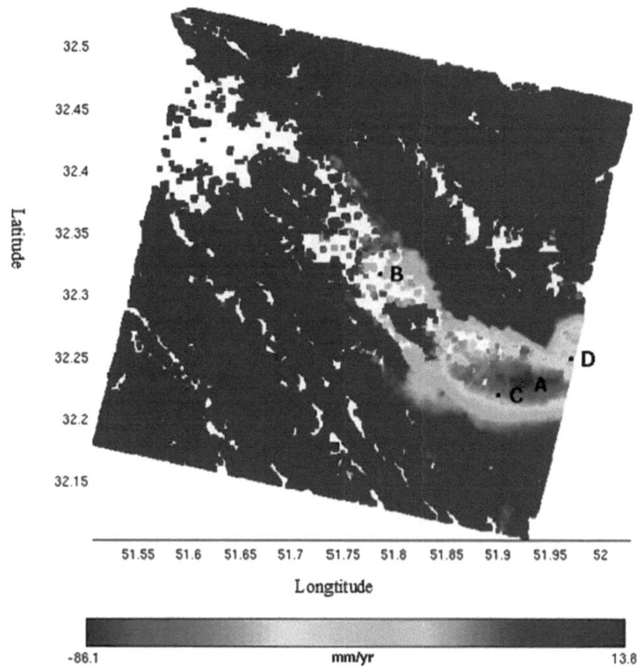

Fig. 4 Mean velocity map of subsidence area in Mahyar

weakness plain in aquifer system control the progress of fissuring (Burbey 2002).

4.3 Piezometric Records

Piezometric record in Mahyar plain shows a correlation between the pattern of land subsidence and the decline in water level. As shown in Fig. 7, the water table in the region has a clear declining trend with a head decline of ~12 m between 1983 and 1995 and only small periodic recoveries, probably caused by seasonal rainfall. Such a decline in water table record suggests that land subsidence in Mahyar plain results from overdrafting of the aquifer system.

Conclusions

This paper has demonstrated the capability of C-band ASAR images for investigating land subsidence in Mahyar plain, an agricultural region south of Isfahan province in Central Iran. Results obtained from the time-series analysis of 15 ASAR images acquired in a descending orbit between 2003 and 2006 indicates a subsidence bowl south of the plain with maximum LOS velocity of ~9 cm/year. Analysis of the piezometric records suggests that subsidence likely results from overdrafting of the aquifer system that has caused an average decline of ~1 m/a in the water table since 1982.

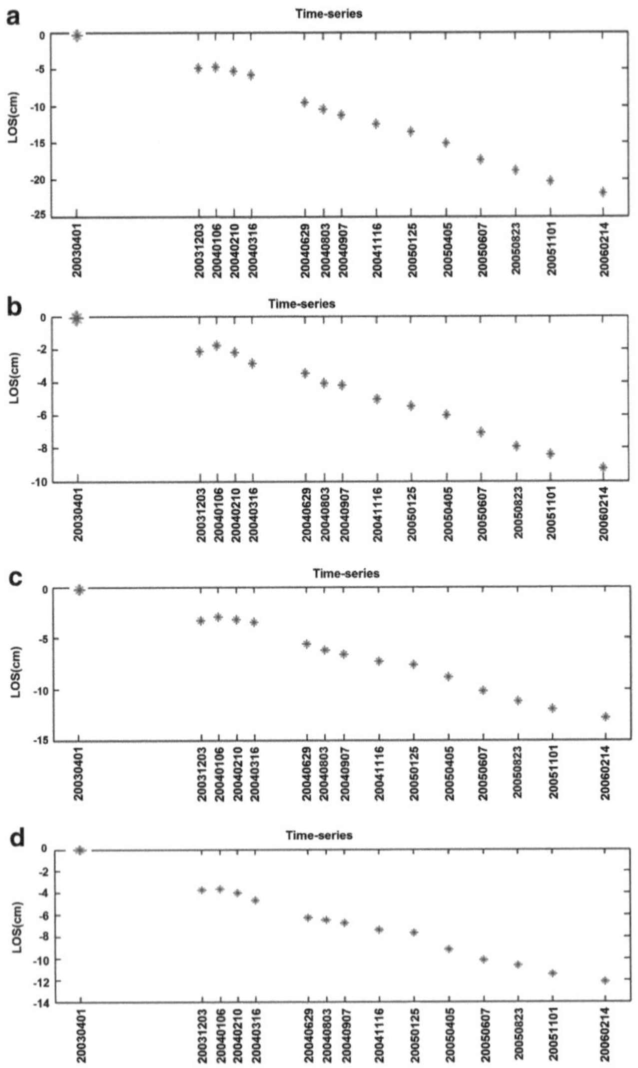

Fig. 5 Examples of time—series results at four selected points. For locations of A, B, C and D please refer to Fig. 4

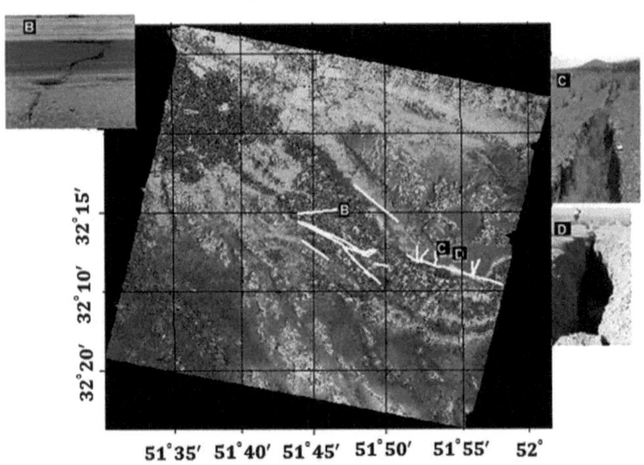

Fig. 6 Fissures detected during field survey of Mahyar plain in 2010. Fissures are overlaid on a descending interferogram. The insets show ground picture of three fissures in Mahyar

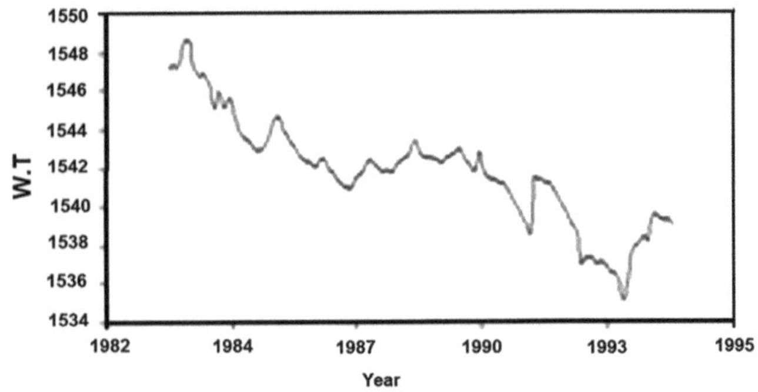

Fig. 7 Average groundwater level decline in Mahyar obtained from piezometric records

Acknowledgments Envisat data used in the study were provided by the European Space Agency (ESA) through proposal AOALO3740. The authors are grateful to an anonymous reviewer for a number of suggestions, which improved the quality of the paper.

References

Ajalloein R, Ghazifard A, Hashemi M, Kamyab E (2006) Effect of stratigraphy on earth fissuring in the northern Mahyar plain, Iran. In: Culshaw MG, Reeves HJ, Jefferson I, Spink TW (eds) Engineering geology for tomorrow's cities. Geological Society, London, Engineering Geology Special Publications, 22 [on CD-ROM insert, paper 596]

Berardino P, Fornaro G, Lanari R, Sansosti E (2002) A new algorithm for surface deformation monitoring based on small baseline differential SAR interferograms. IEEE Trans Geosci Remote Sens 40(11):2375–2383

Burbey TJ (2002) The influence of faults in basin-fill deposits on land subsidence [Las Vegas Valley, Nevada, USA]. Hydrogeol J 10:525–538

Hanssen RF (2001) Radar interferometry: data interpretation and error analysis. ISBN-384 10: 0792369459

Hooper AJ (2008) A multi-temporal InSAR method incorporating both persistent scatterer and small baseline approaches. Geophys Res Lett 35, L16302

Jachens RC, Holzer TL (1982) Differential compaction mechanism for earth fissures near Casa Grande, Arizona. Geol Soc Am Bull 93:998–1012

Motagh M, Walter T, Sharifi MA, Fielding E, Schenk A, Anderssohn J, Zschau J (2008) Land subsidence in Iran caused by widespread water reservoir overexploitation. Geophys Res Lett 35, L16403. doi:10.1029/2008GL033814

Murk WB, Skinner BJ, Porot SC (1995) Environmental geology. Wiley, 535 pp

Waltham AC (1989) Ground subsidence. Chapman and Hall, New York, 202 pp

Recent Impacts of Sensor Network Technology on Engineering Geodesy

O. Heunecke

Abstract

Wireless Sensor Networks (WSN) as an infrastructure comprised of sensing, computing and communication elements are designed for the decentralized recording of environmental information. If the matter of concern is referenced geospatial information this in particular forms a Geo Sensor Network (GSN).

Recently there is an ongoing extensive impact of these new techniques on all geo sciences, whereby it depends on the application which demands with respect to data quality, size and costs of the sensor nodes, number of nodes in the network, requirements of the communication component (coverage, data throughput) and the power management are relevant. In Engineering Geodesy the new possibilities and new ideas emerging from GSN especially concern geo(detic) monitoring of objects like landslides or engineering construction sites. Theory and possibilities of GSN technology as well as some selected aspects for geo monitoring in particular will be discussed. As the early detection of already small variations is essential for Early Warning Systems (EWS) and risk management, data quality and reliability is of utmost importance. Thus, customary utilization of low cost equipment in such a GSN generally requires for calibration procedures and higher sophisticated evaluation concepts in order to provide meaningful results.

Keywords

Engineering Geodesy • Geo monitoring • Geo sensor networks • Low cost GNSS receivers

1 Introduction

According to Stefanidis and Nittel (2005) and Nittel (2009) GSN are specialized applications of WSN technology in geographic space that detect, monitor and track environmental phenomena and spatial-temporal processes in particular.

GSN can be classified in terrestrial ecology observation systems for the modelling of environmental phenomena like weather and climate, real-time event detection, which indicates monitoring tasks in general, and so-called mobile sensor nodes for human, vehicle and animal behaviour tracking and analysis. This itemization concerning mobile nodes should be supplemented by guidance and control, e.g. fleet management, in general.

For catastrophe and disaster management, including force tracking and as well military scenarios, applications can be correlated between these three classes, ranging from small (local) to meso (regional) and finally the global scale. Within a short time, GSNs have already claimed a wide field of applications mainly because the easy available technology of ubiquitous wireless communication, the miniaturization of computing and storage platforms as well as the development of novel and cost effective sensors.

Like the wide-spread use of satellite based positioning techniques, that vary from global to local applications with accuracies spanning the range between several meters to

O. Heunecke (✉)
Institute of Geodesy, University of the Federal Armed Forces Munich, Werner-Heisenberg-Weg 39, 85577 Neubiberg, Germany
e-mail: otto.heunecke@unibw.de

H. Kutterer et al. (eds.), *The 1st International Workshop on the Quality of Geodetic Observation and Monitoring Systems (QuGOMS'11)*, International Association of Geodesy Symposia 140,
DOI 10.1007/978-3-319-10828-5_19,

Table 1 Modelling of spatial-temporal processes (compare Welsch et al. 2000)

	Real object	Model of the object
Geometry domain	Object is a continuum, possibly separated into different regions	Object is dissected by certain number of characteristic points ("discretization in space")
Time domain	Object is more or less permanently in change, in particular deformations	Object is monitored in certain time intervals ("discretization in time")

Table 2 Design parameters and differentiating factors of GSN (compare Heunecke 2008 and Bill 2010)

Design parameter	Properties and distinctions
No. of nodes q	Arbitrary, as a rule $q = 10$–1,000 in small-scale GSN
Node mobility	Static vs. partly in motion (active, passive)
Node design	Homogenous (all identical)/heterogenic
Node autarky	Life cycle of several hours to years
Node deployment	Planed/random (without adaption); one-time/expansion of network (if required)
Data coverage	Scattered/spatially condensed/redundant
Data recording	Permanent/sporadic/event based
Network topology	Infra structural/ad hoc; star/mesh/hybrid topology
Data communication	One-directional/bi-directional; permanent data flow/on request/sporadically
Localization	Planed mounting/position sensor integrated/derived from communication signals

millimetres based on different phase based augmentation techniques, the rapidly ongoing GSN developments arguably make a more detailed discussion comprehensive. The very broad definition of GSNs as specialized applications of WSN lead, on the one side, to a discussion of a large number of tiny, untethered, battery-powered, low cost devices (in the order of a few Euro only or less) deployed in ad hoc fashion without detailed planning as typical (Sohraby et al. 2007; Bill 2010) and, on the other side, to the mention of well-thought-out and sophisticated monitoring systems with expensive high-end measuring units like the German-Indonesia Tsunami Early Warning System (GITEWS; see www.gitews.org) at the same time. Regarding the multiplicity of different scenarios, a universal GSN approach does not exist, but however an all-embracing theory in terms of e.g. terminology, architecture and general properties is developed and available. One ultimate goal for further GSN research is the technical design of such systems in a way known as swarm intelligence in Bionics, however.

Just as the automated data recording with individually laid-out sampling rates has been a standard since many years ("discretization in time") in geo monitoring applications, spatially condensed information clustering ("discretization in space") using GSN technology now becomes possible with passable effort and costs and—which is the main benefit—a more complete picture of the real situation in time and space evolves, see Table 1. Finally, the developing extensively growing data volumes poses new challenges for an ambitious data evaluation and visualization.

All following explanations are focused on small-scale GSN only as they are found in geo and structural health monitoring applications. Such measurements and surveys are carried out to monitor the condition of technical objects (buildings, machinery plants, etc.) and natural objects (embankments, slopes liable to slide, etc.) for the determination of their different kinds of movements and to document the state of the respective objects. Objects equipped with wireless sensors are often termed as 'smart', e.g. smart bridges or smart dykes (Hopman et al. 2011).

2 Geo Sensor Network Technology

2.1 General Statements

A WSN respectively GSN generally consists of the following four basic components (Sohraby et al. 2007):

1. An assembly of distributed sensor nodes;
2. An interconnecting communication network (usually, but not always restricted solely to wireless techniques);
3. A central point of information clustering; and
4. A set of computing resources at the central point or beyond to handle data evaluation, event trending, status querying, etc.

Table 2 lists a few design parameters and technical characteristics which again disclose the omnibus GSN possibilities. Only some of these properties will be discussed in the following. For all geo monitoring applications distinguishing is a well-planned procedure of installing the sensor nodes and the intensive evaluation of data. Thus, a selection of representative recording points well adapted to the object is indispensable instead of a random deployment without adaption which is often discussed in other GSN fields. An expansion of the network planned in advance however can be essential if the detected behaviour of the object is irregular.

Localization in a GSN stands for geo referenced data acquisition in a well defined spatial frame whereas GNSS very often is used in outdoor environments. Using GNSS as a time server allows synchronization of all data to become feasible in an easy way, opposed to the usage of a Real Time Clock (RTC). Positioning, Navigation and Timing (PNT) is generally a key issue for any surveillance of spatial-temporal processes.

With many sensor nodes used in a GSN the price of a single device is an essential parameter for cost effective application. As noted in Table 2 the number of nodes in a geo monitoring network normally is restricted but presumably will increase in future regarding the new possibilities. Projects with about 3,000 sensors and/or targets are already known, especially for the preservation of evidence in venturesome constructions. Additionally, of course the sensitive measurands relevant for the monitored process, whether they are causative forces or geometrical reactions (displacements, inclinations, strain, etc.) at the output side of the investigated process chain, are to be chosen adequately and individually as well as the required measurement uncertainty of the data which generally is ambitious.

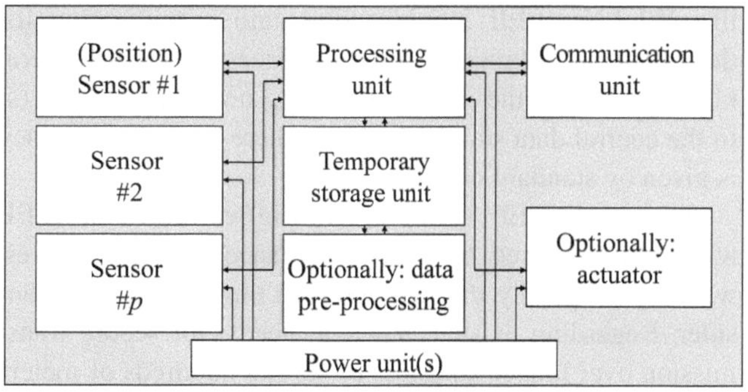

Fig. 1 Typical sensor node design (compare Günther et al. 2008)

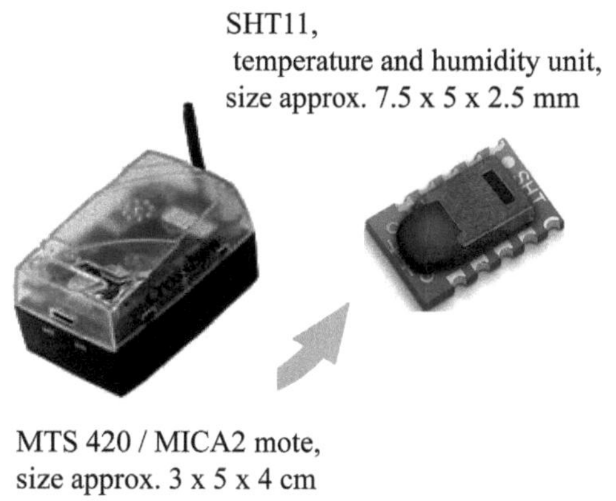

Fig. 2 Crossbow MTS 420/MICA2 mote

2.2 Sensing

Organizationally, a GSN is subdivided into several so-called sensor nodes (number of nodes denoted with q) with distinct addressability, which in general are multifunctional devices (number of sensors at a node denoted with p) in fully autarkic operation. With regard to geo monitoring data quality and secure operation even under harsh conditions are of highest significance.

Although all networked sensors must have the basic availability of localization to attribute the observed information (unless they are mounted fix at a known place) within monitoring applications positioning and changes of positioning (including inclination and distance changes) of the different nodes are of particular significance. This is to be seen along with other data such as trigger factors (rain fall, table water, ...), environmental quantities (temperature, air pressure, ...) and information of the sensor node itself (charge condition, power supply, ...). The previously information may be useful to predict potential failures of a node caused by power shortfalls. The different nodes can be all identical or, which will be the normal case in Engineering Geodesy practise, heterogenic with a certain number of p sensors onboard every multifunctional device. Figure 1 depicts a typical sensor node design. A processing and temporary storage unit makes such a device a micro computer, whereas the possibility of data pre-processing is optional, the so-called decentralized and collaborative spatial computation (Duckham 2008). Finally, with onboard computing a sensor node emerges from a simple data logger to an intelligent platform.

Closed loop applications include also actuators, where a reaction based on a detected situation can be accomplished automatically in real time. Such applications are one of today's Engineering Geodesy challenges and generally termed as 'active monitoring', e.g. hydraulic jacks to compensate settlements or compensation grouting during tunnelling in urban areas. Within such tasks generally size of the nodes etc. is secondary and permanent electric power supply for the devices is presumed, however.

Due to normally restricted resources at the single sensor node, link budgets (reachable ranges and data throughputs of communication), power management of the autarkic units (secure energy supply) and low maintenance (permanent unattended operation) need detailed attention, especially in harsh environmental conditions (e.g. dust, frost, humidity) and for long term monitoring (year round operability). Thus, for steady network operation the robustness of the whole equipment is essential. Using novel technology, e.g. MEMS, today a sensor node has a typical size between a shoe box and a cigarette packet; see Fig. 2, with the prospect of becoming smaller in future. Then often it also would be named a sensor mote.

2.3 Communication

Each sensor node in a GSN has the ability to communicate. The term network initially stands for communication and not spatial relationships of the nodes. Regarding geo monitoring in particular, license free wireless transmission techniques like WLAN (IEEE 802.11) come into consideration for the efficient bridging of huge distances of up to several kilometres from the scattered sensing units via gateway(s) to the central data sink in the field where by default privacy is given by standard encryption.

Bluetooth (IEEE 802.15.1) and ZigBee (based on IEEE 802.15.4, enhanced by ZigBee Alliance) are alternatives, whereby especially the required data throughput is to consider. Regarding existent practical needs, for secure transmission over longer distances of several hundreds of meters or some kilometres free line of sight is necessary to provide adequate connectivity even with suitable external antennas. Repeater nodes may be helpful in topographic restricted vicinities.

Star and mesh network topology with one- and bi-directional data communication is established, both combined is called a hybrid network. Other than at star topology with static and normally well-planned radio connections (infra structural network) by a mesh network different paths to the gateway can be taken by multi-hop fashion in case of obstacles between the transmitter and receiver etc. Here, the messages between the sensor nodes are relayed at short(er) distances until they arrive at the sink. Such non-hierarchical self-healing protocols for multi-hop operating (dynamic routing) which automatically configure themselves (ad hoc) have been developed, for example XMesh by Crossbow (e.g. Nittel 2009), but the presumption always is that there would be sensor nodes available in the achievable range and the cumulated data will not exceed the throughput as well as enough energy at the relay nodes is available. With applications where static radio connections cannot be planned there is no alternative to this approach. At monitoring tasks wireless techniques often are supplemented by wired communication, for instance connecting of subsurface instrumentation in geotechnical applications which as a rule is done by serial cables.

Transition power is a function of distance between the sensor nodes and the gateway and the data volume to be broadcast. As a rule: the shorter the distance the lower the required transition power. Therefore mesh topology with multi-hop offers advantages to simple star networks with longer distances. Each node in the network adds its own data, and the message size increases with each 'hop'. Therefore, nodes close to the gateway need significantly more energy compared to leaf nodes in the network. Once again, it becomes obvious that detailed planning (still) is important for well operating GSNs in practice.

Nittel (2009) points out that sending data consumes about 800 times more energy than computing the same amount of data on a local chip. Thus, minimizing communication is an objective as soon as energy supply is a crucial issue at a sensor node. Optionally, onboard computing and capacity of data buffering (temporary storage) at the sensor nodes can be used to reduce the transmission data volume, to optimize the data packets, to avoid data lost due to communication failures and finally, if necessary, to disburden the master station. These facts assumedly will have impacts on the future design of geo monitoring networks.

2.4 Network Operating

Normally, the central data sink or master station manages all system operations on site, e.g. data collection and controlling of the network. The evaluation of all data in real time normally takes place here and, if necessary, early warnings or even alerts can be given. 'Real time' is given as long as the processed information of the GSN is available for all necessary decision makings and actions. If the processing of the data takes some time, e.g. a few minutes, often this is called 'near real time' but, however, a small delay for data gathering and evaluation must be accepted in any case. Thus, real time always has to be seen regarding the observed process but not the absolute time interval required for the processing in seconds or minutes. The master station is normally linked to a web server for supplementary data archiving and data distribution separated corresponding to access authorisation (esp. administrator, stakeholder, and viewer). Therefore usually there is DSL, UTMS or GSM connectivity.

Energy is possibly the most valuable property at the autarkic sensor nodes and customary communication is the most energy consuming portion of the process. In geo monitoring applications the nodes normally work time based with a dense permanent data recording. When possible the nodes switch over to a sleep mode if the sample rate is low, e.g. a recording hourly or fewer (e.g. Martinez et al. 2009). With temporary onboard data storage, e.g. by embedded boards, broadcast can be reduced to a few orders of events per day. However, this may be restricts real time availability of results.

An event based network operation is optional if a sensor node has the ability of threshold query. Only in case of a threshold exceeding, data would be sent to the master station immediately and the administrator would receive knowledge of the event. But even with optimized operation purely battery-supported operation with respect to long term monitoring without interruptions and unattended running is not possible. Possibilities to be explored for recharge include solar panels, wind generators and, where appropriate, fuel cells. 'Energy harvesting', the process by which energy is for instance derived from thermal environments or kinetic phenomena (e.g. vibrations), may offer new possibilities in future.

So far, proprietary formats—often binary—are exported from the sensors demanding data conversion before handling such information in own evaluations. To enhance data integration the XML-based so-called Sensor Web Enabling (SWE) initiative is proposed to make it easier to share data streams, e.g. the leverage of data from other deployments for own applications (see Bill 2010). Data integration from geo services like official weather stations for example would be much easier with such an interface. Finally, sensor information becomes a part of the omnipresent Spatial Data Infrastructure (SDI) and can be integrated in 'Sensor GIS' applications. These impacts are at the moment beyond the scope of GSN to Engineering Geodesy, however.

3 Selected Aspects

Utilization of low cost equipment, e.g. MEMS, in geo monitoring applications generally requires calibration procedures and/or higher sophisticated evaluation concepts. This shall be explained by two examples.

3.1 Calibration of Motes: An Example

Crossbow Technology Inc. (see www.xbow.com) is one of the leading WSN companies. A modular configured sensor mote of a developer kit equipped with a combined ISM (Industrial, Scientific and Medical bands) radio and power module working with TinyOS (128 kbytes processor, two AA batteries) is the MTS 420/MICA2, see Fig. 2. TinyOS is an operating system designed for extremely restricted devices such as Crossbow motes. The sensor board MTS 420 has $p = 5$ different sensors with in total nine different quantities. There is integrated a GPS receiver LEA-4A (16 channels, L1; additionally an external patch antenna) from ublox (see www.u-blox.com), an accelerometer (from Analog Devices), a light sensor (from TAOS), a barometer (from Intersema) and a temperature and humidity unit SHT 11 (from Sensirion, see www.sensirion.com).

Only the temperature and humidity unit will be discussed in the following. The SHT 11 chip, see Fig. 2, is specified at a range from −40 to +80 °C and an accuracy of ±0.5 °C at 25 °C. For 0 and 50 °C the accuracy specification ±1.0 °C can be found. The humidity transducer is specified with an absolute accuracy of ±3 % of the relative humidity value in a range between 20 and 100 %, which was confirmed by a comparison with reference values in own studies. The complete unit has a size of a few mm and costs of about 25€. The SHT 11 temperature sensor was investigated in a climate chamber at the UniBw between 0 and 35 °C over a time period of 10 h, whereby the reference values can be guaranteed with 0.1 °C. In total, 17 test items were available and tested in two separated runs to check stability over time.

A remarkable offset of several degrees non linear over the examined temperature spectrum exceeding the specified limits by far was to be seen. By polynomial regression analysis (Welsch et al. 2000), individually performed for every test item, calibration functions were established with a remaining rms of better than 0.2 °C. This simple example shows that even with low cost sensors a good data quality can be achieved, provided that a calibration is performed. Increased effort for data handling is requested. Without such investigations and calibration procedures, quality assessments of data may be difficult to accomplish or not possible at all. As quality cannot be compensated by 'mass' (in other words many sensors) at geo monitoring applications such examinations are highly recommended before commencing with a project.

3.2 Carrier Phase Based Positioning Using Simple Navigation Receivers

Geo localization using GNSS techniques is well-known, especially concerning the GPS Standard Positioning Service (SPS). To improve accuracy to cm and mm level augmentation is needed. In general this is done by Phase based Differential positioning (PDGNSS) techniques. Here besides the code, the carrier phase information of the satellites signals is also introduced. Today, even with simple navigation receivers produced for the mass market it is possible to obtain accuracies of coordinate changes in the sub centimetre range if the carrier phase data is available for processing. Many of the simple navigation receivers make use of the carrier phase data for some internal smoothing operations, but do not have the ability of an autonomous phase based positioning like RTK (Real Time Kinematic) rovers.

A procedure to make use of such gear is described in more detail by Günther et al. (2008) and Glabsch et al. (2010).[1] Prerequisite is that a GNSS receiver has not only the ability of carrier phase tracking but also the possibility of read-out these data via a serial interface or USB. By the serial interface or USB the configuration of the receiver also has to be operated. Using commercial off-the-shelf wireless device servers it becomes easily possible to integrate such instruments equipped with serial ports or USB in a WLAN network (Pink 2007). To keep costs low such a sensor node (or sensor node component) in a GSN is just a simple 'data logger' without own processing intelligence.

[1]The research was funded by the German Federal Ministry of Education and Research (BMBF) from 2007 to 2010 in the program 'Geotechnologien'.

The developed monitoring system makes use of a permanent WLAN communication to gather all binary phase and code data from the scattered low cost GNSS receivers at a master station for an attached near real time processing in batch mode. New baseline solutions are computed after a certain time period, e.g. every 15 min, for all involved sensor nodes to detect 3D surface movements. For each epoch an individual solution depending strongly on the actual satellite coverage is obtained. Using the described approach it is necessary to have suitable software which continuously collects the raw data from the *q* sensor nodes, converts the proprietary binary data for the attached baseline processing and hands over the results for the time series configuration, filtering and other analysis tools.

The binary signals transmitted from the low cost receivers to the master station contain not only code and phase information, but also additional data like Signal to Noise ratio (S/N), azimuth and elevation of satellites etc. This meta information can be used to enhance and to optimize the self-designed Near Real Time Processing (NRTP). NRTP distinguishes against RTK but, however, must be considered still 'real time' regarding the monitored process, e.g. a landslide. All results can be stored in a MySQL data base for enclosed (and repeated) evaluations.

The described approach is an alternative to the well-known installation of a total station and offers, under good conditions, similar or even better data quality. However, also a combination of both approaches is technically easily possible. The concept of gathering GNSS raw data over a certain period of time opens all the well-known options of high sophisticated post processing to all kinds of simple receivers with a possibility of read-out of the carrier phase information in a geodetic monitoring network. Of course, this developed technique can be seen just as the position sensor in Fig. 1.

Suitable receivers are available between 300 and 1,000€, for example Novatel Smart V1G Antenna (see Fig. 3) used in a project described in the following. Such a device is called an enclosure with receiver and antenna integrated in one unit. Equipment like the Novatel Smart antenna is all-weathered proofed and temperature resistant, which makes it ideal for year-round operation even in harsh environmental conditions, e.g. landslide or rock fall monitoring in alpine winters. The emerging data amount for such a L1 receiver is about 400 Byte/s depending on the satellite coverage. Standard WLAN is able to meet these volumes even in larger monitoring networks, however.

To test the benefits of low cost receivers several pilot projects are conducted at the UniBw. A time series of 2 months (January and February 2011) from a monitoring project at a lock is presented with Fig. 4. Depicted are plane coordinates X, Y with a used moving average filter of 6 h. Regarding the standard deviation of a single epoch of 15 min

- 14 L1 GPS, 12 L1 Glonass
- 0.15 cm rms accuracy carrier phase
- Power 9-24 V; 1.2 W
- Interfaces RS-232 / RS-422 / USB
- Environmental MIL-STD-810F
- Weight 575 g, size 115 x 90 mm

Fig. 3 Data Novatel Smart V1G antenna

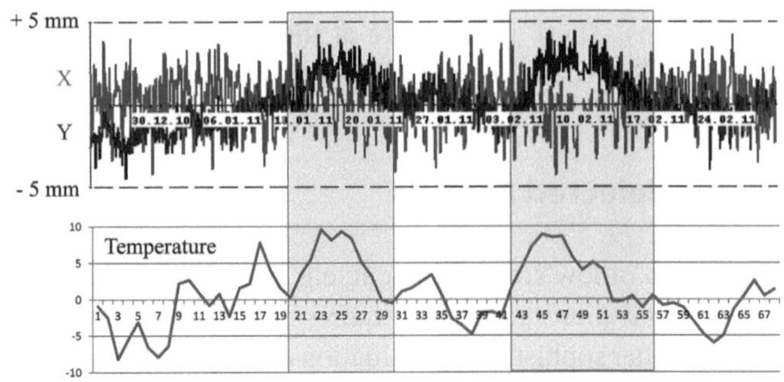

Fig. 4 Example of a time series at a structure obtained by low cost GNSS receivers (time period of 2 months, MA filter of 6 h) versus temperature to demonstrate the potential

Table 3 Disposable and usable base line solutions in a pilot project with free satellite visibility at all sensor nodes

	Basis line solutions		Empirical standard deviation epoch	
Sensor node	Disposable (%)	Usable (%)	X, Y [mm]	H [mm]
# 101	99.9	98.7	3.0	5.3
# 102	99.9	98.7	2.9	5.2
# 103	100	99.4	3.1	4.8
# 104	100	99.4	2.5	4.6

(96 solutions per day) at Table 3 such a solution owns a standard deviation of about 0.6 mm.

Table 3 depicts that there are about 99 % usable epoch solutions in this project where only GPS data is used and the Glonass option still is open. The behaviour of the structure in Y direction with respect to the temperature influence is clearly to be seen in Fig. 4. As consumer grade antennas tend to be sensitive against multipath and other effects (Wieser and Niedermayer 2011), at the moment filter periods of 3–6 h are necessary to obtain reliable results. With a better quality assessment of such antennas further enhancement of results and especially shortening of filter intervals is a challenge.

From a methodical point of view it is conceivable to use the above mentioned low cost GNSS sensor nodes as beacons for a further densification with a class of (cheaper) subordinated sensor nodes not equipped with a receiver (heterogenic design, see Table 2). For instance Bill (2010) discusses such localization approaches with respect to the neighbouring nodes. On the other side it is also possible to have (expensive) high-end receivers on some fiducial points.

This well-known concept from Geodesy, e.g. for a better modelling of troposphere refraction (e.g. Gassner et al. 2002) can be assigned to small scale GSNs as well to improve accuracy and a staged configuration of the monitoring network evolves.

Conclusion and Outlook

In the recent decade, there has been an ongoing pervasive impact of GSN on geo sciences and engineering offering new possibilities for the monitoring of geotechnical objects and civil engineering structures. Avoiding communication and power lines during field installation in addition to cost effective sensing devices and easy installation are some meaningful benefits derived from the new technique. Thus, as an alternative to the exclusive geodetic and geotechnical instrumentation currently available on the market is possible to react appropriately in many applications where cost constraints are dominant and a flexible system set-up is required. As geo monitoring is a multi scale problem concerning dimension of the objects, nature and time dependence of the processes, relevance of the problems etc., GSN is an ideal tool to react flexible and efficiently to different potential situations.

When calibration and high-sophisticated evaluation procedures are used, sufficient data quality for most of the monitoring applications can be obtained even with state of the art low cost equipment like presented in Sect. 3. The essential benefit of the GSN developments is that a densification of characteristic points of an object, the discretization in space according to Table 1, now becomes possible at a higher standard and at lower costs than before. Therefore, a description of spatial-temporal field problems, e.g. in terms of deformation patterns or temperature distributions, can be achieved in a much better way than before.

Because of this, more sophisticated real time capable approximation and interpolation algorithms, among other things, are requested during elevation and processing. Data fusion (compare Arnhardt et al. 2010) stands for the possibility of cross checking data from identical but neighboured sensors by combinations as well as the comparison of different measured quantities (at the same node) and their respective varying rates of change. A better validation of the recorded data evolves from an over determination—more measured quantities than unknowns—of the time depending process. Redundancy, which is well-known and practiced at geodetic networks by means of least squares adjustment for years (see Welsch et al. 2000) will help to detect malfunctions and outliers and other failures as well as an enhancement of data quality, especially with respect to accuracy andreliability. Thus, 'network' furthermore stands not only for communication, but also an integrated data analysis. Bayes-estimation in context with Kalman-filtering, like it is already developed in robotics (Thrun et al. 2005), seems to be a promising tool for prediction and updating the system behaviour. Finally, enhanced data quality management is requested, will lead to better results and also will help to avoid false alarms by better and more reliable decision making, something often lacking in today's acceptability of monitoring systems in practice. However, a lack of suitable and intelligent GSN software to handle the mentioned data fusion as well as the modelling of time variant field problems appropriately including integration of data into models especially for geodetic monitoring purposes is to ascertain at the moment.

References

Arnhardt C, Fernandez-Steeger TM, Azzam R (2010) Sensor fusion in an ad-hoc multi-hop sensor network for real-time monitoring of landslides endangering human infrastructures. In: Geotechnology science report no. 15. Early warning systems for transportation infrastructures, pp 38–49

Bill R (2010) Geosensornetzwerke – neue Technologien und interessante Herausforderungen für die mathematische und datenverarbeitende Geodäsie und Geoinformatik. In: von G. Schmitt (ed) Vernetzt und ausgeglichen, Festschrift zur Verabschiedung. Karlsruhe, pp 41–54. ISBN 978-3-86644-576-5

Duckham M (2008) Ambient spatial intelligence: decentralized computing in geosensor networks. In: EuroSDR & ISPRS Hannover Workshop 2008 on Geosensor Networks, February 20–22

Gassner G, Wieser A, Brunner FK (2002) GPS software development for monitoring of landslides. In: Proceedings FIG XXII Congress, Washington, TS6.4, April 2002

Glabsch J, Heunecke O, Schuhbäck S (2010) Development and testing of a low cost sensor PDGNSS landslide monitoring system using the example of the Aggenalm landslide in the Bavarian Alps. In: Altan O, Backhaus R, Boccardo P, Zlatanova S (eds) Geoinformation for disaster and risk management, examples and best practices, Chapter 11, pp 63–70

Günther J, Heunecke O, Pink S, Schuhbäck S (2008) Developments towards a low-cost GNSS based sensor network for the monitoring of landslides. In: 13th FIG international symposium on deformation measurements and analysis, Lisbon

Heunecke O (2008) Geosensornetze in der Ingenieurvermessung. In: Forum, Zeitschrift des Bundes der Öffentlich bestellten Vermessungsingenieure e. V., vol 2, pp 357–364. ISSN 0342-6165

Hopman V, Kruiver P, Koelewijn A, Peters T (2011) How to create a smart levee. In: 8th symposium on field measurements in geomechanics (FMGM) 2011, Berlin (published on USB-Stick). http://www.fmgm2011.org

Martinez K, Hart J, Ong R (2009) Deploying a wireless sensor network in Iceland. GeoSensor Networks, Lecture Notes in Computer Science 5659:131–137. doi:10.1007/978-3-642-02903-5_13

Nittel S (2009) A survey of geosensor networks: advances in dynamic environmental monitoring. Sensors 9:5664–5678

Pink S (2007) Entwicklung und Erprobung eines multifunktionalen Geosensornetzwerkes für ingenieurgeodätische Überwachungsmessungen. Schriftenreihe im Studiengang für Geodäsie und Geoinformation, Universität der Bundeswehr Munich, vol 83

Sohraby K, Minoli D, Znati T (2007) Wireless sensor networks – technology, protocols and applications. Wiley, New York. ISBN 978-0-471-74300-2

Stefanidis A, Nittel S (eds) (2005) Geosensor networks. CRC Press, Florida. ISBN 0-415-32404-1

Thrun S, Burgard W, Fox D (2005) Probabilistic robotics. The MIT Press, Cambridge

Welsch W, Heunecke O, Kuhlmann H (2000) Handbuch Ingenieurgeodäsie. Auswertung geodätischer Überwachungsmessungen, Wichmann. ISBN 3-8707-295-7

Wieser A, Niedermayer S (2011) Quality assessment of low-cost GPS antenna for use in geodesy. Presentation at the 1st international workshop on the quality of geodetic observation and monitoring system QuGOMS'11, Munich, 14 April 2011

Design of Artificial Neural Networks for Change-Point Detection

H. Neuner

Abstract

An important assumption in the global approach to system identification is the homogeneity of observed time series from statistical point of view. A violation of this assumption leads to biased estimated parameters and a low quality of the model.

This paper addresses the task of change-point detection by means of Artificial Neural Networks (ANN). The focus lies on the appropriate design of ANN by specifying the inputs, outputs and the necessary number of hidden nodes for an error-free classification of the data.

Keywords

Artificial Neural Networks • Change-Point detection • Variance homogeneity

1 Introduction

In a modern perspective deformation analysis deals with the modeling of the whole deformation process which is formed by the causal relationship between acting loads, the monitored object and its deformation. This approach enables the derivation of object properties from the resulting models. The most comprehensive description of the deformation process is performed in dynamic models which express the deformation as a function of acting loads and time (Welsch and Heunecke 2001).

Upon the two distinct strategies of dynamic modeling this study refers mainly to the behavioral approach.

Regardless if the models are linear or non-linear or if the modeling is performed in time or frequency domain a common assumption in this approach is the homogeneity of the statistical parameters of the analyzed data. This refers mainly to the constancy of the mean, of the variance and the dependency of the autocorrelation function on the time lag.

H. Neuner (✉)
Research Group Engineering Geodesy, Department Geodesy and Geoinformation, TU Vienna, Gusshausstrasse 27-29, 1040 Vienna
e-mail: hans.neuner@geo.tuwien.ac.at

The homogeneity assumption is implicit. The linear modeling in time-domain uses the entire data set undifferentiated for the estimation of the weighting coefficients (Neuner et al. 2004). For linear modeling in frequency domain the data is converted using the Fourier-transform, which typically treats the data in a global way (Kuhlmann 1996). Non-linear modeling, i.e. by Artificial Neural Networks (ANN), assumes implicitly homogeneous statistical properties of the data as well. This enables the use of a subset of the data for network training (Miima 2002). In every case the model performance decreases dramatically if the assumptions are violated.

A two-step approach was presented in Neuner (2008) to handle data with heterogeneous mean and/or variance in dynamic deformation modeling. The study refers to piecewise homogeneous series as a special type of heterogeneity. The locations of changes in the statistical parameters are called change-points.

After transforming the time series with a wavelet-transform (Bäni 2005) in a first step, the variance homogeneity is assessed on the basis of the wavelet-coefficients. A main advantage of this approach is the homogeneous mean of these coefficients due to their centering about the zero axis. This allows a separate assessment of mean and variance homogeneity.

H. Kutterer et al. (eds.), *The 1st International Workshop on the Quality of Geodetic Observation and Monitoring Systems (QuGOMS'11)*, International Association of Geodesy Symposia 140,
DOI 10.1007/978-3-319-10828-5_20,

A statistical variance homogeneity test, originally motivated by Durbin (1968) and developed later by Inclán and Tiao (1994), is applied on the series of wavelet coefficients. The test uses a test value based on the centered cumulative sum of squares of the samples (the wavelet coefficients). Its main disadvantage is the availability of probability bounds for the test value only in case of identical and independent distributed (i.i.d.) samples. This is however a very severe restriction which is seldom fulfilled by real-world data. Due to the large variety it is difficult to assess how violations affect the test power and the obtained results in last consequence.

To avoid the abovementioned problems, this paper discusses an alternative method for variance change-point (vcp) detection. The method has a non-statistical background. Its main concept is the treatment of the vcp detection as a pattern recognition task, which is solved by means of ANN. Up to now, ANN were used in engineering surveying in a regression framework to perform non-linear system identification (i.e. Heine 1999; Miima 2002; Neuner 2010).

Different to the regression framework, pattern recognition in the sense it's used in this paper aims the classification of data with respect to a single feature or a set of features. In this case, the feature of interest is the level of variance in parts of the analyzed time series. The outputs of the analysis are labels assigned to the input data according to their membership.

Some advantages of using this ANN-based approach over other vcp detection methods and classification methods, like clustering algorithms, are:

- The method accounts for the alignment in time of the analyzed data. This allows the time localization of identified vcp.
- It is a supervised method. The performance of the model can be controlled by appropriate set up of the model structure (see Sects. 2 and 3) and by special designed training signals.
- The method doesn't assume a specific probability or correlation structure of the analyzed data.

The paper has the following structure: Basics related to ANN are presented in Sect. 2. The theoretical background of the vcp detection method is given in Sect. 3. Obtained results are discussed in Sect. 4. The paper concludes with a summary and an outlook.

2 Artificial Neural Networks

ANN are non-linear model structures that contain units of information processing—the nodes—organized in layers. The interaction between the nodes and their arrangement in layers allowsfor a parallel processing of the information.

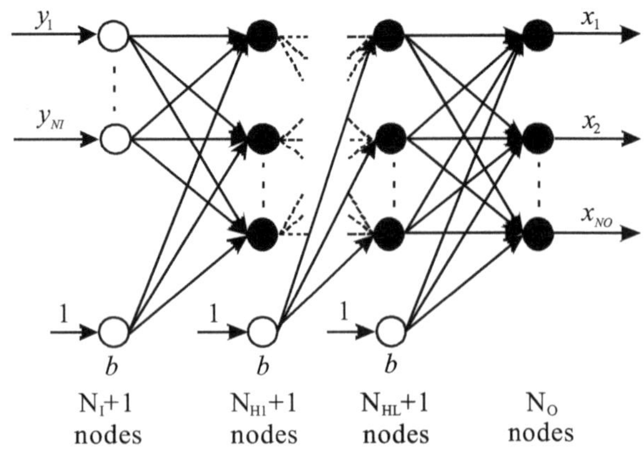

Fig. 1 General structure of an ANN

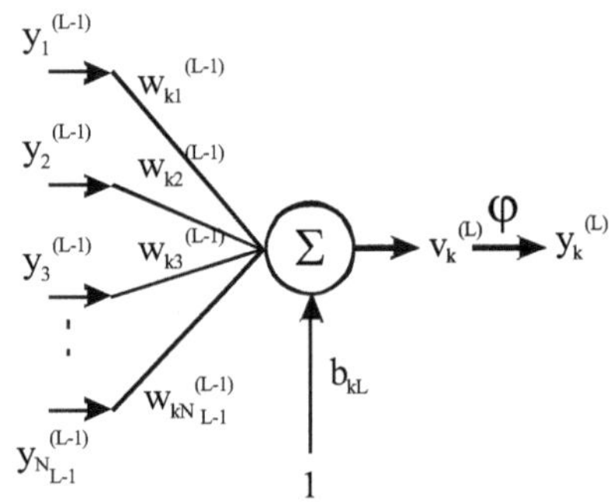

Fig. 2 Information processing at node level

An extensive treatment of ANN is given by Haykin (1999) and Bishop (2005).

The general structure of an ANN is shown in Fig. 1. Starting from left, the first layer contains the input nodes. These are non-processing units that transmit and distribute the input in the network. Usually the number of input nodes N_I equals the number of influencing parameters of the process. The rightmost layer contains the output nodes. These are information processing units which calculate the final result. The number of output nodes N_O corresponds to the number of output parameters of the process. Additionally, the model may contain a variable number of layers situated between the input and the output layer. These are called hidden layers. The number of hidden layers N_H and of nodes included in them is a model selection task that needs to be solved individually in each modeling activity.

The nodes of adjacent layers are completely linked. The strength of the links is expressed by weights. These weights are initially unknown and represent the model parameters that need to be estimated from pairs of input–output data in the training phase.

The information processing at node level is drafted in Fig. 2 and is done according to the formula:

$$y_k^{(L)} = \varphi(v_k) = \varphi\left(\sum_{i=1}^{N_{L-1}} w_{ki}^{(L-1)} \cdot y_i^{(L-1)} + b_{kL}\right), \quad (1)$$

with $k = 1 \ldots N_L$. In (1) $y_k^{(L)}$ is the output from the kth node in layer L, φ is the node specific activation function, $w_{ki}^{(L-1)}$ is the weight corresponding to the link between the kth node in layer L and the ith node in layer (L−1), b_{kL} is a node specific bias term and N_L is the number of nodes in layer L.

Typically, a sigmoidal function, like the tanh, is used for the activation of the hidden nodes and a linear function is used for the activation of the output nodes. This set up was also adopted in the present case.

The resulting non-linear model is fitted to the observed pairs of input and output data. The unknown weights are estimated by minimizing a scope function E, which is built as the sum of squared deviations between the observed and computed values of the output, y_{obs} and y_{ANN} respectively:

$$E = \frac{1}{N}\sum_{i=1}^{N} E_i = \frac{1}{2N}\sum_{i=1}^{N}\sum_{k=1}^{N_O}\left(y_{obs,i}^{(k)} - y_{ANN,i}^{(k)}\right)^2 \rightarrow \min., \quad (2)$$

where N is the number of samples.

The parameter estimation problem was solved in this study with the Levenberg–Marquardt-algorithm (LM-algorithm) which uses a second order approximation of the error surface described by (2). In generally, such an approximation requires the computation of the Hesse–Matrix. This is especially for highly non-linear problems a challenging task. The LM-algorithm circumvents the explicit determination of this matrix by using an approximation which involves the Jacobi-matrix **J**. Thus, starting with an approximate solution $\mathbf{w_0}$ the modification of the estimated parameters $\Delta\mathbf{w}$ is done iteratively:

$$\Delta\mathbf{w} = \left(\mathbf{J}^T\mathbf{J} + \mu\mathbf{I}\right)^{-1} \cdot \nabla\mathbf{E}(\mathbf{w}), \quad (3)$$

where $\nabla\mathbf{E}(\mathbf{w})$ denotes the gradient of the scope function (2), μ a regularisation factor and **I** the identity matrix. The solution (3) is similar to the solution of the Gauss–Markov-model. For $\mu = 0$ the two solutions are identical. For large values of μ the contribution of $\mathbf{J}^T\mathbf{J}$ to the solution is negligible and the solution is equivalent to the one obtained by the steepest-descent.

A main reason for using ANN in vcp detection is their property of universal approximation. As shown by Hornik et al. (1989) an ANN with a single hidden layer of nodes activated by a sigmoidal function and linear activated nodes in the output layer can approximate a continuous function to any degree of accuracy. According to this property the only free structural parameter of the model is the number of nodes included in the hidden layer. The ability of the model structure to approximate complex functional dependencies increases with the number of hidden nodes.

The aimed variance change detection sensitivity is expressed in the ANN model by the number of output nodes. Each output node refers to a specified level of the variance through an assigned numerical label. The higher the detection sensitivity the more complex is the function needed to distinguish between the different variance levels. Therefore, a useful approximation of this function requires an adequate number of hidden nodes. An increase of the number of hidden nodes leads to an increased number of unknown weights that have to be estimated. In last consequence, this affords an increase amount of data in the training. These connections suggest that the aimed detection sensitivity, the number of hidden nodes and the number of training samples should be well-thought-out combined. This aspect is investigated in the next section.

3 Change-Point Detection by ANN

Before starting the processing with ANN the input and output data is usually normalized to a common domain, i.e. [−1, 1]. One reason for this pre-processing step is the use of activation functions, like the tanh, which are S-shaped and have the main transition band in that domain. In order to achieve a good improvement rate of the weights during the iterated estimation it is necessary that the weighted sum of the inputs to a node, which according to (1) represents the argument of the activation function, is situated in this domain. This normalization step is also useful for this study. The transformation of all data independently from their nature, magnitude or source to a common variance level allows an objective assessment of the latter. Thus, a detection sensitivity of the model can be specified according to the specific problem and considered in the design of the ANN. Furthermore, a single ANN structure can be used for the analysis of all influencing parameters and deformation signals if the aimed detection sensitivity is the same.

The input to the network was firstly generated by means of a data window of fixed size. The window was moved over the analysed time series with a stepsize of 1. The data included at a certain position build one epoch of the input to the network. This approach is similar to process control studies (Pugh 1989; Ooh et al. 2005). Based on the analysis of the dependency of the variance of the variance estimation on the number of samples a window size $N_F = 20$ was chosen. This corresponds to an ANN architecture with 20 input nodes. In case of standard normal distributed data the estimation of variance from $N = 20$ samples has a variance of 10 % which was considered to be sufficient for this study.

It turns out that the use of the normalized data in the input leads to poor detection performance. Therefore, in a second approach characteristic measures for the variability of data where used as input to the network. Concretely, the values of the sum of squares and the absolute value of the range

$$\sum_{i=1}^{N_F} x_i^2 \text{ and } \left| \max_{i=1..N_F}(x_i) - \min_{i=1..N_F}(x_i) \right|, \tag{4}$$

where calculated from samples contained in one position of the data window. According to (4), the dimension of the input reduces to two input nodes.

The output from the network are numeric labels assigned to each variance level. These labels characterize only the variance level and are not estimates of it. The number of levels equals the number of output nodes and must be task-oriented predefined. The finer the division of the variance domain the more sensitive must be the detection model. This requires a large amount of free parameters in the model and implicitly large data sets for their estimation. Despite the increase in the computation burden this is not a restriction, due to the fact, that the training of the ANN is based on synthetic signals, as shown in the next section.

The linear activation of the output nodes leads to a continuous output from the network. The nearest label to the computed output determines the assignment to a certain predefined variance level. This corresponds in fact to the use of threshold activation functions for the output nodes.

A main task in the design of ANN vor vcp detection is the specification of the number of hidden nodes. This number determines the structure of the model and thus, the capacity of the network.

A good insight to the notion of capacity is gained by the analysis of an ANN with N_I binary input nodes and one binary output node activated by a threshold function. Without loss of generality the bias term is ignored. There are $2^\wedge(N_I)$ distinct input patterns and each can be classified in 2 different ways by the output node. This leads to $2^\wedge(2^\wedge N_I)$ possibilities in which the N_I inputs can be classified.

The thresholding occurs when the linear output equals to 0. In this case relation (1) can be written for the output node:

$$x_1 = \varphi\left(\sum_{i=1}^{N_I} w_{1i} \cdot y_i\right) = \sum_{i=1}^{N_I} w_{1i} \cdot y_i = 0. \tag{5}$$

For $N_I = 2$ Eq. (5) describes a separating line between two regions of the input space. In this case there exist 4 distinct input patterns $\{(+1, +1); (-1, -1); (+1, -1); (-1, +1)\}$ and 16 possibilities in which they can be classified. Out of these 16 possibilities only 14 can be classified by a linear discriminator. The two unhandled possibilities correspond to the logical XOR-function. The model capacity reflects the number of distinct classification functions that can be implemented by the model.

Note in the abovementioned example, that if the two-dimensional input contains only three input patterns the linear discriminator given by (5) can error-free handle all the eight classification possibilities. This suggests that the capacity of a model should be assessed with respect the number of distinct patterns of the particular input space.

In general, a model of a certain capacity can implement all possible classifications up to a maximum number of distinct patterns N_{max}. In case of the two input linear discriminator $N_{max} = 3$. Beyond this number of samples the model can implement only a fraction of the total amount of possibilities (for $N = 4$ only 14 out of 16 functions can be implemented). The maximum number of distinct samples N_{max} for which a model structure can implement all $2^\wedge(N_{max})$ possible classification functions is called Vapnik–Chervonenkis dimension (VC-dimension) and is a characteristic of the particular (ANN-)model.

Unfortunately, the exact determination of this characteristic measure is very difficult, especially for complex network structures. Up to now a large amount of research concerns the determination of lower and upper bounds for the VC-dimension of various ANN-structures. For the present study the results obtained by Baum and Haussler (1989) are useful. They obtained the following upper bound on the VC-dimension for an ANN that consists of N_K threshold activated nodes and N_W weights:

$$\text{VC-dimension} \leq 2N_W \log_2(eN_K), \tag{6}$$

where e is the base of the natural logarithm. Based on (6) they showed that if out of N patterns, with

$$N \geq \frac{N_W}{\varepsilon} \log_2\left(\frac{N_K}{\varepsilon}\right). \tag{7}$$

a fraction of $1 - (\varepsilon/2)$ is correctly classified by that ANN in the training phase, where $0 < \varepsilon \leq 1/8$, than the network will correctly classify a fraction $1 - \varepsilon$ of future patterns in the so called test phase of the ANN, if these follow the same distribution. This result was used to determine the number of training samples for ANN with a certain number of hidden nodes.

4 Results and Practical Considerations

In a first example the ANN-based change-point detection method is used to detect variance changes in synthetic signals. For the network training $N = 10{,}000$ samples were generated from the standard normal distribution. Three variance change-points of different magnitude were introduced

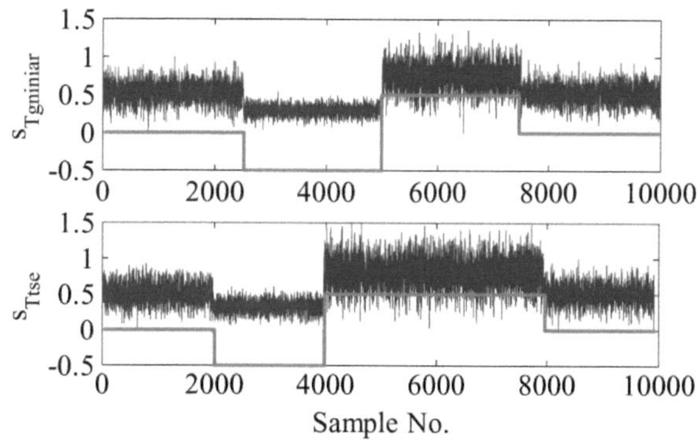

Fig. 3 VCP detection on synthetic signals

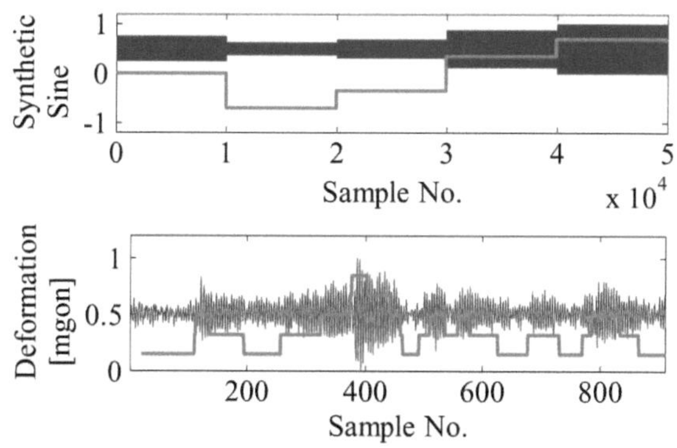

Fig. 4 VCP detection for a deformation signal

at locations of N/4, N/2 and 3 N/4. The magnitude of the changes is expressed with respect to the original variance level of 1, which was not modified in the first part of the series and amounts to: 0.3, 2.0 and 1.0. The synthetic signal is shown in blue in the upper graph of Fig. 3.

The input was generated from the original series using Eq. (4) with $N_F = 20$. The desired output is defined according to the variance level in each of the four sections of the signal. The labels 0, −0.5, 0.5 were chosen to characterize the three distinct variance levels. By this choice one reads in the output signal if the variance increases or decreases with respect to the reference level. Note that these labels are arbitrarily chosen and are not estimates of the variance or change in variance.

The network weights were estimated using the LM-algorithm. The output computed by the network is shown in red in the upper graph of Fig. 3. As can be seen all samples of the synthetic signal were error-free classified. This quote is an exception. In similar studies the classification error amounts to 3-5‰.

To assess the generalisation properties of the network a second synthetic signal consisting of 10,000 samples was generated from the standard normal distribution. The change-points were placed at different locations: N/5, 2 N/5 and 4 N/5. The magnitude of the first 2 change-points was also modified to 0.4 and 2.5 respectively. The structure of this signal is shown in blue in the second graph of Fig. 3.

The already trained network was used to detect the variance changes in this second signal. This corresponds to a run of the model in prediction modus. The classification result is shown in red in the same graph. Again, an error-free classification is attained with that model. This proves its good performance for this test scenario.

In a second application the method was used to detect variance changes in series of wavelet-coefficients of a deformation signal recorded at a wind energy turbine.

The analysed series is shown in blue in the second graph of Fig. 4. The analysed wavelet coefficients contain only the variability information.

Motivated by the fact that many real-world processes contain cyclic components a synthetic cyclic signal was used in the training phase to estimate the weights of the ANN. The periods of the synthetic cyclic signals correspond to the dominant periodicities contained in the analysed series of coefficients. These are identified by spectral analysis.

Five different variance levels were introduced in the synthetic signal by multiplying the unity amplitude with 1.0, 0.5, 0.75, 1.5 and 2.0. The structure of the ANN includes 15 nodes in the hidden layer. According to (7) for an amount of N = 50,000 training patterns the misclassification error of future data is below $\varepsilon = 2.5$ %. Therefore, 1,00,000 samples were generated for each predefined variance level. The use of a synthetic cyclic signal in the training phase is an important advantage of the proposed approach because it allows the generation of sufficiently large data sets in accordance to the desired level of misclassification for future data.

Subsequently, the trained network is used for the variance homogeneity assessment in the series of wavelet coefficients. The output computed by the network in case of the real-world data is shown in red in the second graph of Fig. 4. As can be observed, the model performs fairly good in identifying distinct variance levels of the analysed coefficient series. This result is used to set up a dynamic model which accounts for the different variance levels.

5 Summary and Outlook

This paper deals with the variance change-point detection by means of ANN. The developed method refers to a proper generation of the input and the desired output signals used to train the network. The design of the network architecture is based on the notion of model capacity and related to that, on the VC-dimension of the model structure. Some results from practical applications are discussed in case of synthetic signals and real-world data. Future work is focused on the determination of tight upper bounds for the model capacity and a more thorough theoretical justification of the window size N_F used to generate the input to the network.

References

Bäni W (2005) Wavelets. Oldenbourg, München

Baum EB, Haussler D (1989) What size net gives valid generalization? Neural Comput 1(1):151–160

Bishop CM (2005) Neural networks for pattern recognition. Oxford Press, UK

Durbin J (1968) The probability that the sample distribution function lies between two parallel straight lines. Ann Math Stat 39(2):398–411. Waverly Press, Baltimore

Haykin S (1999) Neural networks: a comprehensive foundation, 2nd edn. Pearson Education, Singapore

Heine K (1999) Beschreibung von Deformationsprozessen durch Volterra- und Fuzzy-Modelle sowie Neuronale Netze. DGK München, Series C, No. 516

Hornik K, Stinchcombe M, White H (1989) Multilayer feedforward networks are universal approximators. Neural Networks 2:359–366. Pergamum Press.

Inclán C, Tiao G (1994) Use of cumulative sums of squares for retrospective detection of changes of variance. J Am Stat Assoc, Theory Methods 89(427):913–923

Kuhlmann H (1996) Ein Beitrag zur Überwachung von Brückenbauwerken mit kontinuierlich registriertenMessungen. PhD-Thesis. Wissenschaftliche Arbeiten der Fachrichtung Vermessungswesen der Universität Hannover, Nr. 218, Hannover

Miima J-B (2002) Artificial neural networks and fuzzy logic techniques for the reconstruction of structural de-formations. PhD-Thesis. Geodätische Schriftenreihe Nr. 18, TU Braunschweig.

Neuner H, Hesse C, Heer R (2004) Kombination verschiedener geodätischer Sensoren zur Überwachung von Kaimauern. In: Schwieger V, Foppe K (eds) Kinematische Messmethoden – Vermessung in Bewegung, vol 58. DVW-Seminar, Stuttgart, Wißner Verlag, Augsburg.

Neuner H (2008) Zur Modellierung und Analyse instationärer Deformationsprozesse. Wissenschaftliche Arbeiten der Fachrichtung Geodäsie und Geoinformatik der Leibniz Universität Hannover, No. 269

Neuner H (2010): Modelling deformations of a lock by means of artificial neural networks. In: Reiterer A, Egly U, Heinert M, Riedel B (eds) Proceedings of the second international workshop application of artificial intelligence and innovations in engineering geodesy. Braunschweig, Germany, pp 32–41

Ooh KJ, Moon MS, Tae YK (2005) Variance change point detection via artificial neural networks for data separation. Neurocomputing 68:239–250

Pugh AG (1989) Synthetic neural networks for process control. Comput Ind Eng 17(1–4):24–26

Welsch W, Heunecke O (2001) Models and terminology for the analysis of geodetic monitoring observations. Official Report of the Ad-Hoc Committee of FIG Working Group 6.1, Copenhagen, Denmark

Spatial and Temporal Kinematics of the Inylchek Glacier in Kyrgyzstan Derived from Landsat and ASTER Imagery

M. Nobakht, M. Motagh, H.U. Wetzel, and M.A. Sharifi

Abstract

Spatio-temporal variations of glacier flow are a key indicator of impact of global warming, as the glaciers react sensitively to change in climate. Satellite remote sensing using optical imagery is an efficient tool for studying ice-velocity fields on mountain glaciers. This study evaluates the potential of Landsat and ASTER imagery to investigate surface velocity field associated with the Inylchek Glacier in Kyrgyzstan. We present a detailed map for the kinematics of Inylchek glacier obtained by cross correlation analysis of Landsat images, acquired between 2000 and 2010, and a pair of Advanced Spaceborne Thermal Emission and Reflection Radiometer (ASTER) images covering the time period of 2007–2008. Our result indicates a high-velocity region in the elevated part of the glacier moving up to a rate of about 0.5 m/day. Time series analysis reveals some annual variations in the mean surface velocity of the Inylchek during 2000–2010.

Keywords

ASTER • Cross-correlation • Glacier dynamics • Inylchek/Kyrgyzstan • Landsat

1 Introduction

One of the greatest concentrations of permanent snow and ice in the mid-latitudes of the Northern Hemisphere is located in glaciated mountains of Central Asia. Tien Shan mountain system hosts some of the largest non-polar glaciers in the world (Mayer et al. 2008). Mountain glaciers are the only renewable fresh water resource in Central Asia, dominated in part by large deserts and arid lowlands with very low precipitation and extremely dry climates (Aizen et al. 2007).

Tien Shan mountain in Central Asia consists of 15953 glaciers with a total area of 15,416 km^2 and a total volume of 1,048 km^3. It is a vital source of water for rivers and lakes in this region, in particular during dry years (Aizen et al. 2007). Inylchek glacier in Kyrgyzstan (Fig. 1) located in Central Tien Shan is the largest glacier of this mountain system with 61 km of length and an area of 567.20 km^2. Its glacial runoff is 54 % of the average annual runoff of Kumarik River (Shen et al. 2006). Inylchek also contains the largest glacial lake of Central Asia, called Lake Merzbacher, located in the conjunction point of northern and southern Inylchek (Wetzel et al. 2005). This ice-dammed lake is founded by an ice dam preventing the northern Inylchek drainage water from direct runoff into the valley (Mayer et al. 2008). Frequent glacier lake outburst floods (GLOFs) of Lake Merzbacher in recent

M. Nobakht • M.A. Sharifi
Department of Surveying and Geomatics Engineering, University College of Engineering, University of Tehran, Tehran, Iran

M. Motagh (✉)
Department of Geography and Environmental Science, The University of Reading, Whiteknights, RG6 6AB Reading, UK

GFZ German Research Centre for Geosciences, Department of Geodesy and Remote Sensing, Potsdam, Germany
e-mail: motagh@gfz-potsdam.de

H.U. Wetzel
GFZ German Research Centre for Geosciences, Department of Geodesy and Remote Sensing, Potsdam, Germany

H. Kutterer et al. (eds.), *The 1st International Workshop on the Quality of Geodetic Observation and Monitoring Systems (QuGOMS'11)*, International Association of Geodesy Symposia 140,
DOI 10.1007/978-3-319-10828-5_21,

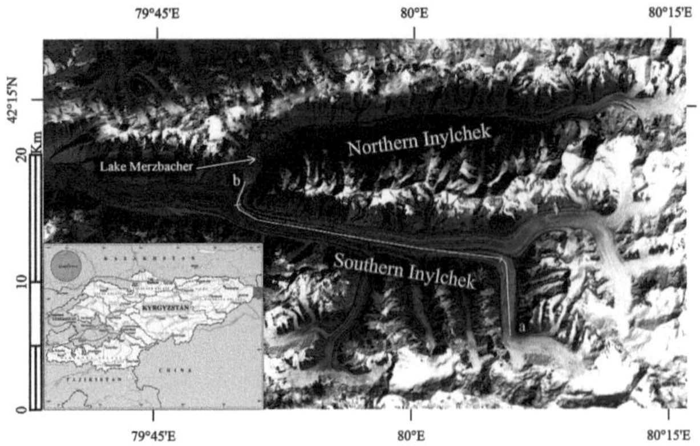

Fig. 1 Position of Inylchek glacier and Merzbacher lake in a Landsat image, acquired 21-aug-2006, bands 4, 3, 2 > RGB

years caused heavy damage to infrastructure along the glacier outflow region.

Valuable information about the dynamics and physical condition of glaciers can be achieved by measurement of surface velocity fields of glaciers. Inylchek glacier is located in a very remote area with an average elevation of 4,000 m above sea level. Performing ground-based in-situ measurement of glacier surface velocity in this area is very costly and extremely time-consuming. Remote sensing measurement using optical imagery provides a valuable tool to study and assess the kinematics of this type of glaciers.

Global coverage, satisfying spatial and temporal resolution (15 m ground pixel size, 16 days revisit period) and low cost of ASTER images provide an opportunity to investigate the dynamic and kinematic of mountain glaciers (Kargel et al. 2005). In comparison with whiskbroom sensors, the use of images such as ASTER obtained from pushbroom sensors provides reliable results, as inherent problems related to effects of attitude variations and inaccurate DEMs can be solved by using raw image metadata. Also, extensive global coverage, relatively good spatial and temporal resolution (30 m ground pixel size, 16 days revisit period) and free availability of Landsat scenes make this type of imagery a viable option among other alternatives for large-scale and long-term monitoring of remote glacial systems. Landsat data has been extensively used in the past three decades to study the kinematics and dynamics behaviour of several glaciers in Antarctic region (Bindschadler 1998; Maas et al. 2008).

By using optical imagery for the kinematics analysis of the glaciers the accuracy of the results greatly depends on several factors including the ground resolution of the images and the ability to precisely co-register consecutive images (Scherler et al. 2008). To obtain an accurate and reliable displacement field, performing precise image to image co-registration is a very critical processing step (Leprince et al. 2007). In this study, several Landsat images and a pair of ASTER images acquired over the Inylchek glacier in the past decade are used to evaluate the temporal and spatial pattern of its surface velocity field.

The paper is arranged as follows. Section 2 presents the characteristics of images and the methodology used for precise co-registration and correlation analysis of the images. In Sect. 3 we discuss the obtained results and perform quality test to determine factors affecting the quality of measurements. We also investigate the annual variations in surface velocity of glacier along a longitudinal profile. Finally we conclude the paper with a summary of our findings in this research in Section "conclusion".

2 Data and Methodology

2.1 ASTER Imagery

ASTER's spectral and geometric characteristics include three bands in VNIR (Visible Near Infra-Red) range with 15 m resolution, six bands in SWIR (Short Wave Infra-Red) with 30 m, five bands in TIR (Thermal Infra-Red) with 90 m, and a 15 m resolution NIR along-track stereo-band looking back 27.6° from nadir. Here we used VNIR band (band 3 N), due to high contrast between features in this frequency band, which facilitates both co-registration and cross-correlation analysis of images. Following the standard procedure described in detail in Leprince et al. (2007) and Scherler et al. (2008), a digital elevation model is used as the global ground truth for orthorectification of ASTER imagery, because the images are not globally georeferenced.

A common problem for both radar-based and optical-based DEMs is the existence of large gaps and voids especially in mountainous terrains. Smaller gaps can be interpolated using the original data but larger gaps and void areas should be interpolated using other data sources. In this study, a 90 m spatial resolution DEM provided by Shuttle Radar Topography Mission (SRTM) is used. The large voids in SRTM are patched with data from topographic maps, freely available from Jonathan de Ferranti (http://www.viewfinderpanoramas.org). Over 10 tie points are selected on stable features between first raw ASTER image of 12 May 2007 and a shaded version of DEM. Selected tie points are locally optimized using sub-pixel correlation and converted into ground control points (GCP) for precise orthorectification. In our analysis an average misregistration of about 50 cm is achieved with a standard deviation of 14 m between ASTER image of 12 May 2007 and hill-shaded DEM. Following this procedure, the other ASTER raw image of 01 July 2008 is orthorectified and co-registered relative to the fist ortho-image with average misregistration of less than 2 cm and standard deviation below 2.5 m (1/6 pixel size). This very precise co-registration and orthorectification is a critical perquisite step for obtaining accurate measurement of surface displacement by cross-correlation of two ortho-images.

To derive accurate and confident velocity field a cross-correlation algorithm in frequency domain, which relies on Fourier shift theorem is applied (Leprince et al. 2007). The relative displacement between two precisely co-registered images can be retrieved from the phase difference of their Fourier transforms, as follows: If $i_1(x, y)$ and $i_2(x, y)$ be two consecutive images that differ only by a displacement of $(\Delta x, \Delta y)$, we have

$$I_2\left(\omega_x, \omega_y\right) = I_1\left(\omega_x, \omega_y\right) e^{-j\left(\omega_x \Delta_x + \omega_y \Delta_y\right)} \quad (1)$$

where I_1 and I_2 denote the Fourier transform of images and (ω_x, ω_y) are the frequency variables in column and row. Normalized cross-spectrum of the images is expressed by

$$C_{i_1 i_2} = \frac{I_1\left(\omega_x, \omega_y\right) I_2^*\left(\omega_x, \omega_y\right)}{\left|I_1\left(\omega_x, \omega_y\right) I_2^*\left(\omega_x, \omega_y\right)\right|} = e^{j\left(\omega_x \Delta_x + \omega_y \Delta_y\right)} \quad (2)$$

where * denotes the complex conjugate. According to cross-spectrum characteristics, relative displacement can thus be recovered by either estimation of the linear phase of the images' cross-spectrum (Van Puymbroeck et al. 2000), or determination of the exact location of the correlation peak (Foroosh et al. 2002). In this study we used a combined method developed by Leprince et al. (2007), using the COSI-Corr module. This process provides us with two correlation images representing horizontal ground displacement in East–West and North–South directions, and a Signal-to-Noise Ratio (SNR) for each measurement, representing the confidence of the results.

2.2 Landsat Imagery

Table 1 lists the images analyzed in this study. Suitable Landsat pairs were selected among more than 70 images acquired in the period 2000–2010 in this region. Many parameters have been considered for selecting these pairs, in particular the presence of strip lines in images, Scan Line Corrector (SLC)-failure in Landsat7 and the similarity of scenes with regards to cloud cover and snow cover. In contrast to ASTER raw images, all Landsat images are accessible in orthorectified and georeferenced format.

Similar to ASTER images, the co-registration of Landsat images is a key step to measure glacier surface velocity fields. It will significantly affect the accuracy of the obtained velocities (Berthier et al. 2003; Leprince et al. 2007). In this study, an automated precise registration and orthorectification package, called AROP, is used for precise co-registration of Landsat Images (Gao et al. 2009). Using this package we achieved co-registered pairs, consistent in the geographic extent, spatial resolution, and projection. Over 300 tie points between two images were generated with the RMSE less than 0.4 pixel size for each pair. Finally, correlation map in sub-pixel accuracy was obtained by applying cross-correlation for a 64 × 64 window as the initial correlation search window and 32 × 32 sliding window as the final correlation window. Final ground resolution of 60 m was obtained by a sliding step of 2 pixels for correlation window.

Table 1 Images used to study Inylchek glacier

Master image	Slave image	Sensor	Separation (day)
13 Sep 2000	16 Sep 2001	Landsat 7	368
13 Sep 2000	01 Jul 2002	Landsat 7	656
16 Sep 2001	01 Jul 2002	Landsat 7	288
23 Feb 2002	14 Mar 2003	Landsat 7	384
21 Aug 2006	24 Aug 2007	Landsat 5	368
12 May 2007	01 Jul 2008	ASTER	416
26 Jun 2009	15 Jul 2010	Landsat 5	384

3 Results

Figure 2 illustrates an example of East–West components of displacement field obtained by sub-pixel correlation of two Landsat images acquired in August 2006 and August 2007 over Inylchek glacier. For a better visualization, a slope masking has been applied using SRTM digital elevation model, and the displacement map has been overlaid on a Landsat image. In this case the slope of all glaciated re is lower than 20°.

As a consistency test three images from the years 2000, 2001 and 2002 are investigated to compare the cumulative displacement of 2000–2001 and 2001–2002 pairs with the one obtained directly from 2000 to 2002 pair (see Table 1). An example of surface displacement along the a-b profile in Fig. 1 is given in Fig. 3. A good agreement over most parts of the profile is observed. There are some discrepancies in the centre of the profile, which is related to the effect of cloud cover and difference in sun azimuth in 2000–2001 and 2000–2002 pairs, as confirmed by visual investigation of images.

Based on our observation along this profile, we can analyse the mean surface velocity in three parts; velocities in a high elevated tributary of glacier (Part I), velocities along the main trunk of southern Inylchek glacier (Part II) and surface velocities right before glacier drainage to the lake Merzbacher (Part III).

Variation of the mean surface velocity along the profile different time period associated with the data in Table 1 is illustrated in Fig. 4. As shown in this figure, the highest mean

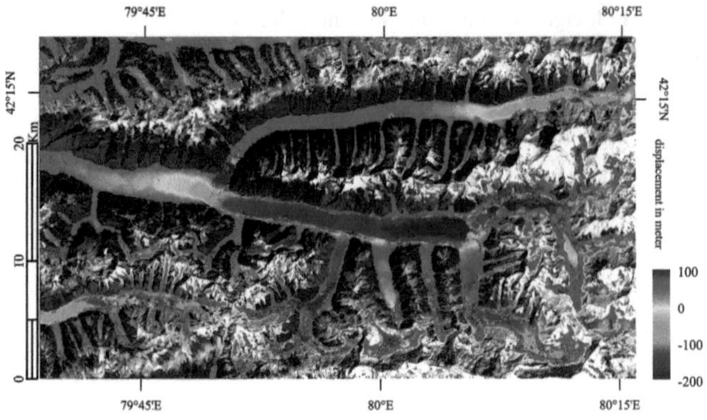

Fig. 2 East–West component of the correlation of two Landsat images over the central Inylchek glacier for the period of 21 Aug 2006 to 24 Aug 2007 overlaid on a Landsat RGB image. Displacements are positive toward the East

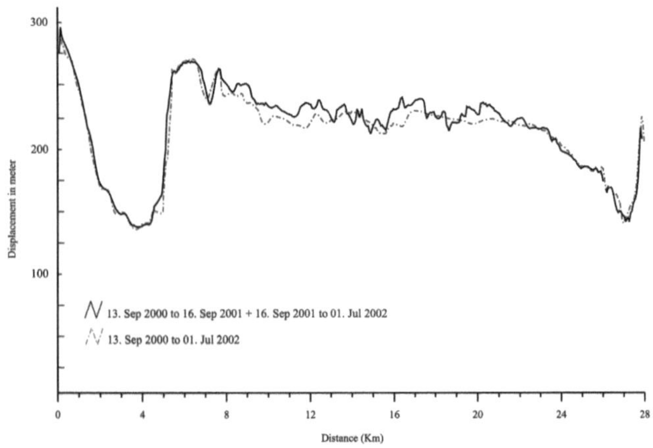

Fig. 3 Surface displacement along the longitudinal profile over Inylchek glacier (see Fig. 1). *Solid* and *dashed lines* depict the total displacement over the time span given in the figure's legend

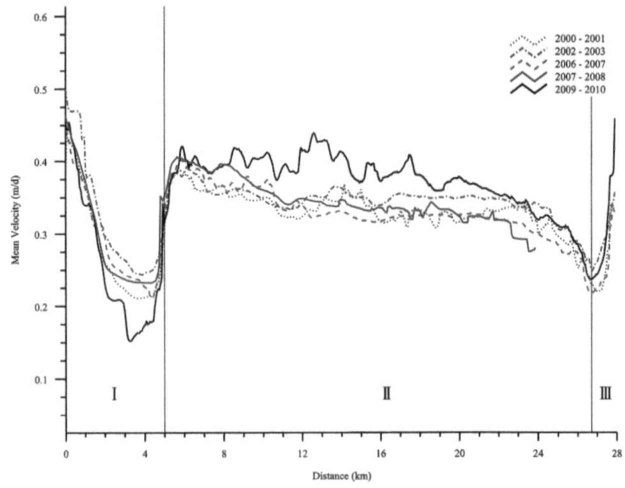

Fig. 4 Surface velocities derived from repeat Landsat and ASTER imagery by cross-correlation analysis along the a–b profile indicated in Fig. 1

surface velocity is occurring at the most elevated part of the profile (Part I), Glacier flows northward in this part and its velocity sharply decreases along the profile. It is interesting to note that a quick change in velocity is happening between Part I and Part II, at the kilometre 5 along the profile a–b. This is due to the difference in velocities between the ablation zone of the glacier and the accumulation zone, as expected. The surface velocity slowly decreases downstream along Part II, as can be expected for parallel ice flow in the ablation zone. This part of the profile shows some fluctuations in the mean surface velocity; the velocity obtained for the 2009–2010 period are about 1.25 times higher than that obtained for the 2006–2007 period.

A major extent of south Inylchek glacier flows toward the Lake Merzbacher (Mayer et al. 2008). Thus, we observe a reduction in the mean velocity exactly before the region of glacier rotation toward the lake. In contrast, close to the ice dam (Part III) a considerable acceleration is observed, which is in accordance with the glacier drainage and glacier calving into the Lake Merzbacher. Our findings in this research agree well with results which have been previously published by Mayer et al. 2008. However they have investigated surface velocities just near the ice dam (Part III) while we present a velocity map for whole glacier extent.

Conclusion

More than 70 Landsat images and 3 ASTER images acquired over the Inylchek glacier in Kyrgyzstan were investigated to derive the temporal and spatial pattern of the surface velocity field of this glacier. Our preliminary results indicate several important features associated with the kinematics of the Inylchek; e.g., a high-velocity region in the elevated part of the glacier is observed moving at a rate of about 0.5 m/day during the 2002–2003 time period. A quick change in velocity is observed between the ablation zone of the glacier and its accumulation zone and also a striking increase in mean surface velocity is observed between 2009 and 2010.

References

Aizen VB, Aizen EM, Kuzmichonok VA (2007) Glaciers and hydrological changes in the Tien Shan: simulation and prediction. Environ Res Lett 2:045019

Berthier E, Raup B, Scambos T (2003) New velocity map and mass-balance estimate of Mertz Glacier, East Antarctica, derived from Landsat sequential imagery. J Glaciol 49:503–511

Bindschadler R (1998) Monitoring ice sheet behavior from space. Rev Geophys 36:79–104

Foroosh H, Zerubia JB, Berthod M (2002) Extension of phase correlation to subpixel registration. IEEE Trans Image Process 11:188–200

Gao F, Masek J, Wolfe RE (2009) Automated registration and orthorectification package for Landsat and Landsat-like data processing. J Appl Remote Sensing 3:033515

Kargel J, Abrams M, Bishop M, Bush A, Hamilton G, Jiskoot H, Kaab A, Kieffer H, Lee E, Paul F (2005) Multispectral imaging contributions to global land ice measurements from space. Remote Sens Environ 99:187–219
Leprince S, Barbot S, Ayoub F (2007) Automatic and precise orthorectification, coregistration, and subpixel correlation of satellite images, application to ground deformation measurements. IEEE T Geosci Remote Sensing 45:1529–1558
Maas HG, Schwalbe E, Dietrich R, Bässler M, Ewert H (2008) Determination of spatio-temporal velocity fields on glaciers in West-Greenland by terrestrial image sequence analysis. Int Arch Photogramm 37:1419–1424
Mayer C, Lambrecht A, Hagg W, Helm A, Scharrer K (2008) Post-drainage Ice Dam response at lake Merzbacher, Inylchek Glacier, Kyrgyzstan. Geogr Ann A 90:87–96
Scherler D, Leprince S, Strecker M (2008) Glacier-surface velocities in alpine terrain from optical satellite imagery-accuracy improvement and quality assessment. Remote Sens Environ 112:3806–3819
Shen YP, Wang SD, Wang GY, Sha C, Mao WY (2006) Response of glacier flash flood to global warming in Tarim River Basin. Adv Climate Change Res 2:32–35
Van Puymbroeck N, Michel R, Binet R, Avouac JP, Taboury J (2000) Measuring earthquakes from optical satellite images. Appl Optics 39:3486–3494
Wetzel HU, Reigber A, Richter A (2005) Gletschermonitoring und Gletscherseebruche am Inyltschik (Zentraler Tienshan)-Interpretation mit optischen und Radarsatelliten. Wirtschaftliche Innovation durch Geodaten, Vorträge Tagungsband, pp 341–350

Response Automation in Geodetic Sensor Networks by Means of Bayesian Networks

S. Horst, H. Alkhatib, and H. Kutterer

Abstract

Today's geodetic sensors allow an almost fully automated data collection. This happens in a previously fixed chain with constant parameters. A reaction to events during the measurement process, for example by adjusting the measurement resolution or a specific control of an actuator, is usually not intended. This lack will be overcome by adaption of new communication techniques with networked sensors and a proper assessment of occurring events. As the basis of such an assessment probabilistic state variables of the processes are introduced. As an analysis method Bayesian networks are used in our study, which are powerful tools to make decisions based on uncertain information. The evidence calculation on the sensor nodes is derived by Kalman filtering and a subsequent compatibility test. The advantages of this method are shown by means of a simulation.

Keywords

Assessment of events • Bayesian networks • Kalman filter • Response automation • Sensor networks

1 Introduction

In today's engineering geodesy applications, a variety of different sensors is used. All these geodetic sensors allow an almost fully automated data collection. For example modern total stations are able to align on different points, laser scanners measure huge point clouds or inclinometer sample in certain intervals. Furthermore, new wireless communication techniques are introduced in geodetic engineering processes. For example in Pink (2007) a wireless sensor network based on GNSS sensors for monitoring measurements is evaluated. Despite these new techniques, all applications have one thing in common: usually each step in the process is performed in a previously fixed chain with constant parameters (e.g. scanning a target point at fixed time intervals). A reaction to events during the measurement process, for example by adjusting the measurement resolution or a specific control of an actuator, is usually not intended. Typically the adaptation of the process is carried out by the operator. Furthermore, the evaluations are usually based on a separate interpretation of the individual sensors. A constant impact of events on different sensors is not considered as preliminary information. The aim of this work is to implement an automatic response capability in engineering survey processes using neighborhood relations. The two main necessary steps for an automatic response are the detection and assessment of an event in the measurement process and a decision and performance of an appropriate response. The realization of these two steps is based on a few key assumptions to facilitate the further development:

- The process consists of a number of sensors to observe certain aspects of the environment.

S. Horst (✉) • H. Alkhatib
Geodätisches Institut Hannover, Nienburger Str. 1,
30167 Hannover, Germany
e-mail: horst@gih.uni-hannover.de

H. Kutterer
Bundesamt für Kartographie und Geodäsie, Richard-Strauss-Allee 11,
60598 Frankfurt am Main, Germany

H. Kutterer et al. (eds.), *The 1st International Workshop on the Quality of Geodetic Observation and Monitoring Systems (QuGOMS'11)*, International Association of Geodesy Symposia 140,
DOI 10.1007/978-3-319-10828-5_22,

- These observations are rather accurate and reliable. Possibly there is not for each state value a sensor available. On the other hand it is possible, that different sensors for the same state value are available.
- Via a communication system, information can be exchanged between the sensor nodes.
- A prediction of future situations is dependent on the subject of quantity and quality of available observations and with a corresponding uncertainty.
- There is some prior information about possible events.

These assumptions are according to the general definition of (geodetic) sensor networks (Sohraby et al. 2007) and make it possible to establish responsiveness.

The paper is organized as follows. Section 2 describes the concept for the description of states and events in geodetic processes by means of state variables. Also a brief overview about two different ways of the assessment of events is given. Section 3 describes the Bayesian networks, which are used as the main tool for the assessment of events in the sensor network. After the implementation of the Bayesian probability in the network Sect. 4 will deal with the evaluation of a probability value on each sensor node by Kalman filtering. In Sect. 5 a simulation of such an assessment of events is presented. Finally, section "Conclusion" summarizes the results and gives an outlook for future work.

2 State Description

The system theory defines a state of a system as the description of a situation, in which an object with respect to certain selected properties remains unchanged (Lunze 2008). States of a system can be described by means of sensor data. Hence, a state transition is a change to an object, where it is transferred from one state to another. Such state transitions are triggered by events. In order to identify and to assess an event the state change must be described. In geodetic processes usually this is carried out by sensor information. Sensor information can be distinguished in raw sensory information (generated by a given sensor data described without any preprocessing), abstract sensor information (mapping the raw sensor information to a standard value range), as well as aggregated information (summarize a number of different observations to aggregate information) and derived information (all information derived by inference from the available information) (Strassberger 2007). It is obvious that, depending on the sensor information, we get different forms of describing variables. Especially if sensor information with boolean return values are used, a safe deduction of the state change is possible and a response to the event can be derived directly. Exemplary total station measurements to reference points will return boolean variables for the availability of the reflectors. These variables will be referred as *hard* variables.

Table 1 Variables of a change in state

Variable	Value
Logical variable (e.g. availability)	Boolean values
Probability of a change in state	$P(event_i)$
Probability of aggregation value of a change in state	$P(event_{agg} \vert event_1, \ldots, event_i)$
Time function of the probability of a change in state	$f(t; P(event_i))$

However, if uncertain or aggregated information is used, the state change can only be described by a probability. In particular the influence of neighboring sensors can only be modeled in this way. The variables used in this work are shown in Table 1.

The straightforward task of describing and assessing state changes based on the presented variables can be divided into two parts. The logical variables can be modeled with simple queries. It is just a question of control engineering where the query variable generates the response directly. This paper will not deal with such scenarios. The second case based on probabilistic variables will be presented by an example of a monitoring task. By using different sensors at several points, neighbourhood relationships have to be taken into account for the entire assessment of the situation. Additionally, preliminary information on the monitoring area can be considered by using probabilistic variables for this assessment. Based on the assessment, an action should be started.

3 Bayesian Networks for the Assessment of Events

In case of raw sensor information and abstract sensor information the uncertainty is limited to measurement inaccuracies. In geodesy, different common evaluation methods are used. However, when aggregated or derived information is used, there is often no safe deduction possible. In general, the uncertainty with respect to the incidence of an event increases during the inference process. Instead of the safe deduction with boolean variables, the inference process is based on statistical knowledge. Such probabilistic inference procedures are particularly advantageous if not all the interdependent aspects of the context are known locally. A common approach to probabilistic inference is the Bayesian network. The subsequent remarks are mainly based on Jensen (2001) and Koch (2007).

A Bayesian network is an acyclic, directed graph whose nodes represent random variables and whose edges represent the direct causal dependencies between these variables.

Further on variables not connected by edges are conditionally independent. The parents of one node X_i are those nodes, which leads an edge to X_i. For each node X_i exists a conditional probability distribution $P(X_i|par(X_i))$, where $par(X_i)$ is the amount of parents. This conditional probability distribution quantifies the influence of the parents on the node. This probability distribution is a multidimensional table [conditional probability table (CPT)] if the random variables take only discrete values. For variables with continuous state space, however, a continuous probability density function is necessary. In our work, we will assume that all the variables are discrete. The conditional probability tables represent the prior information in the sensor network.

In such a probabilistic network the probability of all states of the network can be calculated (joint probability distribution). The joint probability distribution is expressed as a product of distributions over a smaller number of variables, through repeated application of the product rule of probability calculus and by exploiting conditional independence relations described in the graph structure. In the general case of Bayesian networks, consisting of a set of n nodes organized in a directed acyclic graph, where each node X_i has parents, the joint probability distribution is compactly expressed as

$$P(X_1, \dots X_n) = \prod_{i=1}^{n} P(X_i|par(X_i)) \tag{1}$$

An observation on a node leads to a change in the probability distribution of that node. Also, the probability distribution of the directly connected node changes, if they were not yet observed itself. Child nodes are calculated according to the conditional probabilities, parent nodes by the Bayesian rule. In order to make an assessment of events in our monitoring network, we have to calculate the probability that a variable will be in a certain state based on the evidence on other nodes in a current situation. While the prior information is stored in the CPTs the calculated probabilitys are influenced by this information and new evidence of the current situation (Jensen 2001). Bayesian networks support vague, conflicting, and incomplete evidence by allowing one to enter a probability for evidence of a variable being in each state. Therefore various inference algorithms can be used to compute the marginal probabilities for each unobserved node given information on the states of a set of observed nodes. One of the most used methods is the junction tree (Jensen 2001, Chap. 5). Inference in Bayesian networks allows to update the probabilities of the other variables by taking into account any state variable observation. Without any event observation, the computation is only based on a priori probabilities. When observations are given, this knowledge is integrated into the network and all the probabilities are updated accordingly. Knowledge about events is formalized as evidence. We distinguish between hard and soft evidence (refer to Sect. 4) of the random variable. Hard evidence for a node X_i corresponds to safe deduction of the state. Soft evidence corresponds to an update of the prior probability values for the states of this node.

4 Soft Evidence by Kalman Filter

It is obvious that each sensor node generates evidence by its measurements. In the scenario of a sensor network, elementary limitations concerning processing power or limited data storage on the sensor nodes exist. Depending on these limiting preconditions a recursive filtering method should be used for the evaluation of a new system state. The recursion allows for the estimation of a new state by an update of the old state by new measurement data. In addition to the last and current state all previous measurements need not remain in the memory storage of the node. The Kalman filter is such a recursive algorithm to estimate the state of a linear system. The Kalman filter algorithm is shown for example in Caspary and Wichmann (2007) or Welsch et al. (2000) and will not be discussed here. The further remarks refer exclusively to the results of a complete filtering step under the assumption of normally distributed variables.

According to Sect. 2 probabilistic state variables for assessment of an event are needed, which can be included as evidence in the Bayesian network (refer to Sect. 3). Therefore the results of the Kalman filter (the filtered state vector $\hat{\mathbf{x}}_{++}^{(k)}$ and the cofactor matrix of the filtered state vector $\mathbf{Q}_{\hat{x}\hat{x},+}^{(k)}$) cannot be used directly. Hence, the results of the filtering process have to be transformed into probabilistic values according to the desired value (see Fig. 1).

For this task, different approaches are possible. In the first instance the p-value of a compatibility test of the difference of two filter results (epoch k and epoch $k-1$) is used. Equation 2 shows this residual between two filtered measurements of the filtering algorithm.

$$\mathbf{d} = \hat{\mathbf{x}}_{++}^{(k-1)} - \hat{\mathbf{x}}_{++}^{(k)} \tag{2}$$

Beside this residual the variance of $\mathbf{d}$ is needed. It must be considered that the estimated states of two neighboring epochs are not independent. In the usual notation (Kalman gain $\mathbf{K}^{(k)}$, design matrix of the observation model $\mathbf{A}^k$, transition matrix $\Phi^{(k)}$, observation vector $\mathbf{l}^{(k)}$) the states of epoch k and $k-1$ can be written as:

$$\begin{bmatrix} \hat{\mathbf{x}}_{++}^{(k)} \\ \hat{\mathbf{x}}_{++}^{(k-1)} \end{bmatrix} = \begin{bmatrix} (\mathbf{I} - \mathbf{K}^{(k)}\mathbf{A}^k)\Phi^{(k)} & \mathbf{K}^{(k)} \\ \mathbf{I} & \mathbf{0} \end{bmatrix} \begin{bmatrix} \hat{\mathbf{x}}_{++}^{(k-1)} \\ \mathbf{l}^{(k)} \end{bmatrix} \tag{3}$$

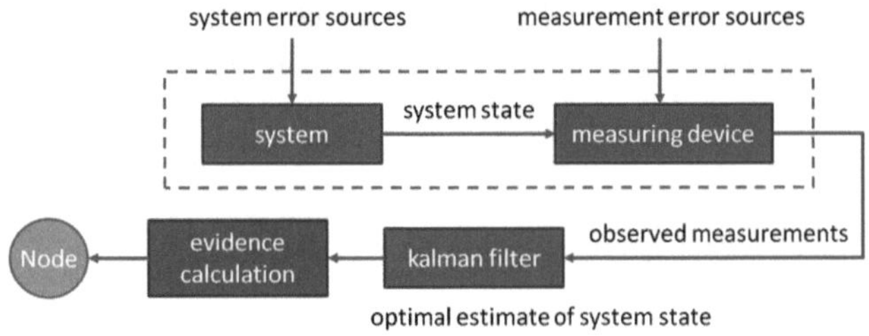

Fig. 1 Soft evidence by local Kalman filtering

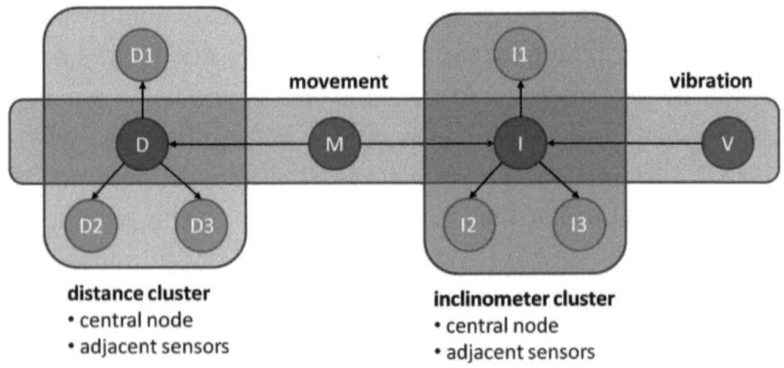

Fig. 2 Bayesian network of a simple monitoring process

The variance-covariance matrix can be obtained by error propagation. With $\mathbf{Q}^{k-1}_{\hat{e}\hat{e},+}$ as the cofactor matrix of the filtering error of epoch $k-1$ the covariance matrix of the two estimated states is shown in Eq. 4.

$$Kov(\hat{\mathbf{x}}^{(k-1)}_{++}; \hat{\mathbf{x}}^{(k)}_{++}) = \mathbf{Q}^{(k)}_{\hat{e}\hat{e},-}((\mathbf{I} - \mathbf{K}^{(k)}\mathbf{A}^{k})\Phi^{(k)})^T \tag{4}$$

Subsequently the difference and the variance of the filtered values are calculated. Under the previously introduced assumption of normally distributed variables, the residual vector will be normally distributed according to

$$\mathbf{d} \sim N(\mathbf{0}; \sigma_0^2 \mathbf{Q}_{dd}) \tag{5}$$

where $\mathbf{Q}_{dd}$ is the cofactor matrix of the residual vector, derived from error propagation.

To introduce a probability of a change in state a compatibility test is performed on the basis of the null hypothesis that the current state is not significantly different from the previous state. As a further simplification the example shown in the following section is calculated only with scalar states. At this point the transition to a multiple test was not investigated yet. Following studies should take this into account. In the special case of a scalar residual vector we get the probability of a change in state directly by calculating the p-value for the hypothesis the residual element $d = 0$.

$$P(|T| = |\frac{d}{\sqrt{\sigma_0^2 q_{dd}}}| \leq z_{1-\frac{\alpha}{2}}|H_0), z \sim N(0,1) \tag{6}$$

This probability can be introduced as soft evidence into the node of the Bayesian network and sent to the evaluation node.

5 Simulation Results

In Bayesian networks only one-directional causal relationships between the variables are defined. Neighboring sensor nodes often have seemingly mutual relationships and the direction of this influence is not clearly determined. In this case, the edges between this correlated variables can not be addressed unambiguously, which would lead directly to an unauthorized cycle in the directed acyclic graph. This fact suggests there are common causes or effects. If this common causes or effects can't be found in the previously identified variables, this indication that not all the relevant variables for the area have been identified (Munkelt 2008). To avoid this problem an additional synthetic hypothesis or rather a synthetic node is introduced, which is equally influenced by the observations of all other nodes. Therefore the network is divided into homogeneous areas where uniform behavior can be assumed. This cluster of small Bayesian networks is combined by superior state variables represented as a central node.

The concept of an automatic response capability within the sensor network based on probabilistic variables will be demonstrated by means of a total station network. The sensor nodes are defined by three reflectors in a homogeneous area. The central node is defined by the total station. On each sensor node we are able to determine a distance measurement by the total station and an acceleration measurement by an accelerometer. As a monitoring scenario, two different events are suspected. On the one hand a point movement of individual sensor nodes is possible, on the other hand partial vibrations of sensor nodes could occur. The homogeneous field in the scenario leads to the assumption that if a single point movement is detected, further points in the field should move also. However, vibrations in the sensor network have only influence on individual nodes. But if on all nodes a shock is detected, it will be presumed that a point motion is present. An illustration of the Bayesian network is shown in Fig. 2. The left rectangle represents the distance cluster of the three reflectors. The right rectangle represents the inclinometer measurements on the three reflector points. Each clustered information is modeled by a central node (for example the total station). The main query about the two states of a movement or vibration in the network is represented by the red rectangle with conditional dependencies to the two sensor clusters.The inference task in the Bayesian network

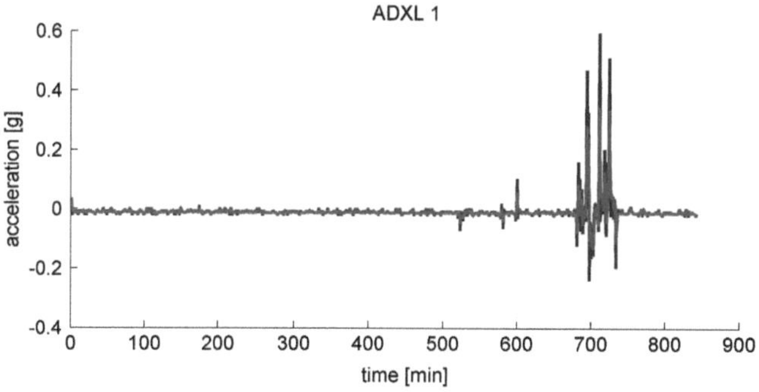

Fig. 3 Raw and filtered data from one accelerometer

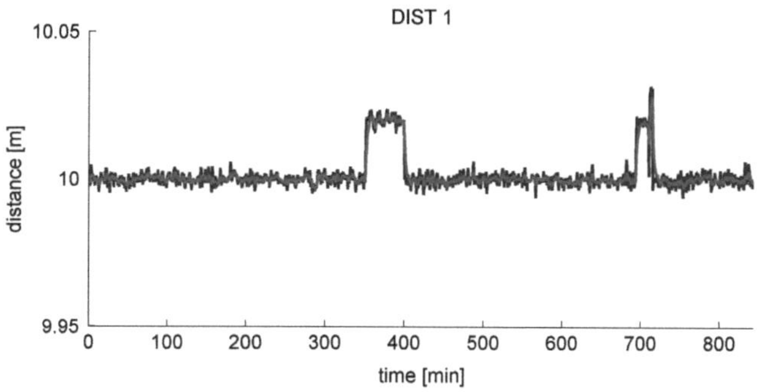

Fig. 4 Raw and filtered data from distance sensor (simulated)

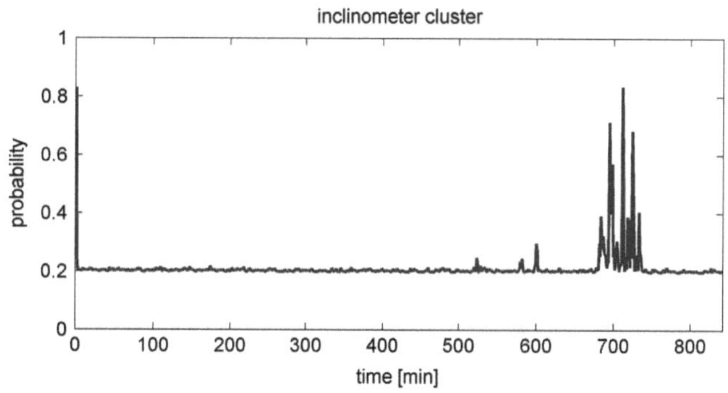

Fig. 5 Calculated p-values for a change in state by one compatibility test

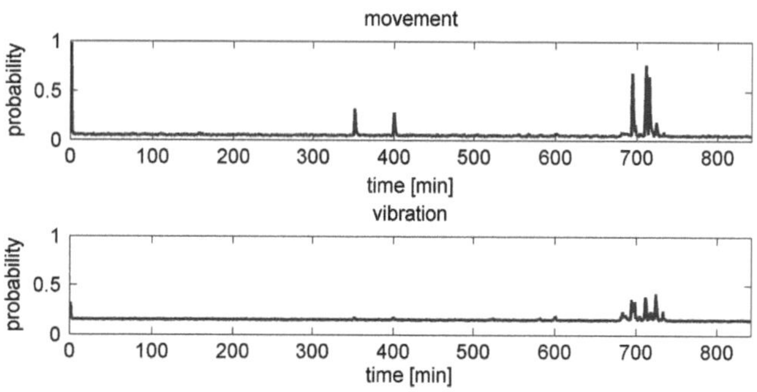

Fig. 6 Probabilities of the global events *move* and *vibration* in the network

is characterized by the calculation of the joint probability of the whole network to determine the actual event of *moving* or *vibrating*.

Referring to Sect. 4 the analysis of sensor information by means of Kalman filter is implemented directly on the node. An exemplary result of one accelerometer is shown in Fig. 3. The noisy line represents the raw accelerations and the smooth line the filtered states. The accelerometer used in this experiment was an low-cost adxl335 mounted on a micro controller (Analog Devices Inc. 2012). For the accelerometer measurements a standard deviation of 0.01 g can be assumed.

Next to the accelerometer measurements simulated distance measurements are introduced into the model. For each node a time series of distance measurements with several changes of the mean was generated. The data was simulated with a normal distribution and standard deviation of 0.002 m. An example of one node is shown in Fig. 4.

According to Fig. 3 the raw measurements and the filtered results can be seen. Furthermore, different changes of the mean are visible in the data. Based on the filtered results the p-value of a change in state is calculated on each sensor node. Exemplary results of such a calculation on one inclinometer node are shown in Fig. 5.

After the p-value of the compatibility test is determined, the value is send to the central node. There, the values obtained are introduced as soft evidence to the Bayesian network. After the update all CPTs of the three sensor nodes, the conditional probabilities of the two events are calculated. Based on real and simulated sensor data a Matlab simulation was calculated for an initial assessment of the concept. The results of this simulation are represented in Fig. 6.

As one can see, the events in the sensor data (different changes of the mean) are clearly visible. By calculating the p-value of the compatibility test, these events can be expressed as probability of a change in state. The graph of this information is similar to the filtered sensor data. Only through the subsequent calculation of the Bayesian network, the probabilities of the global events *move* or *vibration* become visible. It is obvious that the impact of one accelerometer change of the mean is insignificant to the global variables but two or more have a recognizable effect. Only with an additional event in the distance measurements, the probability of a movement in the area rises to one. Based on this assessment it is possible to make an appropriate response. In this case such a response could be to hook up an additional sensor for a more precise observation.

Conclusion

This paper gives an overview about a concept of implementing automatic responses to events in geodetic processes developed at the GIH. We have shown that these events can be modeled by logical and probabilistic state variables. The estimation of the logical state variables and the deduction of responses are straightforward while the calculation of the probabilistic variables and a response based on them are challenging. In our study we used Kalman filter and subsequent p-value calculation of a change in state to determine soft evidence for the Bayesian network. This approach allowed the quantification of existing neighborhood relations and the assessment of occurring events. While one sensor has an insignificant impact on the state of the whole system, two or more have a recognizable effect. The topics of the ongoing research are the improvement of other methods to identify evidence on the different sensor nodes. Referring to Li et al. (2011) Bayesian methods could be used to identify and evaluate changes of the mean in the filtered data. Also the implementation of an appropriate response based on probabilistic assessment should be developed. A possible method for this task is a decision table.

References

Analog Devices Inc. (2012) http://www.analog.com/static/imported-files/data_sheets/ADXL335.pdf. Accessed 03 Sept 2012

Caspary W, Wichmann K (2007) Auswertung von Messdaten, Statistische Methoden für geo- und Ingenieurwissenschaften. Oldenbourg Verlag München, Wien

Jensen FV (2001) Bayesian networks and decision graphs, statistics for engineering and information science. Springer, New York

Koch KR (2007) Introduction to Bayesian statistics. Springer, Berlin/Heidelberg/New York

Li Y, Liu D, Wymeersch H (2011) Bayesian outlier detection in location-aware wireless networks. In: Proceedings Workshop on Positioning, Navigation and Communication (WPNC)

Lunze J (2008) Regelungstechnik 1. Systemtheoretische Grundlagen, Analyse und Entwurf einschleifiger Regelungen. Springer, Berlin/Heidelberg

Munkelt T (2008) Potenzial Bayesscher Netze zur Unterstützung der Produktionsplanung und -steuerung. Dissertation, Technischen Universität Ilmenau: Fakultät für Wirtschaftwissenschaften

Pink S (2007) Geo-Sensornetzwerkes für ingenieurgeodätische Überwachungsmessungen. Dissertation, Universität der Bundeswehr München: Fakultät für Bauingenieur- und Vermessungswesen

Sohraby K, Minoli D, Znati T (2007) Wireless sensor networks. Technology, protocols, and applications. Wiley-interscience, Hoboken

Straßberger M (2007) Kontextbereitstellung in Automobilen Ad-hoc Netzen. Dissertation, LMU München: Fakultät für Mathematik, Informatik und Statistik

Welsch W, Heunecke O, Kuhlmann H (2000) Handbuch Ingenieurgeodäsie, Auswertung geodätischer Überwachungsmessungen. Herbert Wichmann, Heidelberg

Efficiency Optimization of Surveying Processes

I. von Gösseln and H. Kutterer

Abstract

In order to perform an efficiency optimization of surveying processes typical measuring processes can be modeled by using Petri nets. Petri nets are a mathematical and graphical modeling language for the description of concurrent and distributed systems. The modeling allows a simulation and an efficiency optimization of the processes. Simulations of surveying processes can be performed with different input values like the number of staff or the order of activities. The main goals of the optimization are the reduction of cost or the decrease of the required time. Since the exact duration of the individual steps of a measurement task cannot be defined in advance, timed transitions in stochastic Petri nets are selected to introduce the duration of the activities. The presented method is applied to the optimization of a polar network measurement.

Keywords

Efficiency optimization • Modeling and simulation of surveying processes • Petri nets

1 Introduction

In construction processes an ongoing increase of automation is observed. Therefore the compliance with quality becomes more and more important for subsequent tasks. The increasing frequency of measurements requires a very efficient integration of surveying processes into construction processes. Thus, one topic of the DFG[1]-research project EQuiP[2] is the quality assurance. To get more information about the developed quality model see Schweitzer and Schwieger (2011).

Another important task is to improve the efficiency of surveying processes to ensure a good integration of them into construction processes. For this purpose, an efficiency model is set up to evaluate the performance of surveying processes and their interaction with construction processes. The two main efficiency criteria are *cost* and *duration*. The aim is to reach low cost or a short duration.

The optimization of a surveying process is carried out as follows: First, the activities of the process have to be simulated with different input values. Second, the solution with the lowest cost or the shortest duration has to be selected. The input values of the process can be the number of staff or the order of activities. A simulation of various

I. von Gösseln (✉)
Geodätisches Institut, Leibniz Universität Hannover, Nienburger Str. 1, 30167 Hannover, Germany
e-mail: vongoesseln@gih.uni-hannover.de

H. Kutterer
Bundesamt für Kartographie und Geodäsie, Richard-Strauss-Allee 11, 60598 Frankfurt am Main, Germany

[1] German Research Foundation.

[2] German: Effizienzoptimierung und Qualitätssicherung ingenieurgeodätischer Prozesse im Bauwesen, engl.: Optimization of efficiency and quality control of engineering geodesy processes in civil engineering.

H. Kutterer et al. (eds.), *The 1st International Workshop on the Quality of Geodetic Observation and Monitoring Systems (QuGOMS'11)*, International Association of Geodesy Symposia 140,
DOI 10.1007/978-3-319-10828-5_23,

scenarios with different input values requires a timetable for the surveying process. Measuring processes include several different activities that are partly concurrent. Often certain conditions must be fulfilled before an activity can start. For example, the concrete of the foundation slab must be dried before the walls of the building can be staked out.

In particular for complex processes a modeling language is needed that allows computer-based simulation, that can deal with concurrent processes and that can check and take into account the fulfillment of conditions. In our project EQuiP, we have chosen Petri nets because they comply with all these requirements.

2 Process Modeling and Simulation

2.1 Elements of a Petri Net

The modeling of the measuring working steps is realized by a place/transition Petri net, which is one special type of a Petri net. Petri nets are a graphical tool to describe and analyze concurrent and distributed processes. Expressed in mathematical terms a Petri net is a directed bipartite graph in which the two node types are called *places* and *transitions*. They are connected by directed arcs. Arcs run from places to transitions or reverse. Connections between nodes of the same type are not allowed. For more information on Petri nets see Reisig (1985) or Ajmone Marsan et al. (1994).

Figure 1 shows the elements of a place/transition Petri net. In our project EQuiP Petri nets are implemented as a graph in Java.

One of the two node types is called *place*. Places are represented by circles and model the passive components of a process like objects, states or conditions of a process. Places can contain *tokens* which can be represented by small filled circles or numbers. Tokens show the compliance of conditions, the value of an object or the state of the process. In the example in Fig. 2 there are ten "resources" and two members of "staff". The other node type is called *transition*. Transitions are represented by rectangles and model active elements like actions, working steps or events of a process. The main activities of the measurement in Fig. 2 are the "set up of the target points", the "set up of the total station" and the "measurement".

The *directed arcs* represent the relations between places and transitions. An arc from a place to a transition defines the place as an *input place* of the transition. And an arc from a transition to a place defines the place as an *output place* of the transition. If all input places of a transition contain enough tokens, the transition is called *enabled* and it can *fire*. When a transition fires, tokens are removed from each input place and new tokens are created at each output place. After that, subsequent transitions may be enabled and the tokens

Fig. 1 Elements of a place/transition Petri net

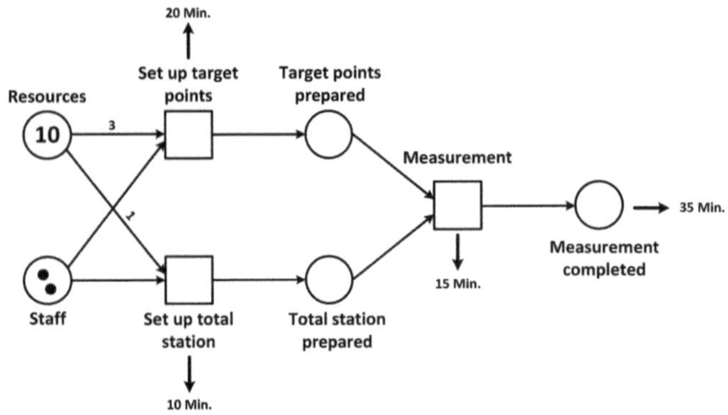

Fig. 2 Place/transition Petri net of a simple measurement

move through the Petri net. Arcs may have an edge weight. The edge weight regulates how many tokens are needed in the input place to enable the transition or how many tokens will be created in the output place after the transition has been fired. In the example there have to be three resources to enable the transition "set up target points". If no edge weight is specified, it is one.

2.2 Timed Petri Nets

To attain a simulation close to reality, timing information has to be assigned to transitions. A concept for the introduction of temporal information into Petri nets was given by (Ramchandani 1974). In his approach each transition t_i gets a firing time τ_i which is the duration of the activity. τ_i is a non-negative bounded real number. Each transition with a firing time is called a *timed transition*. After removing the tokens from the input places the time event starts. After a certain period of time the tokens are created at the output places of the transition and the time event ends. In a simplified assumption a fixed duration can be assigned to each activity. In real-world systems the duration is rather a random quantity that can be approximated by a rectangular or normal distribution in a large number of cases (Ramchandani 1974). This type of Petri net is called a *stochastic Petri net*.

2.3 Run of a Petri Net

Using the example in Fig. 2 one run of the Petri net is explained. The first activity of this measuring process is the "set up of the target points". When the transition fires three tokens from the input place "resources" are removed

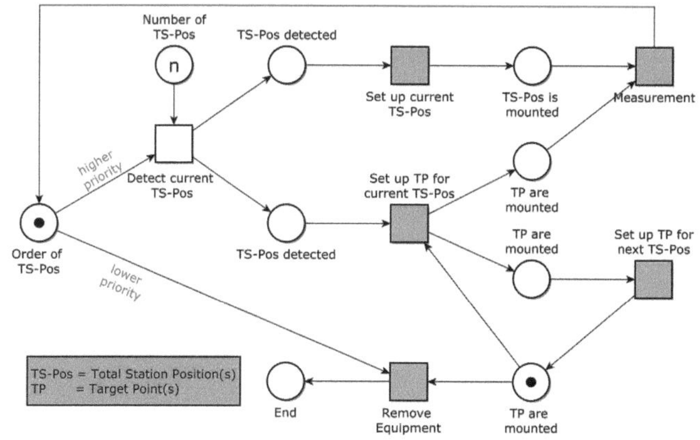

Fig. 3 Place/transition net of a polar network measurement

because the edge weight is three. One token is removed from the other input place "staff". At the same time the surveyor "sets up the total station". When this transition fires, one token is removed from the input place "resources" and one from the other input place "staff". After the set up has been completed, one token is created at the output place "total station prepared". The set up of the target points takes a little bit longer in this example and after its completion one token is created at the output place "target points prepared". Now the "measurement" can be carried out. Both tokens from the input places are removed. After the measurement has been finished, one token is created at the output place "measurement completed". The result of the Petri net run is the elapsed time of the measuring process. This is a very simple model of a measuring process.

A slightly more complex model is the Petri net of a polar network measurement illustrated in Fig. 3. Although the depth of modeling is not very high, all important relations and conditions can be modeled. The input values of the Petri net are a defined "order of total station positions (TS-Pos)" and the number of staff which is not specifically modeled in the Petri net but considered in the simulation. In the first transition the "current TS-Pos is detected" from the given list of total station positions. Then the "set up of the current TS-Pos" by the surveyor and the "set up of the target points (TP) for the current TS-Pos" by the assistants can occur. When the events of both transitions are completed and the tokens at the places "TS-Pos is mounted" and "TP are mounted" are created, the "measurement" can be carried out. At the same time, the assistants can "set up the TP for the next TS-Pos". After the measurement has been finished, the next run of the tokens through the Petri net starts. This is repeated until the place "number of TS-Pos" is empty. Then the last enabled transition is "remove equipment" and after its firing the Petri net has no enabled transitions left and is called *dead*.

The result of the simulation is the duration of the whole network measurement.

3 Optimization

The basis for carrying out the optimization of a network measurement is the definition of the possible input values. In this case the number of staff and the order of the measurements and the set up can vary. The zero to second order design is already set and will not be optimized in the described approach. In the efficiency optimization the different input values can be selected and the simulation of the measurement and thus the calculation of the duration is carried out in the Petri net. Finally, the solution with the shortest duration or lowest cost is selected as the best solution. The output values are the total duration and the total cost of the measurement and the selected input values of the best solution.

Expressed mathematically one is faced with a combinatorial optimization problem. For simple tasks all possible solutions can be evaluated. For complex tasks a good solution has to be found with a search heuristic. Otherwise the computation takes too long. In our project EQuiP genetic algorithms have been selected as a search heuristic. In Rehr et al. (2011) we have described in detail how the optimization of a network measurement with genetic algorithms works. For more information on genetic algorithms see Goldberg (1989).

4 Time Model

As already mentioned, time delays for the transitions of the Petri net are needed for the simulation of the measuring process. The duration can be a fixed value (the expected value) for each activity or a random value that can be approximated by a certain distribution function. Figure 4 shows examples for probability density functions (PDF) of the used input distributions for the activities. The expected value of the activity in this figure is $\tau = 3\,\text{min}$. In addition to the expected value, each activity is associated with an ω. This variable describes the upper and lower limit of the distribution function, thus the variability of the duration. For all activities carried out during a measurement, time values for τ and ω have to be determined. In a first approach, empirical values are used.

In the optimization only the expected values are taken for the Petri net simulation. Otherwise, the optimization process would take too long because for each candidate solution the simulation has to be carried out many times with different random durations for the activities. Using the expected value, the simulation must be performed only once for each candidate solution.

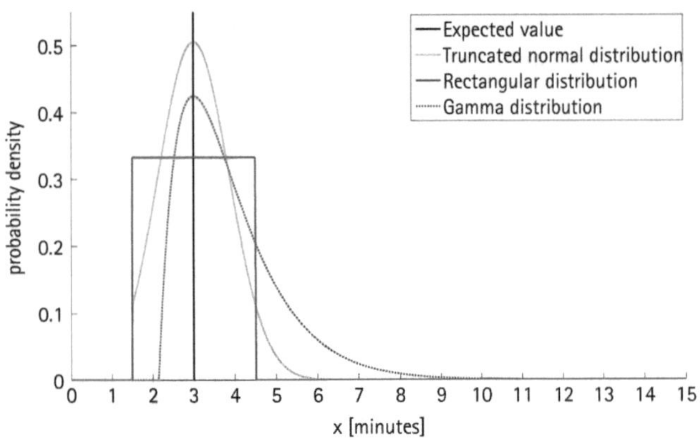

Fig. 4 Input distributions for an activity with the variables $\tau = 3.0$ min and $\omega = 1.5$ min

After the optimization, the process is simulated many times with the input values of the best solution. In this case 10,000 iterations are used, because with this number the repeatability of the result (mean value and variance) is given in the range of a few seconds. The duration of each working step is a random value that follows one of the distributions shown in Fig. 4. Different distributions are tested to analyze the effect of their choice for the whole process. It is conceivable that different activities has to be approximated by different distributions in reality. For example the unload of the equipment might follow a normal or a rectangular distribution. It is equally probable that the duration is shorter or longer than the expected duration. Other activities like measuring sets of angles have a specific duration at the minimum. In general it is more probable that the activity lasts longer than shorter than the expected value. Such activities are often done automatically or semi-automatically. They might be approximated by a gamma distribution or another unbalanced distribution.

All PDFs of the distributions illustrated in Fig. 4 can be expressed with the variables τ and ω. The notation of the rectangular distribution with the lower bound $a = \tau - \omega$ and the upper bound $b = \tau + \omega$ is:

$$X \sim U(a,b) \Rightarrow X \sim U(\tau - \omega, \tau + \omega) \tag{1}$$

The PDF of the rectangular distribution is:

$$f(x) = \begin{cases} \frac{1}{2\omega}, & \text{for } -\omega \le x - \tau \le \omega \\ 0, & \text{otherwise} \end{cases} \tag{2}$$

The normal distributions expressed with the mean value $\mu = \tau$ and variance $\sigma^2 = \frac{\omega^2}{3}$ is:

$$X \sim N(\mu, \sigma^2) \Rightarrow X \sim N\left(\tau, \frac{\omega^2}{3}\right) \tag{3}$$

The PDF of the normal distribution is truncated at the bottom to prevent values appearing below a lower bound $a = \tau - \omega$ (see Robert 1995):

$$f(x) = \begin{cases} \dfrac{\frac{1}{\sqrt{2\pi\sigma^2}} \cdot \exp\left(-\frac{(x-\tau)^2}{2\sigma^2}\right)}{\frac{1}{2} \cdot \left(1 - erf\left(\frac{a-\tau}{\sqrt{2\sigma^2}}\right)\right)}, & \text{for } x \ge a \\ 0, & \text{for } x < a \end{cases} \tag{4}$$

The notation of the gamma distribution with the shape parameter $k = 2$ and scale parameter $\theta = \frac{\omega}{\sqrt{3}}$ is:

$$X \sim \Gamma(k, \theta) \Rightarrow X \sim \Gamma\left(2, \frac{\omega}{\sqrt{3}}\right) \tag{5}$$

The PDF is shifted to the right, with the result that the x-value of the PDF maximum is the excepted value τ. Then the function is:

$$f(x) = \begin{cases} \frac{(x-\tau+\theta)}{\theta^2} \cdot \exp\left(-\frac{x-\tau+\theta}{\theta}\right), & \text{for } x \ge \tau - \theta \\ 0, & \text{for } x < \tau - \theta \end{cases} \tag{6}$$

5 Numerical Study and Results

Figure 5 shows the network that should be measured and optimized beforehand. The optimization is performed only for the sequence of activities and the number of staff. The observation network and the positions of the points remain unchanged in this first approach. The network of paths with travel times from point to point must be known in advance, too.

The polar network measurement has been selected as an example because each point of the network must be visited repeatedly to set up and orientate the prisms to the current total station position. That implies a high logistic effort. The solution of an extended multiple traveling salesman problem is required for planning the optimal order of total station positions and the routes of all participants. One is faced with a very complex optimization problem.

The optimization is carried out with genetic algorithms. The fitness value that is used to compare the different solutions is calculated by using the Petri net simulation which model is illustrated in Fig. 3.

As described, the Petri net simulation in the optimization is carried out with the expected values for the duration of the activities. The best result of the optimization is a solution that takes $x_{Opt} = 4.03$ h. Subsequent to the optimization the temporal variation of this best solution is computed. Therefore, this variant is simulated 10,000 times with random values for each activity. Successively, the three distribution functions were used that were described in Sect. 5. The results of this

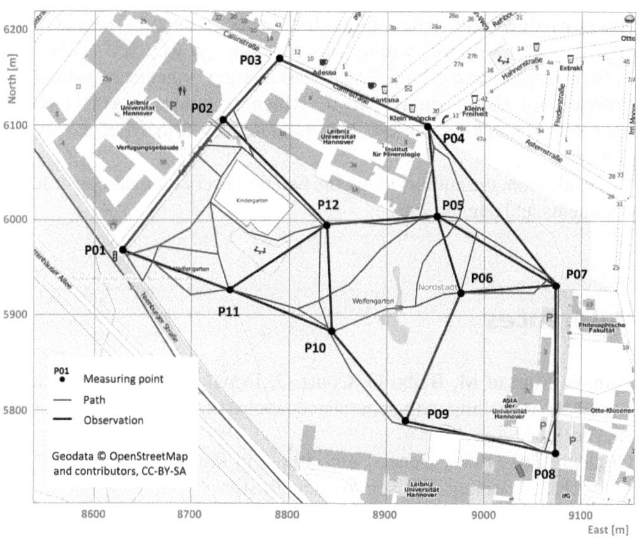

Fig. 5 Welfengarten network

computation is shown in Fig. 6. The doted line in all subfigures is the result of the optimization calculated with the expected values.

For the first calculation of the temporal variation, normal distributions were assumed for all activities (Fig. 6a). The temporal variation of the result seems to follow a normal distribution, too, but the mean value ($\overline{x}_{ND} = 4.07\,\text{h}$) is slightly shifted to the right, compared with the result of the optimization. The reason is probably, that the input normal distribution is cut on the left side and a few more random values are bigger than the expected value.

The next tested distribution was the rectangular distribution for all activities (Fig. 6b). The result seems to follow a normal distribution again. The mean value ($\overline{x}_{RD} = 4.05\,\text{h}$) is approximately the same as the result of the optimization. Such a distribution has been expected because the number of random values to the right and left side of the expected value are the same.

In the last simulation the gamma distribution was chosen to describe the duration of the activities (Fig. 6c). Once more the temporal variation seems to follow a normal distribution. But in this case the mean value ($\overline{x}_{GD} = 4.45\,\text{h}$) is strongly shifted to the right compared with the other output distributions and the result of the optimization. Furthermore, the variance of the result is larger than in the other two examples. The input gamma distribution has a steep ascent close to the expected value and then decreases slowly. The majority of the random values is bigger than the expected value. In addition, the range of random values is larger than the range of the other two input distributions. This explains the right shift and the higher variance of the output distribution, but such a strong shift was not expected ahead of the tests.

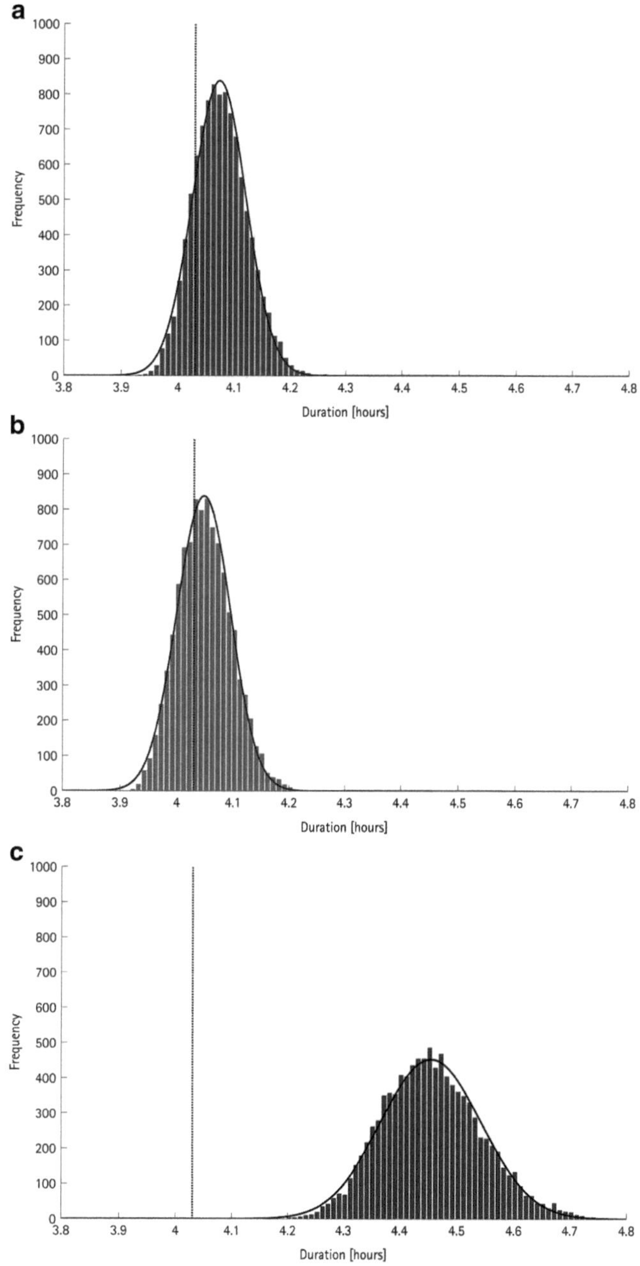

Fig. 6 Temporal variation of the network measurement calculated with a random duration for the activities (10,000 iterations). (**a**) Input distribution: truncated normal distribution, (**b**) input distribution: rectangular distribution, (**c**) input distribution: gamma distribution

If the gamma distribution turns out to be the most appropriate distribution function to approximate the duration, we have to find a way to consider this strong right shift already in the optimization to get realistic results.

In Fig. 6 the effect of the selected distribution becomes clear. But it remains the open question which input distribution is the closest to reality.

6 Summary and Outlook

Petri nets are qualified for modeling and simulation of surveying processes. With their assistance it is possible to test different input values in an optimization process and to select the best of all solutions. After the optimization the temporal variation of this solution can be computed to get a simulation closer to reality.

In future work a characteristic distribution for each activity has to be found. Probably different activities have to be approximated by different distributions. This requires that the durations are recorded during a network measurement. A recording of the duration with a stopwatch would be expensive and the logging of values would take time, too. Ideally, the duration is recorded automatically and does not obstruct the measurement staff in their work. An execution of the simulated network measurement in the Welfengarten is planned. To derive the durations of the activities, the measurement should be monitored by GPS loggers.

Furthermore, it is necessary to find out if it is really the best way to carry out the optimization only with the expected value. If the gamma distribution is used to model the time delay or even different input distributions, it might be better to simulate a certain number of each scenario to get a more realistic total duration.

For further developments, an integration of the quality parameters (accuracy and reliability) into the optimization process is aimed.

Acknowledgements The presented paper shows results and approaches developed within the research project KU 1250/10-1 "Effizienzoptimierung und Qualitätssicherung ingenieurgeodätischer Prozesse im Bauwesen (EQuiP)", which is funded by the German Research Foundation (DFG). This is gratefully acknowledged by the authors.

In addition, the authors thank the two reviewers for their constructive comments and suggestions.

References

Ajmone Marsan M, Balbo G, Conte G, Donatelli S, Franceschinis G (1994) Modelling with generalized stochastic Petri nets. Wiley, New York

Goldberg DE (1989) Genetic algorithms in search, optimization, and machine learning. Addison-Wesley, Boston

Ramchandani C (1974) Analysis of asynchronous concurrent systems by Petri nets. Ph.D. Thesis. Massachusetts Institute of Technology, Project MAC, Cambridge

Rehr I, Rinke N, Kutterer H, Berkhahn V (2011) Maßnahmen zur Effizienzsteigerung bei der Durchführung tachymetrischer Netzmessungen. AVN 1:2–13

Reisig W (1985) Petri nets. An introduction. Springer, Berlin (EATCS monographs on theoretical computer science, 4)

Robert CP (1995) Simulation of truncated normal variables. Stat Comput 5:121–125

Schweitzer J, Schwieger V (2011) Modeling of quality for engineering geodesy processes in civil engineering. J Appl Geodesy 1:13–22

Modeling and Propagation of Quality Parameters in Engineering Geodesy Processes in Civil Engineering

Jürgen Schweitzer and Volker Schwieger

Abstract

Quality assurance in civil engineering is a complex and multifaceted field. Especially for successful automation it plays an important role. One aspect of the quality assurance relates to the geometry of a building. In order to determine and control geometric elements, measurement and evaluation processes of engineering geodesy have to be integrated into the construction processes. So the task of engineering geodesy is to create the basis for bringing the planned building geometry in quality-assured reality.

One way to describe the quality is to define a quality model, which describes the quality on the basis of characteristics, which are substantiated by parameters. In generally the characteristics and the parameters are derived from the requirements.

For engineering geodesy processes in civil engineering a process- and product-oriented quality model consisting of the characteristics "accuracy", "correctness", "completeness", "reliability" and "timelessness" was build. These five characteristics are substantiated by altogether ten parameters. In addition to the well-known parameters like "standard deviation" in geodesy and "tolerance" in civil engineering, other parameters like "number of missing elements" and "condition density" help to have a complete and detailed description of the quality of the geometry of a building and the related processes. The parameters can be differentiated in process- and product-related parameters. Finally the quality parameters can be analyzed to get a significant statement about the actual reached quality within the process.

Keywords

Engineering Geodesy • Monte-Carlo-Method • Quality Model

1 Introduction

In the standard (DIN EN ISO 9000) quality is defined as "degree to which a set of inherent characteristics fulfils requirements". A characteristic is defined as a distinguishable feature. In general one can say that quality is the fulfillment of requirements of a product or process. Quality in construction, respectively in civil engineering is generally measured by the adherence to guidelines or standards, such as the "recognized rules of technology", DIN/ISO standards, association guidelines as well as individually contracted specifications.

In this article the considered product is the geometry of a high-rise building in shell construction. So the main focus lies on the engineering geodesy processes, since these, among others, are responsible for delivering geometry on a high quality level.

The complexity of the quality is considered by a quality model that describes the quality on the basis of characteristics which are substantiated by parameters. The quality

J. Schweitzer • V. Schwieger (✉)
Institute of Engineering Geodesy, University of Stuttgart, Geschwister-Scholl-Str. 24D, 70174 Stuttgart, Germany
e-mail: volker.schwieger@ingeo.uni-stuttgart.de

H. Kutterer et al. (eds.), *The 1st International Workshop on the Quality of Geodetic Observation and Monitoring Systems (QuGOMS'11)*, International Association of Geodesy Symposia 140,
DOI 10.1007/978-3-319-10828-5_24,

model is developed within the project "Optimization of Efficiency and Quality Control of Engineering Geodesy Processes in Civil Engineering" [*German*: Effizienzoptimierung und Qualitätssicherung ingenieurgeodätischer Prozesse im Bauwesen (EQuiP)] which is supported by the DFG (German Research Foundation). In contrast to most quality models which are only product-orientated, this quality model is product- and process-orientated.

In this article the basics of the quality model are firstly explained and in the following the propagation of the quality parameters of the quality model for engineering geodesy processes in civil engineering are presented.

2 Quality Model

2.1 Fundamentals

A quality model is a conceptual framework in which the abstract term of quality is gradually resolved into individual aspects and so the abstract term is substantiated. Using a quality model, the quality of a product or process should be completely describable and comparable.

A quality model typically consists of characteristics and parameters. A quality characteristic is an inherent feature of a product or process, related to a requirement. The quality characteristics form the first hierarchical level. Each characteristic may be described by a number of parameters. The parameters substantiate the characteristics. Each parameter may be quantified with a specific (measurable) value (e.g.: standard deviation $\sigma = 1$ cm). Details about quality models in the area of software development and Transport Telematics can be found in Boehm et al. (1978), ISO 1926 (2001) and Wiltschko (2004). On the base of this information, a quality model for engineering geodesy processes in civil engineering is developed (see Table 1).

Details about the derivation and the development of the quality model can be found in Schweitzer and Schwieger (2011).

2.2 Exemplary Parameters

Exemplary, two parameters of the quality model from Table 1 are chosen and explained in detail:

The **standard deviation** is a parameter for the characteristic accuracy. It is derived from the random scatter of the measurements x_i of a random variable X around the expected value $E(X) = \mu$. The standard deviation σ_x is defined as the square root of the variance Var(X) and can be calculated from repetitions [see Eq. (1)] or derived from accuracy information of e.g. the measurement devices. To ensure accuracy of two or three dimensional variables, such as points, a covariance matrix has to be used [for detailed information see Niemeier (2002) or Teunissen (2003)].

Table 1 Characteristics and parameters of the quality model

Parameters	Characteristics
Standard deviation	Accuracy
Tolerance correctness Topological correctness	Correctness
Number of missing/odd elements Adherence to the plan	Completeness
Condition density Minimal detectable error (MDE) Impact of MDE on parameters Vulnerability to failures	Reliability
Time delay	Timeliness

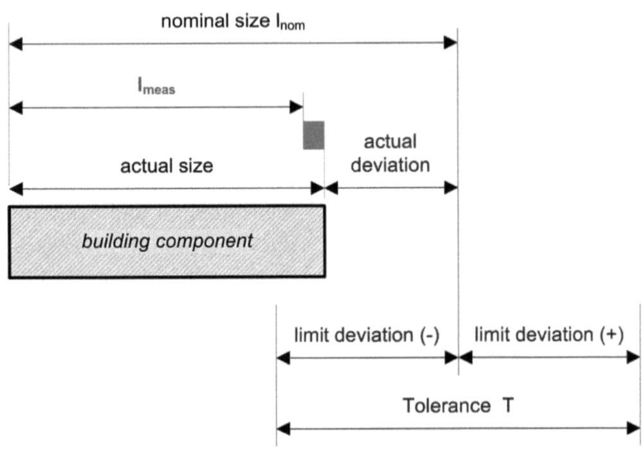

Fig. 1 Use of terms in the field of building tolerances [translated and modified from (DIN 18202)]

$$\sigma_x = \sqrt{\frac{\sum (x_i - \mu)^2}{n}} \tag{1}$$

The **tolerance correctness** t_c is a parameter for the characteristic correctness. It is an aggregated value that delivers a statement of compliance with the required tolerance. Further information can be found in Schweitzer and Schwieger (2011). It often relies on the measured length of a building component. In some cases angles and other features may play a role, but for the applications discussed in this paper, lengths are important. If t_c is greater or equal than zero, the tolerance is met. If t_c is negative, the tolerance is not met [see Eq. (2)]

$$t_c = 0.5\sqrt{T^2 - T_M^2} - |l_{meas} - l_{nom}| \tag{2}$$

The tolerance T is a specified value, which can be taken for example from standards like (DIN 18202). l_{meas} is the measured size and l_{nom} is the nominal size of a building component (see Fig. 1). The surveying tolerance T_M is a tolerance value, which is derived from the uncertainty of the measuring results. T_M is derived from the confidence interval of the standard deviation of the measuring process [see Eq.(3)]

$$T_M = 2k_{1-\alpha/2}\sigma_{measure} \quad (3)$$

The factor $k_{1-\alpha/2}$ is regarded as a quantile of the corresponding distribution function with the given error probability α. The surveying tolerance is equal to the length of the confidence interval $[C_l, C_u]$. Further information on this topic can be found in Schweitzer and Schwieger (2011).

Considering the Guide to the Expression of Uncertainty in Measurement (GUM 1995) the standard deviation in Eq. (1) corresponds to the uncertainty u (type-A). The tolerance T_M corresponds to the extended uncertainty U which is derived from the standard deviation $\sigma_{measure}$. If $\sigma_{measure}$ is derived from device uncertainties the parameter is of type-B. Parameters of type-A are determined by traditional statistical methods, while parameters from type-B are evaluated by other means like experience with prior knowledge about the measuring device.

The problems regarding the systematic and random components of the parameters will not be vanished. Further details on this can be found e.g. in Koch (2008).

3 Quality Propagation Methods

3.1 General

The quality parameters described in Sects. 2.1 and 2.2 are related to single products or processes. To describe the quality of several processes or finished products, an appropriate computational procedure to propagate the quality parameters through the process has to be developed for each parameter. The exemplary parameters from Sect. 2.2 are taken to explain the propagation procedure. For the parameter standard deviation the Monte Carlo Method or the variance covariance propagation law can be used. For the parameter tolerance correctness the quadratic tolerance propagation law is used. Correlations can be neglected, since no stochastic relationship between the two tolerances T and T_M exists.

3.2 Standard Deviation

To propagate the standard deviation on the one hand you can use an empirical method, the **Monte Carlo Method** (MCM). This is a numerical method to propagate random variables through a process or a system. A large number m of scattered observations are generated computer-based in a "virtual

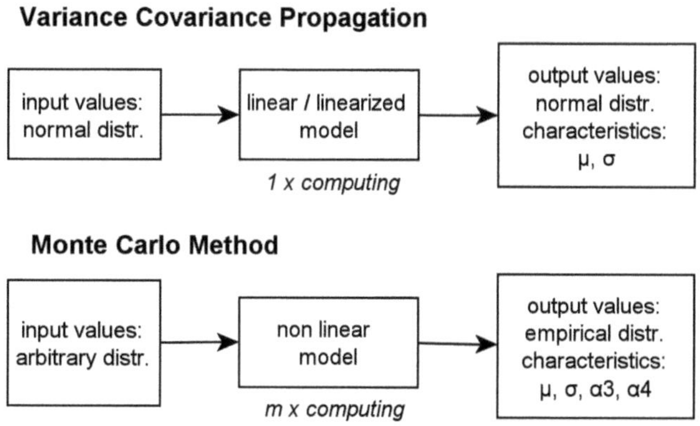

Fig. 2 Difference between the Monte Carlo method and the variance covariance propagation

experiment" whose impact on the outcome is determined. It also ensures that each scatter of the input variables $X_1 \ldots X_l$ suffice its statistical probability distribution. With each sample, the functional model $f(X_1 \ldots X_l)$ is going through the calculation and thus gives results $y = f(X)$. On the basis of the m results, a statistical analysis is performed. Important statistical parameters like the expected value μ, the standard deviation σ, confidence interval $[C_l, C_u]$, kurtosis α_4 and skewness α_3 can be determined empirically. The kurtosis is a measure for the peakedness and the skewness is a measure for the asymmetry of a probability distribution of a real-valued random variable

$$\alpha_3 = m^{-1}\sigma^{-3}\sum(\mu - x_i)^3 \quad (4)$$

$$\alpha_4 = m^{-1}\sigma^{-4}\sum(\mu - x_i)^4 - 3 \quad (5)$$

For further information, see Koch (2008) or Binder (1979). This procedure is also accepted by the GUM, written down in Supplement 1 to the GUM (JCGM 2008).

On the other hand, the **variance covariance propagation** law can be used. This is a statistical method to propagate the standard deviation of a random variable. The basis is a linear or linearised model and normal distributed input variables. The statistical parameters skewness and kurtosis of a probability density function (PDF) of a normal distributed random variable are always zero. The normal distribution is always completely defined by the expected value and the variance. For further information see e.g. Niemeier (2002).

In Fig. 2 the main differences between the two propagation methods are shown. For linear processes and normal distributed random variables the variance covariance propagation law is a reliable tool. Only in the case of non-linear processes and non-gaussian distributions, the MCM has to be considered. One expressive property of the MCM is the computational cost, because the functional model has to be computed m times.

3.3 Tolerance Correctness

The tolerance correctness parameter cannot be propagated directly. Here, the particular parts of Eq. (2) have to be regarded separately: The tolerances T and T_M [Eq. (3)] are propagated by the quadratic tolerance propagation law, where each of the tolerances is added quadratically [see Ertl ((2006)]. The other parameters l_{meas} and l_{nom} may be propagated in a linear way. This propagation method is only applicable on tolerances from building components, which are in a straight line. Further information is given in Schweitzer and Schwieger (2011).

4 Simulation Results

4.1 Example: Basement Walls

As an example for the propagation of the quality parameters, the process "building of the basement walls" of a high-rise building in civil engineering is regarded. In Fig. 3 the process is shown in a flowchart represented by a petri-net. The propagation of the quality parameters relies on the first two processes (1.1 stationing and 1.2 align formwork) only.

The first process "stationing" describes the free stationing of a total station. The input parameters of the first process are the polar elements (measured with a total station), the coordinates of the control points, and all related standard deviations. The output parameters are the station coordinates (x, y, z) with their standard deviations. The functional model is a free stationing, which is realized by a Helmert Transformation [see for example (Niemeier 2002)].

The second process "align formwork" describes the alignment of a formwork, which has to be adjusted in x- and z-direction see Fig. 4.

The input parameters are the station coordinates from the first process including standard deviations. The output parameters are the coordinates of the formwork corners the respective standard deviations and the tolerance correctness, relying on the vertical and horizontal position of the formwork.

Following the GUM (1995) standard deviations should be considered as uncertainties independent of their construction as type A or B or even combined.

The functional model is a polar survey of the corner marks 1, 2, 3 and 4 of the formwork and the subsequent calculation of the distance among these points.

4.2 Different Simulation Scenarios

As a numerical example the Monte Carlo Method and the variance covariance propagation law are chosen as propagation methods for the standard deviation. It is focused on the first two processes, which are described in Fig. 3.

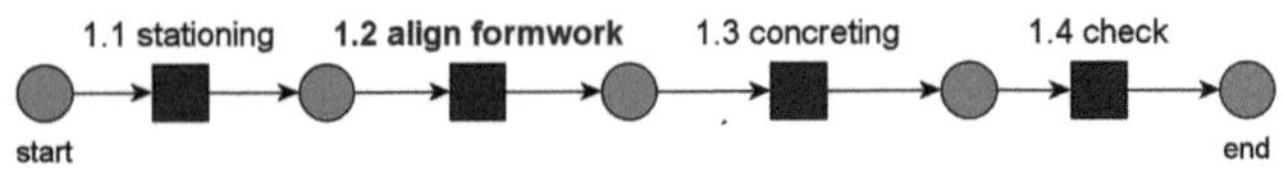

Fig. 3 Process-building of the basement walls

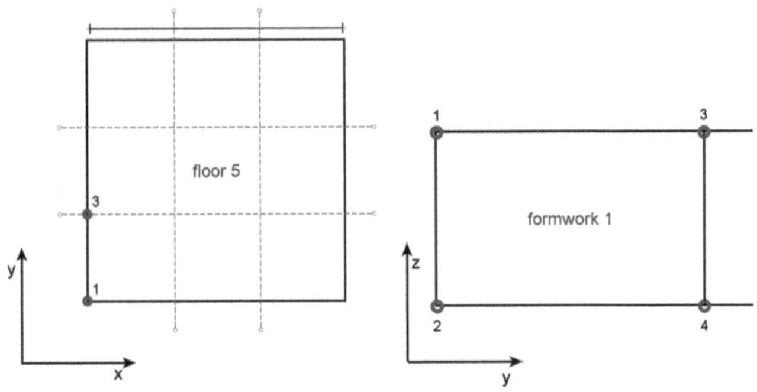

Fig. 4 Ground plan (*left*), vertical section of the formwork (*right*)

Table 2 Input and output parameters of the processes

Processes	Input	Output
1.1	$3\,x\,(\sigma_{hz},\sigma_v,\sigma_s,\mu_{hz},\mu_v,\mu_s)$	*emp.* $PDF_{X,Y,Z}$
	$3\,x\,(\sigma_{x,y,z},\mu_{x,y,z})$,	
1.2	*emp.* $PDF_{X,Y,Z}$	$4x\,(\sigma_{xp,yp,zp}\,,\mu_{xp,yp,zp})$
	$4\,x\,(\sigma_{hz},\sigma_v,\sigma_s,\mu_{hz},\mu_v,\mu_s)$	$4x\,(\sigma_{d,},\mu_{d,})$
		$4x\ t_c$

In Table 2 the input and output parameters of the processes "1.1 stationing" using three reference points and "1.2 align formwork" using the four points shown in Fig. 4 are presented.

The polar elements are introduced with the standard deviation $\sigma_{hz} = \sigma_v = 0.0008$ gon and $\sigma_s = 0.01$ m and the control points are introduced with σ_x, σ_y, $\sigma_z = 0.005$ m. The number of random samples m for the MCM is 100,000 [recommended by Koch (2008)]. For this example three control points are chosen which are located within a distance of about 100 to 200 m away from the station. For the evaluation of the probability density function (PDF) resulting from MCM the standard deviation σ, skewness α_3 [Eq. (4)] and kurtosis α_4 [Eq. (5)] are used.

The Monte Carlo Method is performed by using two different variations. The first one is done where all input variables are normal distributed, the second variant the polar elements are uniform distributed (interval $\left[-\sigma\sqrt{3};\ \sigma\sqrt{3}\right]$) and the coordinates of the control points are normal distributed.

The results for the free stationing step are shown in Table 3.

There is no significant difference between the standard deviation of the variance covariance propagation law and both Monte Carlo Method variations.

Table 3 Results of the process 1.1 free stationing

	VCP	MCM(1)	MCM(2)
$\sigma_{x\,[mm]}$	5.2	5.1	4.9
$\sigma_{y\,[mm]}$	5.5	5.6	5.6
α_4 (x/y)	0	−0.03/0.01	−0.17/−0.11
α_3 (x/y)	0	−0.004/−0.01	0.01/−0.01

Table 4 Results of the process 1.2 align formwork

	VCP	MCM(1)	MCM(2)
1.3 $\sigma_{x\,[mm]}$	6.3	6.5	6.1
1.3 $\sigma_{y\,[mm]}$	11.0	11.1	11.0
σ_d	7.7	7.7	7.7
t_c [mm]	8	10	14
α_4 (d)	0	0.08	−0.88
α_3 (d)	0	−0.03	0.02

And there is also no significant difference between the PDFs of the MCM(1) and the variance covariance propagation (VCP). This is underlined by the significance test for a_3 and a_4 with respect to the theoretical values ($E(a_3) = \alpha_3 = 0$, $E(a_4) = \alpha_4 = 0$) assuming normal distribution. With a significance level of $\alpha = 5$ % and inequalities 6 and 7 it can be shown that for the values $a_3 = -0.01$ and $a_4 = -0.03$ we get the test quantities of $y_d(\alpha_3) = 1.29$ and $y_d(\alpha_4) = 1.93$ as a result.

$$y_d\,(\alpha_3) = |a_3| \cdot \left(6 \cdot m^{-1}\right)^{-1/2} < y_{1-\alpha/2} = 1.96 \quad (6)$$

$$y_d\,(\alpha_4) = |a_4| \cdot 2\left(6 \cdot m^{-1}\right)^{-1/2} < y_{1-\alpha/2} = 1.96 \quad (7)$$

Both test quantities are smaller than the threshold of the normal distribution, so a_3 and a_4 do not differ significantly from zero.

But a small difference between the PDFs of the second MCM variation and other two can be seen while the hypothesis is discarded, because the test value $y_d = 3.4$ according to the kurtosis $\alpha_4 = -0.17$ is greater than the threshold of 1.96. In this process this has no effect on the standard deviation, but only on the confidence interval [C_l, C_u] and so the tolerance correctness of the following processes.

For the second process "align formwork" the results from the first process are introduced. For the MCM the complete result values from the x- and y-coordinate are introduced. In Table 4 one can see the results of the second process, where only two of the four points are chosen. σ_d describes the standard deviation of the exemplary distance $d = \overline{13}$. It is computed in the first variant (VCP) with full covariance information and the second and third variant (MCM) using the MCM.

Again there is no significant difference between the standard deviation σ_d of the three variants. At the second variation of the MCM, the kurtosis of the empirical PDF differs significantly from the other two. This can also be seen in Figs. 5 and 6 where the two PDFs of the Monte Carlo Method variants are plotted together with the normal distribution. It can be seen that the second PDF [MCM(2)] is narrower than the first one. This has no influence on the standard deviations but again on the confidence intervals. The confidence intervals are empirically computed by the Monte Carlo Method. The interval between the maximum and the minimum of the m random results is divided into small intervals, see Koch (2008), and relative frequencies are computed. Then the confidence interval [C_l, C_u] can be determined by adding the probabilities at both ends until the probability $\alpha/2 = 0.025$ is reached. The confidence interval resulting from MCM(2) is [199.986; 200.0135] (shown by the red speckled lines in Fig. 6) in contrast to the confidence interval resulting from VCP [199,98; 200,02] (shown by the blue lines in Fig. 6). The different confidence intervals have an influence on the surveying tolerance T_M [compare Eq. (3)] which can be calculated by $T_M = (C_u - C_l)$. For the calculation of t_c in Table 4 the values l_{meas} and l_{nom} are assumed as equal, because there are no real measurements available for the

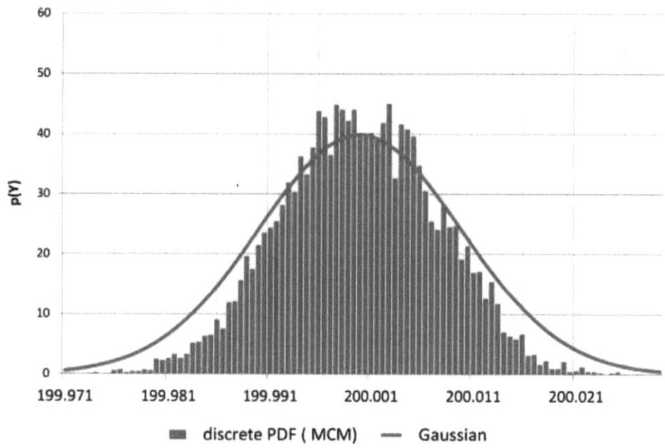

Fig. 5 PDF's of the resulting distance d [VCP and MCM(1)]

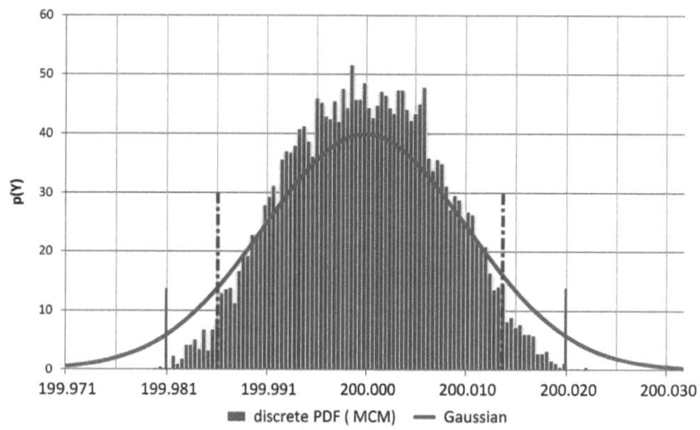

Fig. 6 PDF's with confidence intervals of the resulting distance d [VCP *blue lines* and MCM(2) *red speckled lines*]

statistical simulation case. The numeric values for t_c in Table 4 show the maximum difference between l_{nom} and l_{meas} that would deliver a positive t_c, if real measurements would be available.

In this case it can be concluded, that different input distributions have no influence on the standard deviations but on the tolerance correctness. The influence on other quality parameters e.g. for reliability are not discussed in this paper.

Conclusions

A quality model for engineering geodesy processes in civil engineering, which can be used to describe the quality of the geometry of a building is presented in this paper. Additionally, computational procedures are presented to propagate the quality parameters through the processes. The Monte Carlo Method and the variance covariance propagation law are chosen here for propagating parameters with different input distributions.

For one application case it is shown, that different distributions (Gaussian and uniform) of input values have a very small influence on the standard deviation but a decisive on the confidence intervals and so on the tolerance correctness. This was numerically demonstrated for the alignment of a formwork of a high rise building. The tolerance correctness of the MCM with partly uniform distributed input values is 4 mm greater than the tolerance correctness of MCM with Gaussian input values and 6 mm greater than the VCP result.

Because of the high requirements to the tolerances in building construction, correct knowledge about the measurement processes has to be available in construction phase. In the demonstrated case the correct propagation by MCM leads to a more relaxed decision (2 mm in comparison to VCP). This is due to non-linearity in the model. If uniform distributed observations are assumed, the decision is even more relaxed (6 mm). Generally it can be shown that non-linearity and non-normal distributions require MCM for correct uncertainty propagation through the process.

Acknowledgment The investigations published in this article are granted by the German Research Foundation (DFG) under the sign SCHW 828/3-1. The authors cordially thank the funding agency.

References

Binder K (1979) Monte Carlo methods in statistical physics. Springer, Berlin

Boehm BW, Brown JR, Kaspar H, Lipow M, Macload GJ, Merrit MJ (1978) Characteristics of software quality. North-Holland, Amsterdam

DIN 18202 (2005) Tolerances in building construction – Buildings

Ertl R (2006) Toleranzen im Hochbau. RudolfMüller, Köln

DIN EN ISO 9000 (2005) Quality management systems – fundamentals and vocabulary

ISO, Guide to the Expression of Uncertainty in Measurement (GUM), First edition, 1993, corrected and reprinted 1995, International Organisation for Standardization (ISO), Geneva, 1993 and 1995

JCGM (2008) Evaluation of Measurement Data – Supplement 1 to the "Guide to the Expression of Uncertainty in Measurement" – Propagation of distributions using a Monte Carlo method. JCGM 101:2008. Joint Committee for Guides in Metrology. www.bipm.org/en/publications/guides/gum.html

Koch KR (2008) Evaluation of uncertainties in measurements by Monte Carlo simulations with an application for laserscanning. J Appl Geodesy 2(2):67–77

Niemeier W (2002) Ausgleichungsrechnung. Walter de Gruyter, Berlin

Teunissen PJG (2003) Adjustment theory. Delft University Press, Delft

Schweitzer J, Schwieger V (2011) Modeling of quality for engineering geodesy processes in civil engineering. J Appl Geodesy 5(1):13–22

Wiltschko T (2004) Sichere Informationen durch infrastrukturgestützte Fahrerassistenzsysteme zur Steigerung der Verkehrssicherheit an Straßenknotenpunkten. PhD thesis, University of Stuttgart, 2004. Available at: http://elib.uni-stutttart.de/opus/volltexte/2004/1929/pdf/Dissertation_ThomasWiltschko.pdf (last accessed Sep 2010)

Part V

Multi-Mission Approaches with View to Physical Processes in the Earth System

Completion of Band-Limited Data Sets on the Sphere

W.-D. Schuh, S. Müller, and J.M. Brockmann

Abstract

In this study we propose the complementation of satellite-only gravity field models by additional a priori information to obtain a *complete model.* While the accepted gravity field models are restricted to a sub-domain of the frequency space, the complete models form a complete basis in the entire space, which can be represented in the frequency domain (spherical harmonics) as well as in the space domain (data grids). The additional information is obtained by the smoothness of the potential field. Using this a priori knowledge, a stochastic process on the sphere is established as a background model. The measurements of satellite-only models are assimilated to this background model by a subdivision into the commission, transition and omission sub-domain. Complete models can be used for a rigorous fusion of complementary data sets in a multi-mission approach and guarantee also, as stand-alone gravity-field models, full-rank variance/covariance matrices for all vector-valued, linearly independent functionals.

Keywords

Complete models • Covariance functions • Smoothness of potential fields • Spherical harmonics • Stochastic processes on the sphere • Variance/covariance estimation

1 Introduction

The observation of the Earth's system is one of the most important research topics of these days. Although huge sets of data are already available, additional measurements are still necessary to obtain a better understanding of the Earth's processes. Measurements from satellites have the great advantage of delivering homogeneous data sets over large areas (oceans or continents) or covering the whole globe. Due to special measurement conditions with respect to the satellite orbits as well as the observation technique, such data sets often have a band-limited spatial or spectral resolution. For example, gravity field model have, due to the downward continuation process, only a restricted frequency resolution. The global behavior of long-wavelength information is well determined, whereas the short-wavelength content is only weakly determined or not estimable. To avoid unstable systems gravity field models are often restricted to a special sub-domain, e.g. spherical harmonics up to a maximal degree. Users of such models must keep in mind that these models only describe a part (commission domain) of the real phenomena and that also the omission domain has to be taken into account. This fact is well known and was often discussed (cf. e.g. Losch et al. 2002). One typical example is the determination of the mean dynamic ocean topography as the difference between the mean sea surface height and the geoid height (Becker et al. 2011). The trace-wise spatially highly resolved altimetric measurements have to be combined with the spectral band-limited information of the gravity field and computed on a predefined grid of an ocean circulation model. Beside the different representations

W.-D. Schuh • S. Müller (✉) • J.M. Brockmann
Institute of Geodesy and Geoinformation, Nussallee 17, 53115 Bonn, Germany
e-mail: schuh@geod.uni-bonn.de; schuh@uni-bonn.de; mueller@geod.uni.bonn; brockmann@geod.uni-bonn.de

H. Kutterer et al. (eds.), *The 1st International Workshop on the Quality of Geodetic Observation and Monitoring Systems (QuGOMS'11)*, International Association of Geodesy Symposia 140,
DOI 10.1007/978-3-319-10828-5_25,

of the data, especially the different information content is crucial for the modeling. To overcome the shortcomings of band-limitation and incomplete information representation two approaches are possible: A restriction of the model to a least common sub-domain or an extension to the entire space by additional a priori information.

Special filter processes are usually introduced to homogenize all the available information with respect to a least common subspace (Jekeli 1981, 1996; Wahr et al. 1998; Swenson and Wahr 2006). It goes without saying that with this filter approach also valuable information of the signal is filtered out. To avoid this drawback of the filter processes, a rigorous fusion of the gravity field and a priori smoothness conditions is here proposed to form a complete model. Kusche (2007) showed an appropriate way to adapted smoothness conditions regarding preprocessed unconstrained gravity field models. In contrast to Kusche (2007), we extend our model to the complete space including the commission as well as the omission space.

The concept of our approach is characterized by the following items. As background model we introduce the Hilbert space H^1_Γ, as a complete and separable space for continuous functions on the sphere Γ, with square integrable first derivatives. We use an isotropic stationary stochastic process on the sphere to represent this background model. This process can be represented in form of random coefficients as well as in form of a covariance function. This flexibility in representing the stochastic process allows one now to treat this information individually for the commission and omission space. The spherical harmonics as base functions allow us, due to the orthogonality relations, to split up the Hilbert space into sub-domains. With respect to the gravity field models we divide the space into three sub-domains: commission, transition and omission domain. The commission sub-domain is mainly fixed by the real measurements (e.g. satellite-to satellite tracking data, gravity gradient measurements, ...). In the transition zone the information of measurements is supported by the stochastic background process in form of random coefficients, which are modeled by their expectations and variances. This additional a priori information prevents the well-known over-estimation of information content for high frequencies. And finally, the omission domain up to infinity is dominated only by the a priori knowledge about the smoothness of the potential field modeled in form of covariance functions. The shape of the covariance function is given by the theoretical assumptions on the smoothness or by other a priori information. For degrees up to 360, 720 or 2160 the knowledge about the smoothness can be supported by high-resolution gravity field models (e.g. EGM96, EGM08), but up to infinity only theoretical assumptions, like smoothness conditions with respect to the Hilbert space H^1_Γ can be used (Meissl 1971; Schuh and Becker 2010).

The main point is now that we construct with this approach a complete model which enables us to express any functionals and also their variance/covariance information in a consistent way. All functionals of this model are unbiased estimable functionals and can therefore be used without filter processes for data assimilation as well as for hypothesis testing. Because of the completeness with respect to the frequency domain, all vector-valued functions with linear independent functionals (e.g. profiles, gridded data,...) possess an invertible variance/covariance matrix.

This paper is organized in the following way. The construction of stochastic processes on the sphere is discussed in Sect. 2 before we introduce the model building process and the separation into sub-domains in Sect. 3. An example with a GRACE and combined GRACE/GOCE field provides a proof of concept in the final section. A summary concludes this article.

2 Stochastic Processes on the Sphere

Consider an arbitrary square integrable function $u(\vartheta, \lambda)$ on the unit sphere Γ. Because of the completeness of the orthonormalized Laplace's surface spherical harmonics $\bar{C}_{\ell m}(\vartheta, \lambda)$ and $\bar{S}_{\ell m}(\vartheta, \lambda)$ this function can be written in form of a spherical harmonic synthesis

$$u(\vartheta, \lambda) = \sum_{\ell=0}^{\infty} \sum_{m=0}^{\ell} \left[\bar{c}_{\ell m} \bar{C}_{\ell m}(\vartheta, \lambda) + \bar{s}_{\ell m} \bar{S}_{\ell m}(\vartheta, \lambda) \right] . \quad (1)$$

Treating the spherical harmonic coefficients $\bar{c}_{\ell m}$ and $\bar{s}_{\ell m}$ as random variables $\bar{\mathcal{C}}_{\ell m}$ and $\bar{\mathcal{S}}_{\ell m}$ the function $u(\vartheta, \lambda)$ becomes a random process $\mathcal{U}(\vartheta, \lambda)$ on the sphere,

$$\mathcal{U}(\vartheta, \lambda) = \sum_{\ell=0}^{\infty} \sum_{m=0}^{\ell} \left[\bar{\mathcal{C}}_{\ell m} \bar{C}_{\ell m}(\vartheta, \lambda) + \bar{\mathcal{S}}_{\ell m} \bar{S}_{\ell m}(\vartheta, \lambda) \right] . \quad (2)$$

We will assume now that the random variables have zero expectations, i.e.

$$E\left\{\bar{\mathcal{C}}_{\ell m}\right\} = E\left\{\bar{\mathcal{S}}_{\ell m}\right\} = 0 \qquad \ell = 0, \ldots, \infty, m = 0, \ldots, \ell \quad (3)$$

are mutually uncorrelated, i.e.

$$\left.\begin{aligned} \Sigma\left\{\bar{\mathcal{C}}_{\ell m}, \bar{\mathcal{C}}_{sr}\right\} &= 0 \\ \Sigma\left\{\bar{\mathcal{S}}_{\ell m}, \bar{\mathcal{S}}_{sr}\right\} &= 0 \end{aligned}\right\} \text{ if } s \neq \ell \text{ or } r \neq m \text{ or both}$$
$$\Sigma\left\{\bar{\mathcal{C}}_{\ell m}, \bar{\mathcal{S}}_{sr}\right\} = 0 \text{ in any case} \quad (4)$$

and have constant variances per degree, i.e.

$$\Sigma\left\{\bar{\mathcal{C}}_{\ell m}\right\} = \Sigma\left\{\bar{\mathcal{S}}_{\ell m}\right\} := \frac{1}{2\ell+1}\Sigma\left\{\mathcal{S}_\ell\right\} \tag{5}$$

where $\Sigma\{\mathcal{S}_\ell\} = E\left\{\mathcal{S}_\ell^2\right\}$ with the random variable

$$\mathcal{S}_\ell^2 = \bar{\mathcal{C}}_{\ell 0}^2 + \sum_{m=1}^{\ell} \bar{\mathcal{C}}_{\ell m}^2 + \bar{\mathcal{S}}_{\ell m}^2 \,. \tag{6}$$

The stochastic process (2) can be characterized by the expectation $E\{\mathcal{U}(\vartheta,\lambda)\} = 0$ and the stationary, isotropic covariance function

$$\mathrm{Cov}\left\{\mathcal{U}(\vartheta,\lambda);\, \mathcal{U}(\vartheta',\lambda')\right\} = \sum_{\ell=0}^{\infty} \Sigma\{\mathcal{S}_\ell\} P_\ell(\cos\psi) \ , \tag{7}$$

which depends only on Legendre polynomials P_ℓ as functions of the spherical distance ψ between (ϑ,λ) and (ϑ',λ') (cf. e.g. Meissl 1971, Sect. 4; Moritz 1980, Sect. 34). In many applications the estimated degree variances

$$\bar{\sigma}_l^2 = \bar{c}_{\ell 0}^2 + \sum_{m=1}^{\ell} \bar{c}_{\ell m}^2 + \bar{s}_{\ell m}^2 \tag{8}$$

as a single realization of the stochastic variable $\mathcal{S}_\ell^2$ is used to fix $E\left\{\mathcal{S}_\ell^2\right\} = \Sigma\{\mathcal{S}_\ell\}$, or $\Sigma\{\mathcal{S}_\ell\}$ is substituted by deterministic quantities.

In the following it will be shown that a stochastic process on the sphere can also be formulated for deterministically defined degree variances by utilizing the amplitude and phase formulation. Recalling the relations between the Laplace's surface spherical harmonics $\bar{C}_{\ell m}$ and $\bar{S}_{\ell m}$

$$\begin{aligned} \bar{C}_{\ell m}(\vartheta,\lambda) &= \bar{P}_{\ell m}(\cos\vartheta)\cos m\lambda \\ \bar{S}_{\ell m}(\vartheta,\lambda) &= \bar{P}_{\ell m}(\cos\vartheta)\sin m\lambda \end{aligned} \tag{9}$$

with the fully normalized Legendre functions $\bar{P}_{\ell m}$ we can express (1) using the amplitude $A_{\ell m}$ and phase $\Phi_{\ell m}$ notation

$$u(\vartheta,\lambda) = \sum_{\ell=0}^{\infty}\sum_{m=0}^{\ell} A_{\ell m}\bar{P}_{\ell m}(\cos\vartheta)\cos\left(m\lambda + \Phi_{\ell m}\right) \ . \tag{10}$$

The connections between the parameters $\bar{c}_{\ell m}$, $\bar{s}_{\ell m}$ and $A_{\ell m}$, $\Phi_{\ell m}$ are given by

$$\left.\begin{aligned} \bar{c}_{\ell m} &= A_{\ell m}\cos\Phi_{\ell m} \\ \bar{s}_{\ell m} &= -A_{\ell m}\sin\Phi_{\ell m} \end{aligned}\right\} \Leftrightarrow \left\{\begin{aligned} A_{\ell m} &= \sqrt{\bar{c}_{\ell m}^2 + \bar{s}_{\ell m}^2} \\ \Phi_{\ell m} &= \tan^{-1}\tfrac{-\bar{s}_{\ell m}}{\bar{c}_{\ell m}} \ . \end{aligned}\right. \tag{11}$$

We generate now a stochastic process on the sphere by introducing the following conditions. The amplitudes $A_{\ell m}$ are constant quantities depending on the degree variances σ_ℓ^2 of degree ℓ

$$A_{\ell m}^2 = \frac{2}{2\ell+1}\,\sigma_\ell^2 \ . \tag{12}$$

The phases $\Phi_{\ell m}$ are mapped into a vector $\boldsymbol{\Phi}$ by an appropriate numbering scheme. The random counterpart is defined by $\mathcal{P}_{\ell m}$ and $\boldsymbol{\mathcal{P}}$, respectively, where the probability density function $f_{\mathcal{P}}(\mathbf{p})$ is defined by

$$f_{\mathcal{P}}(\mathbf{p}) = \prod_{\ell=0}^{\infty}\prod_{m=0}^{\ell} f_{\mathcal{P}_{\ell m}}(p_{\ell m})$$

$$f_{\mathcal{P}_{\ell m}}(p_{\ell m}) = \begin{cases} 0 & p_{\ell m} \le -\pi \\ \frac{1}{2\pi} & -\pi \le p_{\ell m} \le \ \pi \\ 0 & \pi \le p_{\ell m} \ . \end{cases} \tag{13}$$

This means we have a uniformly distributed phase, for each degree ℓ and order m, and phases for different degrees/orders are mutually independent. The stochastic process $\mathcal{U}(\vartheta,\lambda)$ on the sphere is defined by

$$\mathcal{U}(\vartheta,\lambda) = \sum_{\ell=0}^{\infty}\sum_{m=0}^{\ell} A_{\ell m}\bar{P}_{\ell m}(\cos\vartheta)\cos\left(m\lambda + \mathcal{P}_{\ell m}\right) \ . \tag{14}$$

The expectation

$$E\{\mathcal{U}(\vartheta,\lambda)\} = \int_{-\infty}^{\infty} u(\vartheta,\lambda) f_{\mathcal{P}}(\mathbf{p})d\mathbf{P} \tag{15}$$

is fixed again by $E\{\mathcal{U}(\vartheta,\lambda)\} = 0$, and the covariance

$$\mathrm{Cov}\left\{\mathcal{U}(\vartheta,\lambda),\mathcal{U}(\vartheta',\lambda')\right\} = \int_{-\infty}^{\infty} u(\vartheta,\lambda)\, u(\vartheta',\lambda')\ f_{\mathcal{P}}(\mathbf{p})d\mathbf{P} \tag{16}$$

of this process, can be expressed again by the isotropic covariance function

$$\begin{aligned} \mathrm{Cov}\left\{\mathcal{U}(\vartheta,\lambda),\mathcal{U}(\vartheta',\lambda')\right\} &= \sum_{\ell=0}^{\infty}\sigma_\ell^2\ P_\ell(\cos\psi) \\ &=: \mathrm{cov}(\psi,\sigma_\ell^2) \end{aligned} \tag{17}$$

and depends only on the degree variances σ_ℓ^2 and the spherical distance ψ (cf. appendix for a detailed derivation).

We see that the definition of a uniformly distributed phase, independent for each degree and order, yields an isotropic stochastic process. However, now the degree variances are not represented as second moments of a stochastic variable, in contrast to the first stochastic process (7). The degree variances are defined by the amplitudes, are deterministic quantities, and can be fixed arbitrarily. This allows us to distinguish rigorously between stochastic quantities, realizations and deterministic quantities. As shown in Schuh and Becker (2010), Table 1, the smoothness conditions for the Hilbert spaces H^p_Γ are formulated as asymptotic restrictions $\sigma^2_\ell < \frac{c}{\ell^{2p-1}}$ on the degree variances. This deterministic approach coincides with the amplitude/phase formulation (10) of the stochastic process.

Nevertheless, both representations of the stochastic process in the spectral domain result in a stationary isotropic covariance function. This allows for a representation of the processes in the spectral domain by (2) and (14) as well as in the space domain by (7) and (17). Due to the orthogonality relations of the spherical harmonics (Moritz 1980, p. 21 (3–16)) the infinite Hilbert space can be separated into subspaces, and in each subspace we can express the stochastic process in an individually appropriate way.

3 Model Building and Separability

The Hilbert space H^1_Γ is separated into different subspaces. We divide the space H^1_Γ into the commission space $(H^{(c)})^1_\Gamma$, a transition space $(H^{(t)})^1_\Gamma$, and into the omission space $(H^{(o)})^1_\Gamma$. While the content of the commission space is dominated by the measurements, the content of the omission space reflects basically the weak knowledge of the a priori model.

In the transition space the observations are supported by the a priori model, represented by a stochastic process in terms of randomized spherical harmonic coefficients as documented in (2). The degree variances of the corresponding covariance function (7) are introduced as random variables defined by the spherical harmonic coefficients with $E\left\{\bar{C}_{\ell m}\right\} = E\left\{\bar{S}_{\ell m}\right\} = 0$ and the variances

$$\Sigma\left\{\bar{C}_{\ell m}\right\} = \Sigma\left\{\bar{S}_{\ell m}\right\} = \frac{10^{-10}}{\ell^4}, \quad \begin{matrix} \ell_{t_{min}}, \dots, \ell_{t_{max}} \\ m = 0, \dots, \ell \end{matrix} \tag{18}$$

according to Kaula's rule (Kaula 1966, p. 98 (5.51))

$$\Sigma\left\{\bar{S}_\ell\right\} = \frac{10^{-10}(2l+1)}{\ell^4} \tag{19}$$

considering again the relation (5). $\ell_{t_{min}}$ and $\ell_{t_{max}}$ represent the degree range of the transition zone.

In contrast to random coefficients as prior information in the transition sub-domain, the prior information in the omission sub-domain is defined by the covariance function (17) of the stochastic process. The degree variances are represented as constant quantities. The sizes of these quantities are fixed with respect to Kaula's rule

$$\sigma^2_\ell = \frac{10^{-10}(2\ell+1)}{\ell^4}, \qquad \ell = \ell_{o_{min}}, \dots, \ell_{o_{max}} \tag{20}$$

where $\ell_{o_{min}}$ and $\ell_{o_{max}}$ defines the range of the omission sub-domain.[1]

Due to the orthogonality relations (Moritz 1980, p. 21 (3–16)) of the spherical harmonics, the subspaces are orthogonal to each other for continuous sampling. In our case the discrete satellite measurements have a very uniform distribution of data points over the sphere and therefore our subspaces are almost orthogonal, as will be seen later on.

4 Application and Simulations

To study the performance of complete models with state-of-the-art gravity field models we assimilate the normal equations of two gravity field models with the a priori model: the static solution of the *ITG-Grace2010s* model (Mayer-Gürr et al. 2010) and a combined *GRACE/GOCE* model including 7 months of GOCE measurements (Pail et al. 2011). Many tests show that the exact definition of the ranges for the commission, transition and omission domain is not crucial. For the computation we used the following partitioning strategies. In the case of the *ITG-Grace2010s* the transition sub-domain is defined in the range between degree 151 up to 180, and in case of the combined *GRACE/GOCE* model we choose 171 up to 250.

To assess the behavior of the different models we implemented an error propagation into height anomalies with the variance/covariance information of the particular gravity field models. Tables 1 and 2 summarize the characteristic values. For the *ITG-Grace2010s* (*GRACE/GOCE*) model the standard deviation of the complete model of ±41.7 cm

[1] For practical reasons we work with $\ell_{o_{max}} = 18,000$. It is well known that the variance $\text{cov}(0, \sigma^2_l)$ of the stochastic process using Kaula's degree variances (20) is finite and given by

$$\begin{aligned} \text{cov}(0, \sigma^2_\ell) &= 10^{-10} \sum_{\ell=1}^{\infty} \frac{2\ell+1}{\ell^4} \\ &= 10^{-10}\left(2\sum_{\ell=1}^{\infty}\frac{1}{l^3} + \sum_{\ell=1}^{\infty}\frac{1}{l^4}\right) \\ &= 10^{-10}\left(2\zeta(3) + \zeta(4)\right), \end{aligned} \tag{21}$$

where $\zeta(3)$ and $\zeta(4)$ denote the function values of Riemann's zeta function. R. Apéry proved in 1977 that $\zeta(3)$ is irrational with a value of $\zeta(3) = 1.20205690315959\dots$ (Hata 2000). Euler (1740) p. 133, §18 already derived $\zeta(4) = \frac{\pi^4}{90}$. These constants can be used to compute the relative approximation error for the finite summation up to 18,000 with $1 \cdot 10^{-4}$ $(2 \cdot 10^{-4})$ starting the omission space at 181 (251).

Table 1 Characteristics of the accuracies of height anomalies for a point value of a complete model with commission subspace 2–150 (2–170), transition space 151–180 (171–250) and omission space

ITG-Grace2010s	Commission (2–150)	Transition (151–180)	‖Mixed‖	Model (2–180)	Omission (181–∞)	All (2–∞)
Without a priori information						
std. dev. (m)	0.029	0.220	0.009	0.221	0.353	0.417
With a priori information in the range from 151–180						
std. dev. (m)	0.026	0.121	0.007	0.124	0.353	0.375
GRACE/GOCE	Commission (2–170)	Transition (171–250)	‖Mixed‖	Model (2–250)	Omission (251–∞)	All (2–∞)
Without a priori information						
std. dev. (m)	0.026	0.342	0.011	0.343	0.255	0.427
With a priori information in the range from 171–250						
std. dev. (m)	0.026	0.158	0.007	0.160	0.255	0.301

The standard deviations (std. dev.) result from an error propagation using the model *ITG-Grace2010s* and the combined *GRACE/GOCE* model respectively. All values are averaged over the region $-5°$ to $5°$ latitude and $-5°$ and $5°$ longitude with a grid spacing of $0.1°$, but to be precise, these values reflect the mean behavior of the point information and not mean values over a specific region

Table 2 Characteristics of the accuracy of the geoid undulations for mean values in a specific square area

	ITG-Grace2010s			GRACE/GOCE		
Side length of the area	Model (2–180)	Omission (181–∞)	All (2–∞)	Model (2–250)	Omission (251–∞)	All (2–∞)
0.000	0.124	0.353	0.375	0.160	0.255	0.301
0.500	0.112	0.265	0.288	0.137	0.157	0.208
1.000	0.076	0.123	0.145	0.069	0.040	0.079
2.000	0.011	0.031	0.033	0.019	0.019	0.027
4.000	0.007	0.013	0.015	0.009	0.005	0.010

($\pm$42.7 cm) can be split up into the deviations $\pm$2.9 cm ($\pm$2.6 cm) in the commission zone, $\pm$22.0 cm ($\pm$34.2 cm) in the transition zone and $\pm$35.5 cm ($\pm$25.5 cm) in the omission zone. As a first point we can see that the introduction of a priori information in the transition zone reduces the standard deviation of the height anomalies to $\pm$37.5 cm ($\pm$30.1 cm). We can state that the standard deviation up to 180 (250) is reduced by the a priori information from $\pm$22 to $\pm$12 cm ($\pm$34 to $\pm$16 cm), but the additional omission error of $\pm$35 cm ($\pm$25 cm) yields again a very inferior overall performance. Note also that the overall standard deviation of the combined *GRACE/GOCE* model is larger than that of the *ITG-Grace2010s* before adding a priori information. This is a clear indication for the over-parameterization of high frequencies in the transition zone for the *GRACE/GOCE* model if no prior information is introduced. This is also reflected by Fig. 1 by the increasing degree variances in the transition zone. The mixed term defined by the norm of the covariances between commission and transition space is very small. As mentioned above, these sub-domains are almost orthogonal due to the regular data distribution on the sphere within the gravity field models.

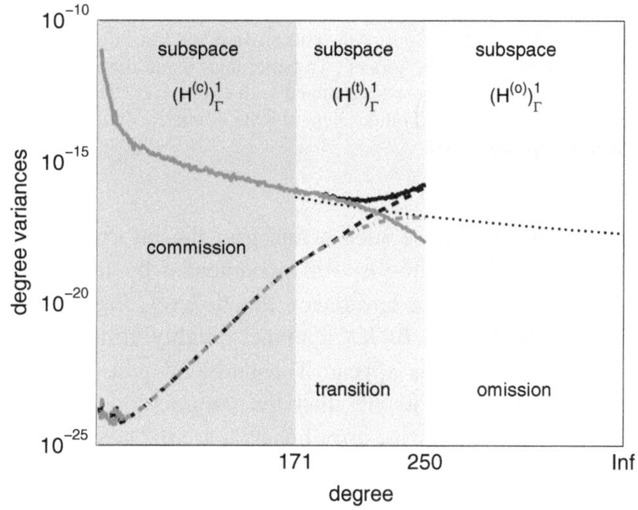

Fig. 1 Complete model: commission–transition–omission subspaces. The *solid lines* mark the degree variances of the signal in the combined *GRACE/GOCE* model (*black*) in contrast to the *complete model* (*gray*). The *dashed lines* reflect the accuracy of the signals. The *dotted line* marks Kaula's rule for the transition and omission sub-domain

As expected the point values perform very poorly (cf. first row in Table 2 with area-size 0.000) because the request of a point value cannot be answered satisfyingly by the band-limited gravity field information. But if we ask for the standard deviation of averaged height anomalies in a quadratic region we may expect accurate values. For this experiment we compute point values on a $0.1°$ grid with variances and covariances, and compute the moving average over a square area with side lengths of $0.5°$, $1°$, $2°$

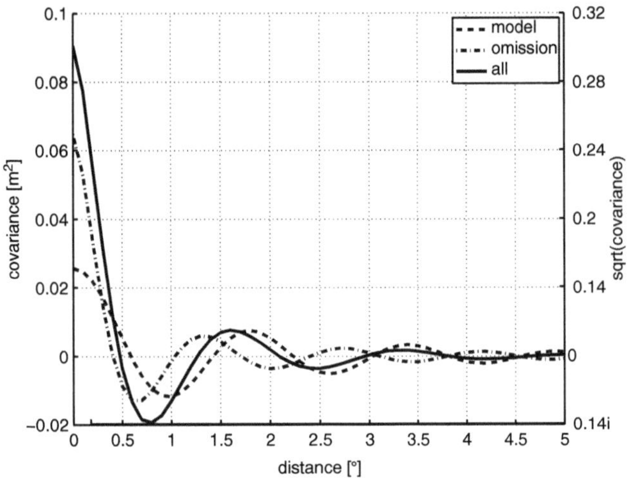

Fig. 2 Covariance information for height anomalies from the combined *GRACE/GOCE* model computed on a profile at the equator ($\phi = 0°, \lambda = 0° \ldots 5°$). The model is structured in a commission space (2–170), a transition space (171–250) and an omission space (251–∞). The contributions of the commission and transition space is summarized by the term 'model'. To better understand the magnitude of the variances this figure is equipped with two y-axes. The *left one* reflects the variances and on the *right one* the square root values of the left scalar are displayed

and 4°. If we choose such a fine grid the information is highly correlated. This is well documented by the shape of the corresponding covariance function (cf. Fig. 2). For the combined *GRACE/GOCE* model roughly up to degree 3 strong correlations appear. The omission part shows a smaller correlation length than the commission part. Of particular interest for the error behavior are the positive and negative slopes. In positive areas an averaging process does not gain high accuracy, because the positive correlations counteract. By contrast, the negative parts of the covariance function accelerate the benefits of an averaging process, and the accuracy decreases disproportionately. The numbers in Table 2 reflect exactly this behavior and give a comparison of the resolutions and expected accuracies for the two models.

Summary and Conclusions

In this article the construction and advantages of *complete models* are illustrated. A background stochastic process forms the basis for a complete model. A priori information defines the amplitude and smoothness of this background field in terms of degree variances. This background model forms a complete base which means that each square integrable function is a member of this infinite space. The stochastic process can be described equivalently in the space domain by covariance functions and in the frequency domain by spherical harmonic coefficients. Into this background model the band-limited data sets from satellite missions are assimilated in a sequential least squares approach. The resulting model is complete again and reflects exactly the strengths and weaknesses of the involved information. In addition it has the following properties:

- The variance/covariance matrices of all vector-valued functions with linearly independent functionals (e.g. for geoid undulations on a grid) have full rank, i.e. their inverses exist. This allows for a rigorous assimilation of gravity information into other Earth system models with an arbitrary grid (e.g. ocean circulation models).
- All linear functionals are unbiased estimable functions, because the complete model spans the entire space and no nullspace is left. No additional computations are necessary to prove that the linear functional is a member of the commission sub-domain. This enables us to define arbitrary problem dependent functions in the space domain and compute the variance/covariance information rigorously (e.g. the mass variation over Greenland).

Complete models are universally applicable as stand-alone models as well as for assimilation purposes.

Acknowledgements This work was funded within the DFG priority program SPP 1257 'Mass transport and mass distribution in the system Earth'. The authors acknowledge the European Space Agency for the provision of the GOCE data and the GOCO-group for providing the normal equations of the *GOCE-TIM2.0* and the *ITG-Grace2010s* model.

Appendix

Expectation and Variance of a Stochastic Process in Amplitude/Phase Notation on the Sphere

The stochastic process $\mathcal{U}(\vartheta, \lambda)$ on the sphere is defined by

$$\mathcal{U}(\vartheta, \lambda) = \sum_{\ell=0}^{\infty} \sum_{m=0}^{\ell} A_{\ell m} \bar{P}_{\ell m}(\cos\vartheta) \cos(m\lambda + \mathcal{P}_{\ell m}) \quad (22)$$

where the phases constitute random variables. The distribution is defined by

$$f_{\mathcal{P}}(\mathbf{p}) = \prod_{\ell=0}^{\infty} \prod_{m=0}^{\ell} f_{\mathcal{P}_{\ell m}}(p_{\ell m})$$

$$\text{with}\, f_{\mathcal{P}_{\ell m}}(p_{\ell m}) = \begin{cases} 0 & p_{\ell m} \leq -\pi \\ \frac{1}{2\pi} & -\pi \leq p_{\ell m} \leq \pi \\ 0 & \pi \leq p_{\ell m}\ . \end{cases} \quad (23)$$

This means we have uniformly distributed phases, for each degree ℓ and order m, and phases for different degrees/orders are mutually independent.

We are interested in the expectation

$$E\left\{\mathcal{U}(\vartheta,\lambda)\right\} = \int_{-\infty}^{\infty} u(\vartheta,\lambda)\, f_{\mathcal{P}}(\mathbf{p}) d\mathbf{P}$$

$$= \int_{-\infty}^{\infty} \sum_{\ell=0}^{\infty}\sum_{m=0}^{\ell} A_{\ell m}\bar{P}_{\ell m}(\cos\vartheta)\cos\left(m\lambda + \mathcal{P}_{\ell m}\right) f_{\mathcal{P}}(\mathbf{p})\, d\mathbf{P}\,.$$

Due to the independence of the phases the integral can be rewritten as

$$\int_{-\infty}^{\infty} \sum_{\ell=0}^{\infty}\sum_{m=0}^{\ell} A_{\ell m}\bar{P}_{\ell m}(\cos\vartheta)\cos\left(m\lambda + \mathcal{P}_{\ell m}\right) f_{\mathcal{P}_{\ell m}}(p_{\ell m}) dp_{\ell m}\,.$$

If we now interchange integration and summation and introduce the individual distribution (23) we get

$$E\left\{\mathcal{U}(\vartheta,\lambda)\right\} = \sum_{\ell=0}^{\infty}\sum_{m=0}^{\ell} A_{\ell m}\bar{P}_{\ell m}(\cos\vartheta) \times \frac{1}{2\pi}\int_{-\pi}^{\pi}\cos\left(m\lambda + p_{\ell m}\right) dp_{\ell m}$$

$$= \sum_{\ell=0}^{\infty}\sum_{m=0}^{\ell} A_{\ell m}\bar{P}_{\ell m}(\cos\vartheta)\,\frac{1}{2\pi}\big(-\sin m\lambda + \sin m\lambda\big)\,.$$

Finally we see that

$$E\left\{\mathcal{U}(\vartheta,\lambda)\right\} = 0\,. \tag{24}$$

The variance of this process is given by

$$\operatorname{Cov}\left\{\mathcal{U}(\vartheta,\lambda),\mathcal{U}(\vartheta',\lambda')\right\} = \int_{-\infty}^{\infty} u(\vartheta,\lambda)\, u(\vartheta',\lambda')\, f_{\mathcal{P}}(\mathbf{p}) d\mathbf{P}\,. \tag{25}$$

Because of the independence of the random variables this can be written as

$$\operatorname{Cov}\left\{\mathcal{U}(\vartheta,\lambda),\mathcal{U}(\vartheta',\lambda')\right\} = \sum_{\ell=0}^{\infty}\sum_{m=0}^{\ell} A_{\ell m}^2\bar{P}_{\ell m}(\cos\vartheta)\bar{P}_{\ell m}(\cos\vartheta') \times \int_{-\infty}^{\infty}\cos\left(m\lambda + p_{\ell m}\right)\cos\left(m\lambda' + p_{\ell m}\right) \times f_{\mathcal{P}_{\ell m}}(p_{\ell m})\, dp_{\ell m}\,. \tag{26}$$

If we extend the first cosine term in the integral to $(m(\lambda-\lambda') + m\lambda' + p_{\ell m})$, use the relation

$$cos(x+y)\cos(x-y) = \frac{1}{2}(\cos 2x + \cos 2y)$$

and substitute

$$x = \frac{1}{2}m(\lambda-\lambda') + m\lambda' + p_{\ell m}$$
$$y = \frac{1}{2}m(\lambda-\lambda')$$

the integral can be solved and yields

$$\frac{1}{2}\sin\left(m(\lambda-\lambda') + 2m\lambda' + 2p_{\ell m}\right)\Big|_{p_{\ell m}=-\pi}^{p_{\ell m}=\pi} + \frac{1}{2}\cos\left(m(\lambda-\lambda')\right) p_{\ell m}\Big|_{p_{\ell m}=-\pi}^{p_{\ell m}=\pi}\,.$$

The first term vanishes because of the skew symmetry of the sine and only the cosine term is relevant. Substituting this result into (26) yields

$$\operatorname{Cov}\left\{\mathcal{U}(\vartheta,\lambda),\mathcal{U}(\vartheta',\lambda')\right\} = \sum_{\ell=0}^{\infty}\sum_{m=0}^{\ell}\frac{1}{2}A_{\ell m}^2\bar{P}_{\ell m}(\cos\vartheta)\bar{P}_{\ell m}(\cos\vartheta')\cos\left(m\lambda - m\lambda'\right)\,.$$

Applying the addition theorem

$$\cos\left(m\lambda - m\lambda'\right) = \cos m\lambda\cos m\lambda' + \sin m\lambda\sin m\lambda'$$

and recalling the definition of Laplace's surface spherical harmonics (9) the right hand side can be reformulated as

$$\sum_{\ell=0}^{\infty}\sum_{m=0}^{\ell}\frac{1}{2}A_{\ell m}^2\left(\bar{C}_{\ell m}(\vartheta,\lambda)\bar{C}_{\ell m}(\vartheta',\lambda') + \bar{S}_{\ell m}(\vartheta,\lambda)\bar{S}_{\ell m}(\vartheta',\lambda')\right)\,.$$

Introducing now the amplitudes defined in (12) we get

$$\sum_{\ell=0}^{\infty}\sum_{m=0}^{\ell}\frac{1}{2\ell+1}\sigma_\ell^2\left(\bar{C}_{\ell m}(\vartheta,\lambda)\bar{C}_{\ell m}(\vartheta',\lambda') + \bar{S}_{\ell m}(\vartheta,\lambda)\bar{S}_{\ell m}(\vartheta',\lambda')\right)\,.$$

The decomposition formula or addition theorem for spherical harmonics (cf. e.g. Moritz 1980, p. 23 (3–30))

$$P_\ell\big((\vartheta,\lambda);\ (\vartheta',\lambda')\big) =$$

$$\frac{1}{2\ell+1}\sum_{m=0}^{\ell}\left(\bar{C}_{\ell m}(\vartheta,\lambda)\bar{C}_{\ell m}(\vartheta',\lambda') + \bar{S}_{\ell m}(\vartheta,\lambda)\bar{S}_{\ell m}(\vartheta',\lambda')\right)$$

allows the for further simplification

$$\mathrm{Cov}\left\{\mathcal{U}(\vartheta,\lambda),\mathcal{U}(\vartheta',\lambda')\right\} = \sum_{\ell=0}^{\infty}\sigma_\ell^2 P_\ell\big((\vartheta,\lambda);\ (\vartheta',\lambda')\big)$$

where the function value of the Legendre polynomial $P_\ell\big((\vartheta,\lambda);\ (\vartheta',\lambda')\big)$ depends only on the spherical distance $\cos\psi$ between (ϑ,λ) and (ϑ',λ')

$$\cos\psi = \cos\vartheta\cos\vartheta' + \sin\vartheta\sin\vartheta'\cos(\lambda-\lambda')\,. \qquad (27)$$

Finally this results in

$$\mathrm{Cov}\left\{\mathcal{U}(\vartheta,\lambda),\mathcal{U}(\vartheta',\lambda')\right\} = \sum_{\ell=0}^{\infty}\sigma_\ell^2\ P_\ell(\cos\psi) = \mathrm{cov}(\psi,\sigma_\ell^2).$$

References

Becker S, Freiwald G, Losch M, Schuh WD (2011) Rigorous fusion of gravity field, altimetry and stationary ocean models. J Geodyn (online first). doi:10.1016/j.jog.2011.07.006

Euler L (1740) De summis serierum reciprocarum. Commentarii academiae scientiarum Petropolitanae 7:123–134. http://eulerarchive.maa.org/pages/E041.html

Hata M (2000) A new irrationality measure for $\zeta(3)$. Acta Arithmetica 92(1):47–57

Jekeli C (1981) Alternative methods to smooth the Earth's gravity field. Reports of the Department of Geodetic Science, Ohio State University (OSU), Ohio, No. 327

Jekeli C (1996) Spherical harmonic analysis, aliasing, and filtering. J. Geod 70:214–223. doi:10.1007/BF00873702

Kaula W (1966) Theory of satellite geodesy. Blaisdell Publishing Company, Toronto

Kusche J (2007) Approximate decorrelation and non-isotropic smoothing of time-variable GRACE-type gravity field models. J Geod 81:733–749. doi:10.1007/s00190-007-0143-3

Losch M, Sloyan B, Schöter J, Sneeuw, N (2002) Box inverse models, altimetry and the geoid: problems with the omission error. J Geophys Res 107(C7):3078. doi:10.1029/2001JC000855

Mayer-Gürr T, Kurtenbach E, Eicker A (2010) ITG-Grace2010 gravity field model. University Bonn, Institute of Geodesy and Geoinformation. http://www.igg.uni-bonn.de/apmg/index.phg?id=itg-grace2010

Meissl P (1971) A study of covariance functions related to the earth's disturbing potential. Reports of the Department of Geodetic Science, Ohio State University (OSU), Ohio, No. 151

Moritz H (1980) Advanced physical geodesy. Wichmann, Karlsruhe

Pail R, Bruinsma S, Miggliaccio F, Förste C, Goiginger H, Schuh WD, Höck E, Reguzzoni M, Brockmann J, Abrikosov O, Veicherts M, Fecher T, Mayrhofer R, Krasbutter I, Sansó F, Tscherning C (2011) First GOCE gravity field models derived by three different approaches. J Geod 85:819–843. doi:10.1007/s00190-011-0467-x

Schuh WD, Becker S (2010) Potential field and smoothness conditions. In: Contadakis M, Kaltsikis C, Spatalas S, Tokmakidis K, Tziavos I (eds) The apple of knowledge - in honour of Prof. N. arabelos, pp 237–250. University of Thessaloniki, AUTH - Faculty of Rural and Surveying Engineering

Swenson S, Wahr J (2006) Post-processing removal of correlated errors in GRACE data. Geophys Res Lett 33:L0840211. doi: 10.1029/2005GL025285

Wahr J, Molenaar M, Bryan F (1998) Time variability of the Earth's gravity field: hydrological and oceanic effects and their possible detection using GRACE. J Geophys Res 103:30205–30299. doi:10.1029/98JB02844

List of Reviewers

Hamza Alkhatib
Orhan Akyilmaz
Markus Antoni
Alireza A. Ardalan
Oliver Baur
Matthias Becker
Mathis Bloßfeld
Klaus Börger
Xavier Collilieux
Andreas Eichhorn
Annette Eicker
Xing Fang
Karl Foppe
Christian Gerlach
Engin Gülal
Andrea Heiker
Hans Heister
Katja Heine
Maria Hennes
Otto Heunecke
Boris Kargoll
Mahmut O. Karslioglu
Tobias Kersten
Thomas Kersten
Jürgen Kusche
Hansjörg Kutterer
Michael Meindl
Frank Neitzel
Wolfgang Niemeier
Werner Lienhart
Jens-Andre Paffenholz
Alexander Reiterer
Björn Riedel
Roelof Rietbroek
Michael Schmidt
Steffen Schön
Wolf-Dieter Schuh
Volker Schwieger
Willfried Schwarz
Florian Seitz
Peter Steigenberger
Werner Stempfhuber
Daniela Thaller
Torsten Mayer-Gürr
Markus Vennebusch
Matthias Weigelt
Andreas Wieser
Thomas Wunderlich

H. Kutterer et al. (eds.), *The 1st International Workshop on the Quality of Geodetic Observation and Monitoring Systems (QuGOMS'11)*, International Association of Geodesy Symposia 140,
DOI 10.1007/978-3-319-10828-5, © Springer International Publishing Switzerland 2015

Author Index

H. Kutterer et al. (eds.), *The 1st International Workshop on the Quality of Geodetic Observation and Monitoring Systems (QuGOMS'11)*, International Association of Geodesy Symposia 140,
DOI 10.1007/978-3-319-10828-5, © Springer International Publishing Switzerland 2015

MIX
Papier aus verantwortungsvollen Quellen
Paper from responsible sources
FSC® C105338

If you have any concerns about our products,
you can contact us on
ProductSafety@springernature.com

In case Publisher is established outside the EU,
the EU authorized representative is:
Springer Nature Customer Service Center GmbH
Europaplatz 3, 69115 Heidelberg, Germany

Printed by Libri Plureos GmbH
in Hamburg, Germany